Biocybernetics of Vision

Integrative Mechanisms and Cognitive Processes

Istituto Italiano per gli Studi Filosofici

Series on Biophysics and Biocybernetics

Coordinating Editor: C. Taddei-Ferretti

Vol. 1: Biophysics of Photoreception: Molecular and Phototransductive Events
edited by: C. Taddei-Ferretti

Forthcoming volumes:

Vol. 3: High Dilution Effects on Cells and Integrated Systems
edited by: C. Taddei-Ferretti and P. Marotta

Vol. 4: Macromolecular Interplay in Brain Associative Mechanisms
edited by: A. Neugebauer

ISTITUTO ITALIANO PER GLI STUDI FILOSOFICI
SERIES ON BIOPHYSICS AND BIOCYBERNETICS
Vol. 2 – Biocybernetics

Istituto Italiano per gli Studi Filosofici
Gruppo Nazionale di Cibernetica e Biofisica, CNR

Biocybernetics of Vision

Integrative Mechanisms and Cognitive Processes

Proceedings of the International School of Biocybernetics
Casamicciola, Napoli, Italy, 16–22 October 1994

Edited by

C. Taddei-Ferretti
Istituto di Cibernetica, CNR, Arco Felice, Napoli, Italy

World Scientific
Singapore • New Jersey • London • Hong Kong

Published by

World Scientific Publishing Co Pte Ltd
5 Toh Tuck Link, Singapore 596224
USA office: 27 Warren Street, Suite 401-402, Hackensack, NJ 07601
UK office: 57 Shelton Street, Covent Garden, London WC2H 9HE

British Library Cataloguing-in-Publication Data
A catalogue record for this book is available from the British Library.

BIOCYBERNETICS OF VISION: INTEGRATIVE MECHANISMS AND COGNITIVE PROCESSES

ISBN-13 978-981-02-3229-0
ISBN-10 981-02-3229-2

PREFACE

This is the second volume of the Series on Biophysics and Biocybernetics promoted by the Istituto Italiano per gli Studi Filosofici.

It appears as the Proceedings of the "International School E.R. Caianiello on Biocybernetics of Vision: Integrative Mechanisms and Cognitive Processes", which was inaugurated at the site of the Istituto, Palazzo Serra di Cassano, Naples, Italy, and was held at Casamicciola on the isle of Ischia, Italy, on October 16-22, 1994, under the direction of this volume's Editor.

This School was promoted and supported jointly by the Istituto Italiano per gli Studi Filosofici and by the Gruppo Nazionale di Cibernetica e Biofisica of the Italian Consiglio Nazionale delle Ricerche (CNR).

The School has been organized by the Istituto di Cibernetica, CNR, in the 25th year of its foundation: for this reason it has been dedicated to the memory of Prof. E.R. Caianiello, who was the founder of the Institute.

The topics covered ranged from "Integrative Mechanisms of Vision", to "Visual Perception and Cognitive Processes" and to "(From Animal Vision to) Machine Vision".

The topics are related to visual cognition, which is an important area of biocybernetics.

It ranges from the filtering processes of early vision, to the structural and functional organization of the visual centres, as well as, in higher animals, to the neuronal plasticity, the decision-making rules, the effect of noise, the role of attention, the ambiguity of patterns and the time dimension.

All those and other factors contribute to the cognitive interpretation of visual sensation that takes place in visual perception.

A side field of research is that of machine vision, in which the signal processing known from animal vision is applied to the construction of mobile robots responding to light stimulation.

I would like to thank the members of the School Advisory Board for their fruitful advice and suggestions, as well as for their helpful cooperation also as reviewers of the participants' communications: L. Cordella (I), I. De Lotto (I), A. Fiorentini (I), N. Franceschini (F), K.G. Götz (D), S. Hochstein (IL), E. Mingolla (USA), T. Radil (CZ).

I would also acknowledge the partial financial support of the Istituto di Cibernetica, CNR.

Furthermore, only due to the help of the Istituto Italiano per gli Studi Filosofici it has been possible to award grants to several deserving participants needing financial help.

I wish to thank all the scientists who agreed to lecture, who contributed to the School with their discussions and who favoured, toghether with all the participants, a charming atmosphere in a highly stimulating scientific milieu.

Finally, I thank the members of the local organizing committee: A. Cotugno, M. Izzo, A.M. Mazzarella, C. Musio, S. Piantedosi, S. Santillo, of the Istituto di Cibernetica, CNR; without their work the School could not have been realized, nor its cordial atmosphere obtained.

The beauty of the isle of Ischia and the courtesy of the staff of the Hotel Gran Paradiso at Casamicciola, where the School was held, completed the pleasantness of the environment.

Cloe Taddei-Ferretti

CONTENTS

PARTICIPANTS

IN MEMORY OF E.R. CAIANIELLO

EDUARDO R. CAIANIELLO: *IN MEMORIAM* (*)

Maria Marinaro

Facoltà di Scienze, MFN, Università di Salerno, Salerno, Italy

Eduardo R. Caianiello, the brilliant outstanding scientist whose scientific contributions have had a great impact both in theoretical physics and in cybernetics, died October 22, 1993, in Naples, Italy.

The unique character of his scientific personality manifested itself in all the work he made. His high general culture, his unpredictable way of thinking, and his wide spectrum of interests always generated admiration in all who approached him. He was a great "maestro" ever sensitive to the problem of the young people, whom he always tried to help and understand. He shared his ideas generously with his students and colleagues and was gratified whenever they succeeded.

In the fifties his formulation of Feynman graphs in terms of permanents, determinants, pfaffians, and hafnians, and the consequent mathematically consistent treatment of the renormalization in quantum field theory was the starting point for the rigorous formulation of the renormalization group and of the Callan-Symanzik equation, which was formulated in the seventies. His seminal contributions in the field of neural nets, hierarchical systems, composite sets (c-calculus) are well known in the world scientific community.

In the work "Outline of a Theory of Thought Processes and Thinking Machines", published in the *Journal of Theoretical Biology* in April 1961, a complete mathematical formulation of the behaviour of a neural system appeared for the first time.

The neuronic equations that describe the behaviour of the net with fixed connections; the mememonic equations that account for the change of the coupling among the neurons; and the adiabatic learning hypothesis reported in the literature as Caianiellello's equations (Encyclopaedic Dictionary of Mathematics, MIT Press, 1974), have been the basis for many subsequent contributions on associative memories and learning rules.

The composite set calculus (c-calculus), a new algebraic method introduced by Caianiello as a general technique for the analysis of complex syastems, was applied with success in automatic image analysis, particularly in pattern recognition, filtering, contour extraction, image refinement, and texture analysis. c-Calculus has given a unified methodology for the study of different aspects of automatic analysis.

His studies on the structure of language and of biological and social systems convinced him to the relevance of hierarchical structure in the study of complex systems, and led him to the analysis of modular systems that he applied to the monetary, urban, and other complex systems.

(*) Reprinted from *Neural Networks*, Vol. 7, No. 2, p. xiii, 1994, with kind permission from Elsevier Science Ltd, The Boulevard, Langford Lane, Kidlington 0X5 1GB, UK.

These contributions do not represent a complete picture of the activity of Eduardo R. Caianiello but are only a few aspects of his many accomplishments. The best way to comment on his scientific activity is to recall what he used to tell his students and colleagues: "What matters is not the problem one is facing, but the scientific method which must be the guide of every research". Scientific method and intellectual integrity constantly characterised all his activities as a scientist.

Dr. Caianiello was born in Naples in 1921. After his "laurea" degree in Physics at the University of Naples, Caianiello moved to United States, where he received his PhD in Physics at the University of Rochester in 1950. He returned to Italy: to Turin and Rome; and in 1956 to Naples, as a full Professor of Theoretical Physics, in the chair that had been that of E. Majorana. In 1957, he founded the Institute of Theoretical Physics of Naples, which very soon became one of the most active places in Italy, where physicists from all over the world met to exchange ideas and propose new theories. The presence of Eduardo Caianiello and the stimulating atmosphere of the Institute under his guidance was a very fruitful experience for young Italian students. In 1968, he founded the Laboratory of Cybernetics at Arco Felice, which soon became one of the leading institutions in areas like pattern recognition, analysis of form and language, and neural nets. Eduardo Caianiello was a man continuously attracted by new challenges, he could not be quiet for very long. In 1972, he moved to the University of Salerno, where he founded the Science Faculty. Salerno thus became one of the four Italian universities offering a laurea degree in the new science of Informatics.

His last gift to thc scientific community was the foundation of the International Institute for Advanced Scientific Studies in Vietri sul Mare (SA), now the seat of SIREN (Italian Society of Neural Nets). He dedicated the last years of his life to the development of this institution that today has gained a relevent position for research in neural nets and its applications.

Caianiello's equations and cognitive processes

Ernesto Burattini
Istituto di Cibernetica - CNR
Arco Felice - Napoli

It is a great honor for me to give this talk in memory of Eduardo Caianiello. I partly feel it as an obligation as well, in view of the intellectual and human debts that I carry towards him. It is, unfortunately, an obligation I will not able to comply with in the ways deserved by the man this school is dedicated to, whose scientific achievements have been vividly reminded to many of us by distinguished lecturers on various special meetings held in his memory during the period of time separating us from his sudden and untimely death.

This school, as everyone here knows, is organized by the Istituto di Cibernetica, a scientific institution that Eduardo Caianiello founded twenty-five years ago overcoming, as everyone here can very well imagine, formidable practical problems. He was also the founder of the Istituto di Fisica Teorica of the Università di Napoli, the Facoltà di Scienze dell'informazione of the Università di Salerno and, more recently, the Istituto Internazionale per gli Alti Studi Scientifici.

These permanent marks of his thorough dedication to the scientific pursuit still provide highly valuable forums of scientific interchange and fertile grounds for significant scientific work.

It is this work initiated by him, rather than his personal merits, that Eduardo Caianiello would wish us to concentrate on today. And it is a very small portion of this work that I wish to focus on in my talk, with the intention of showing that some of his ideas on intelligent systems formulated in the early 1960's still carry today the seeds of interesting scientific developments.

When I first met Eduardo Caianiello in 1968, his celebrated neural equations and model were well known to the international scientific community, and much work on them had already been done.

And before breaching new ground in Cybernetics with his equations, Eduardo Caianiello had given proof of his outstanding scientific quality with his work in physical science.
Thus, when I met him in 1968, he could have happily rested on his laurels. Instead, he felt he was only at the beginning of a long and engrossing scientific undertaking. His enthusiastic and unfaltering commitment to the ideal of scientific knowledge enabled him to overcome the unavoidable frustrations involved in pursuing overarching and ambitious research goals, and to look beyond the limitations affecting the technology available in the early 1970's. Indeed, some of the results he was looking for could be obtained only with the much more powerful computing technologies available today.

My talk is directly concerned with this latter point. What I would like to show is the fact that computational elements directly inspired by the biological neuron, and taking the form of the abstract neuron model adopted by Eduardo Caianiello, can provide an adequate basis for the simulation of some forms of high-level intelligent behavior. More specifically, suitably organized networks of such elements provide a basis for performing intellectual tasks which fall in the domain of investigation of an important area of Artificial Intelligence: expert system research.

This area of AI is concerned with the study and simulation of problem-solving activities requiring the knowledge of highly trained human professionals and involving sophisticated forms of stepwise reasoning. Networks of relatively simple, biologically inspired, computational elements are becoming very valuable tools even in this discipline, which has been for a long time programmatically inattentive to the biological bases of our thinking.

Everybody here knows what a neuron is in terms of its biological structure and its chemical and electrical behaviour. Even since McCulloch and Pitts, a very rough mathematical model was introduced with the hope that even an elementary model of the neuronal cell may help one to design nets having behaviours or functions similar to those expressed by the human brain.

Eduardo Caianiello in his fundamental work, published in 1961 on the Journal of Theoretical Biology (Outline of a Theory of Thought-Processes and Thinking Machines, 1961 2,204-235), claimed:

"Our main guiding principle has been the conviction, ..., that the human brain, tremendous in its complexity, yet obeys, if one looks at the operation of individual neurons, dynamical laws that are not necessarily complicated; and that these laws are such as to engender in large neuronal assemblies collective modes of behavior, to which thought-processes are correlated".

Then the basic hypothesis was that a very simple model might be useful to mimic at least a partial activity of the human brain.

This model captures, in a very simplified manner, the electrical functioning of the neuronal cell, suggesting that it consists in a system which sums a lot of electrical excitatory impulses originating in other elements of the same sort. Those impulses are excitatory or inhibitory and through the links connecting each neuron to the other ones, may be amplified or changed in sign. The total amount of excitation supplied to the input of each element is compared to a value, named threshold, which may differ from one neuron to another. If the sum is greater than the threshold then the neuron fires, that is, an excitation is sent to all other neurons connected to it.

The neuronic equations, based on the previous considerations, were stated by Caianiello as follows:

$$u_h(t+\tau) = \mathbf{1}\left[\sum_{k,r} a_{h,k} \cdot u_k(t-r\tau) - s_h\right] \qquad (1)$$

where $u_h(t)$ is the state (1 or 0) of the neuron h at time i

$a_{h,k}$ is the coupling coefficient that transfers the pulse from neuron k to neuron h

τ is the delay time after which the neuron may or not be excited r is a constant that determines the refractoriness of each neuron, i.e., the time interval for which, after a firing, the neuron cannot be activated

s_h is the threshold of the neuron h, such a neuron fires at time t+τ

$\mathbb{1}[x]$ is $\begin{cases} 1 \text{ if } x > 0 \\ 0 \text{ if } x \le 0 \end{cases}$

Another set of equations was introduced by Caianiello: The mnemonic equations.

While the neuronic equations have constant coefficients and determine the instantaneous behavior of the system, the mnemonic equations take into account the semi-permanent or permanent changes in the structure of the system caused by its past operation. We don't report here the matematical model of the mnemonic equations since they are not involved in the following and moreover their structure is quite complicated to be briefly explained. We remark only that while the neuronic equations have been studied in depth by many authors, only few people have investigated the mnemmonic ones and this may be an interesting challenge for researchers of neural nets.

After introducing equations (1) many authors developed theories and obtained results. Two main problems arise in the study of Caianiello's equations: the analysis and the synthesis of a neuronic net.

The first problem may be stated as follows: given a well defined net, i.e., assigned coupling coefficients, thresholds and refractoriness time, forecast its behavior in time, emphasizing periodical cycles (reverberations) and transients. This problem was addressed by the Caianiello's works and by many other authors, first of all by people coming from the japanese school, and some satisfactory answers were given.

A more complicated problem is the second one. In this case, in fact, assigned a desired behavior, in terms of reverberations and transients, one must find the structure of the net, i.e., the values for the coupling coefficients, the thresholds and the refractoriness such that the required behavior is realized. This is a very significant problem both from the biological point of view and from the computational one. In the first case observing the electrophysiological output of very limited portion of animal brains, if we were able to obtain the same behavior using a neuronic net, we may verify the structural similarities or dissimilarities between the biological and the mathematical nets. Few results have been obtained in this direction.

The second case is also very interesting since if we were able to design a net having a preassigned behavior then we might store a quantity of information exponential with respect to the number of neurons of the net. Unfortunately this kind of results has not yet been obtained even though Eduardo Caianiello was very confident in the prospect of finding a solution.

From the 1970's until 1981 many people unsuccessfully worked on such matters, and the fascination with neural net studies decreased. A new impulse to such researches was given by the work of Hopfield, Ruhmelart et al. who paved the way to approaching and solving the fundamental problem of learning in neural nets. Their connectionist approach is quite different from that of Eduardo Caianiello. They in fact introduce a strong modification in the threshold function converting it into a continuous non decreasing function. In other words, they considerably changed the basic neuron model overcoming in such a way the mathematical difficulties arising with the previous model.

On this approach the synthesis problem using learning algorithms based on the possibility of changing the coupling coefficients during a training period which stops only when a satisfactory behavior has been reached, that is, until the prescribed input/output relationships have been obtained. Connectionist nets are formed by a certain number of layers, many of which have no biological or semantic justification: they are introduced only to accomplish the prescribed behavior. In other words, if we are unable to satisfy our goal with a given number of neurons then we increase such a number and run again the learning procedure. It is evident that such a net cannot satisfy the Eduardo Caianiello criteria: Caianiello in fact firmly believed that only simplicity and a clear mathematical formalization may help solving the problem of forcing a net into a prescribed behavior.

Our work has been conducted more in the spirit of Caianiello's requirements than adopting the now prevailing connectionists methodologies.
These investigations in AI and ES show that in certain well restricted domains it is possible to emulate human behavior in terms of reasoning and knowledge.

The question to which we tried to give an answer was: *Which parts of expert knowledge and reasoning can be adequately and efficiently captured by neural networks, and by which types of such networks?*
This is the pertinent question to be addressed when exploring the possible uses of neural elements in the design of an expert system. A standard answer to this question is that neural nets are unsuitable for stepwise expert reasoning, possibly involving subgoaling and interactions with the environment, as well as for the related task of providing informative explanation facilities. In contrast, the learning and generalization capacities of neural nets can be profitably employed, e.g., for handling noisy or incomplete data, acquiring expert knowledge by examples, solving one-step classification problems.

This popular answer is partially misleading, because many types of non-connectionist networks can be used to carry out elaborate forms of stepwise reasoning (including justification). Indeed, representing the knowledge bases of most expert systems does not require an essential use of a first-order language, for these are supposed to capture relations over a finite domain of elements (components and (ab)normal states of a class of objects, and explanatory hypotheses for the observable anomalies). In turn, such knowledge bases can be codified by means of (weighed) production rules over propositional literals, which are neurally representable using a localist semantic interpretation: individual neurons in the network are interpreted as literals occurring in the rules and their relationships can be expressed by means of neural connections. Under this approach, individual neurons are given a precise semantic interpretation in terms of the atomic elements in the domain of the expert system. Moreover, hidden layers of neurons and learning algorithms are not necessarily involved in constructing the neural analog of the rule interpreter.

This approach falls squarely outside the connectionist paradigm: there is, e.g., a manifest conflict with the distributed representation principles presented by Smolensky. One can, however, amalgamate in a principled way symbolic neural processing and connectionist methodologies in *totally neural* systems for diagnostic reasoning.

The localist semantic interpretation and non-learning based, symbolic neural inferences are appropriate for most diagnostic tasks, i.e., whenever the system is supposed to carry out stepwise inferential processes and to provide informative explanations for the conclusions that have been reached, whereas layers of neurons governed by learning algorithms can be introduced whenever the robustness and learning capacities of neural nets can be profitably put to work.

This is our short answer to the initial question: expert knowledge and reasoning can be simulated, in many significant cases, by means of totally neural systems suitably intertwining symbolic neural processing and connectionist methodologies.

A most appealing feature of totally neural expert systems is the possibility of fully exploiting the parallel processing potentialities of neural systems: in view of the possibilities offered by new parallel architectures, this design choice may have significant practical consequences whenever reaction times are decisive parameters for evaluating performances. Diagnoses of malfunctioning in industrial environments, such as chemical or nuclear power plants, are cases in point. However, parallel processing of information has to be appropriately regimented in diagnostic tasks: a diagnostic process takes place in a certain number of *sequential* steps, while a *parallel* processing of the available information can be performed within each step.

Let me now recall that a production system consists of a *knowledge base* containing *rules*, a *data base* containing *facts*, and a *rule interpreter* to control the inference process.

The *rules* we shall consider here are conditional expressions in the Horn clause form

$$p_1 \wedge \ldots \wedge p_k \rightarrow c$$

with the additional constraint that $p_1, \ldots, p_k$ and c are propositional literals. These restrictions are immaterial in a wide variety of expert system domains, in view of the considerable expressive power of (propositional) Horn clause logic [Kowalski 1979] and the fact that the knowledge base of most diagnostic systems is supposed to capture relations over a finite domain of elements.

The *facts data base* of a production system is a collection of facts, whether asserted or inferred by applying the rules. The *rule interpreter* works iteratively in recognize-and-act cycles, which can be used to implement various kinds of searches. *Forward* and *backward chaining* are basic inferential strategies in production systems, which may also be amalgamated to obtain mixed strategies. For example, given facts and rules, forward chaining may be used initially in order to infer a certain goal and then backward chaining can be applied to find other facts that support the same goal.

To represent a production rule we used the Caianiello neural model.
A rule r

$$p_1 \wedge \ldots \wedge p_k \rightarrow h$$

can be represented as a net having k neurons $p_1, \ldots, p_k$ connected to a neuron $\boldsymbol{h}$ (see fig.1) with the following settings:

$$\begin{aligned} &\forall j \quad a_{j,\boldsymbol{h}} = 1 \quad (1 \le j \le k) \\ &s_{\boldsymbol{h}} = k - \varepsilon \quad (0 < \varepsilon < 1) \\ &r = 0 \\ &\tau = 1 \end{aligned} \qquad (2)$$

By (1) and the settings in (2), one has that:

$$u_h(1) = \mathbf{1}\left[\sum_{j=1}^{k} u_{p_j}(0) - (k - \varepsilon)\right].$$

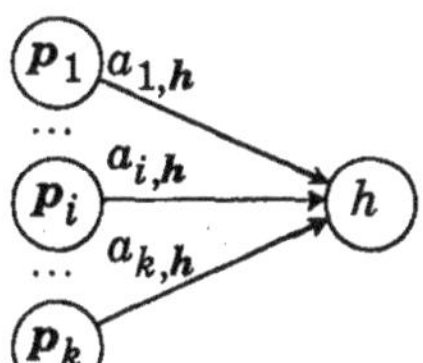

fig.1: *a neural rule model*

Thus,

$$u_h(1) = 1 \quad \text{iff} \quad \forall j\, u_{p_j}(0) = 1$$

that is, the neuron h is active (its state is 1) at time $t = 1$ if and only if all $p_1, \ldots, p_k$ are simultaneously active at time $t = 0$. More in general,

$$u_h(t+1) = 1 \quad \text{iff} \quad \forall j\, u_{p_j}(t) = 1.$$

Thus, the behaviour of the net formed by the neurons $p_1, \ldots, p_k$ and h reflects faithfully the behaviour of a rule interpreter applied to rule r. Using this representation of rules as basic building block, one can design a neural production system, organized into five different layers of neurons, capable of carrying out forward search on a knowledge base of facts and production rules. A specific example of such system, for a set of four rules, is presented in fig.2.

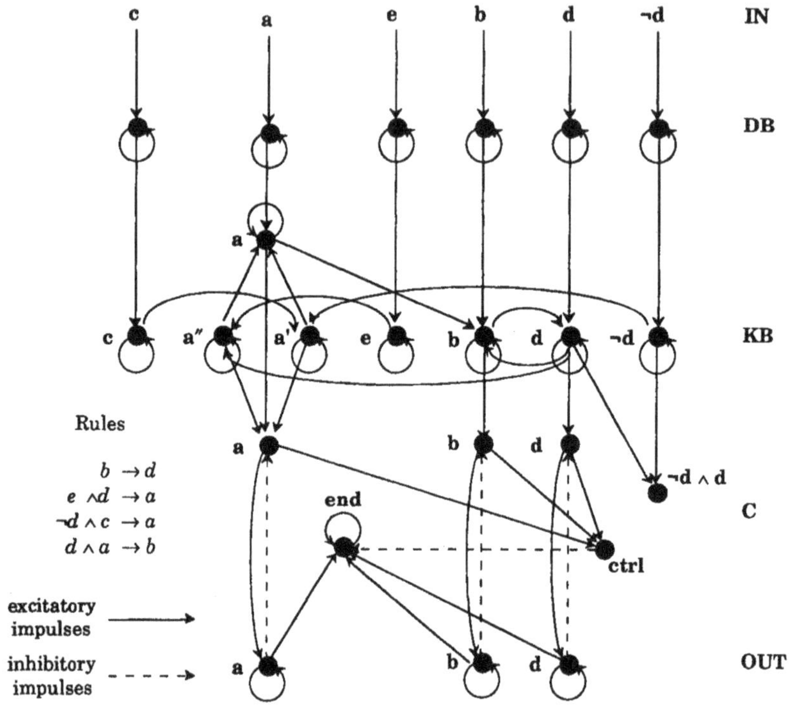

fig.2: *neural forward chaining*

We have seen that many symbolic, rule-based reasoning processes can be emulated, and in several cases efficiently so, by means of neural networks utilizing a localist semantics and fully symbolic neural processing. These non adaptive networks retain the reasoning and justification capacities of traditional symbolic systems, while sharing the parallel processing of neural production systems which make use of distributed representations and hidden layers of neurons governed by learning algorithms. They are affected, however, by some central limitations of traditional symbolic systems: brittleness and no increase in performance with experience. Thus, injecting *in a principled way* learning capacities in our networks would be highly desirable.

Which parts of diagnostic reasoning can be *better* emulated exploiting the learning capacities of neural systems? This is the central question underlying the problem of amalgamating in a principled way symbolic neural processing and connectionist methodologies in totally neural diagnostic systems. A similar question is addressed in [Sun 1991], who argues for connectionist models of rule-based reasoning on the basis of commonsense reasoning protocols presented in [Collins and Michalski 1990] that are difficult to deal with in symbolic models. But this sort of commonsense reasoning is hardly ever encountered in diagnostic domains, and thus more specific arguments are needed to make a substantive case for connectionist techniques in diagnostic processes. The expert systems currently under development or revision in our research programme for totally neural diagnostic systems provide interesting clues.

The first diagnostic system we have provided with a totally neural architecture operates in the medical domain of paediatric gastroenterology, duplicating the performance of a traditional symbolic expert system [Burattini *et al.* 1988]. The selection of candidate diagnoses and appropriate queries in this system is not context-sensitive: a given class of initial data will always induce the same plausibility ordering of pre-diagnostic hypotheses and the same priority ordering for hypothesis-driven queries. Although this rigidity is immaterial with respect to the final diagnosis, taking into account contextual information may improve the efficiency of the system. For example, queries concerning infectious diseases should be given a different priority order according to whether, e.g., the system operates in a northern or tropical country.

Neural net learning may be exploited to make this system context-sensitive: a learning network can be used to monitor the input-output relationships of the system's modules in a given context, detecting the regular association of certain initial data and final diagnosis, through pre-diagnosis and hypothesis-driven query; in turn, this information can be used to modify the plausibility ordering of pre-diagnostic hypotheses and the priority ordering for queries as a function of the most commonly encountered pathological situations and diseases in the given context. Neural nets would assume here the role of a *metarule* construction tool in diagnostic, rule-based systems.

A problem with this proposal is that the system would accept a metarule on the basis of the network's response, although the latter may not capture a correct contextual generalization.
This is, in our view, a general problem towards the use of learning by examples in diagnostic systems: how can one ever be sure that an uncontrolled use of training sets leads to correct generalizations?
In view of this difficulty, it seems more sensible to use networks as an *interactive* knowledge acquisition tool: the metarules isolated by means of the network can be added to the system's knowledge base only under the experts' approval. In this mode, neural networks would provide a facility to assist experts on the task of building and augmenting a diagnostic system rather than an automatic knowledge acquisition system.

We would outline that we were able not only or reproducing the deductive reasoning but also the causal one and introducing in our nets the more classical approach to the uncertainty traitments, that is, the bayesian and the certainty factors.

Its likely taht a satisfactory answer to the knowledge acquisition problem (learning) might be given by the application of menmonic equations proposed by Eduardo Caianiello in 1961.

In conclusion I belive that while the neuronic equations enabled us to implement the knowledge and the reasoning performed by human experts, the menmonic equations could provide a partial answer to another fundamental problem of cognitive processes: learning.

Let me conclude by quoting once again Eduardo:
"If ever there was a field (Cybernetics) of research in which the true qualities of a scientist should be displayed, we think this is the one; for a scientist is not a man who knows, but a man who knows that he does not know, and tries to find out, with caution and humility"

INTEGRATIVE MECHANISMS OF VISION

PROCESSING VISUAL INFORMATION IN VERTEBRATE RETINAE

Silvana Vallerga
Istituto di Cibernetica e Biofisica del CNR, Via De Marini, 6 - 16149 Genova, and International Marine Centre, Lungomare Eleonora D'Arborea 22 - 09072 Oristano.

Introduction

The retina is essentially a part of the central nervous system. It is remarkably similar in all vertebrate species. It is composed by a receptor surface apposed onto highly organised neural layers. The photoreceptors lie in the innermost part, thus the light crosses the whole retina before being absorbed by the visual pigment molecules. Photon catch initiates a cascade that results in a variation of the membrane potential of the receptor cell. The visual signal is processed through interactions in the retinal neural network, up to ganglion cells, whose axons convey to brain the signal coding the visual scene [2, 10, 60, 63].

The retina utilises 'parallel' and 'serial' pathways for information processing. A direct pathway links photoreceptors to ganglion cells, through bipolar cells. Two lateral pathways, photoreceptors-horizontal-bipolar cells and bipolar-amacrine-ganglion cells, mediate processing over a large area. Feedback circuits are active locally between neighbouring cells. Other circuits carry back to the retina information processed in higher centres[17, 19, 21].

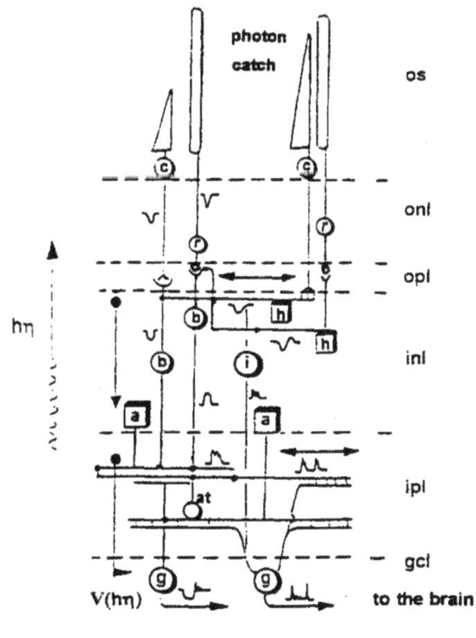

Schematic diagram of the morphological structure and information flow in vertebrate retina. os=outer segments of photoreceptors; onl=outer nuclear layer; c=cone cells; r=rod cells; opl=outer plexiform layer; inl=inner nuclear layer; h=horizontal cells; b=bipolar cells; a=amacrine cells; i=interplexiform cell; ipl=inner plexiform layer; g=ganglion cells; gcl=ganglion cell layer.

1. Processing in the retina of lower and higher vertebrates

The complexity of the visual processing in the retina decreases from lower to higher vertebrates, along with the increase in central processing [40, 60]. Fish retina encodes luminosity, colour, shape, movement and direction of objects in the visual scene. In primate visual system the complex processing resulting in the perception of direction of movement is a task of the cortex [21, 32]. The fish retina is a mini brain performing virtually all visual tasks required for the fish survival. The primate retina is the sensory input to a complex visual system, which is able to codify a tridimensional world and to clearly perceive objects close by and landscapes far away [48].

The functional complexity of the retina of lower vertebrates is supported by a higher morphological variety of cell subtypes. Photoreceptors in lower vertebrates are well discriminated both morphologically and physiologically, whilst in primate retina the distinction between rod and cone cells require careful morphological examination. Neural cells have a large morphological diversity in fish than in mammalian. In the cat retina there are three morphological subtypes of amacrine cells differing for dendritic architecture. In fish retina there are five different classes of dendritic architecture in amacrine cells[57]. Birds that are among the vertebrates with most acute vision have the largest variety of subtypes of retinal neurones [12, 57].

2. Outer retina

Luminosity, colour and position in space of each element of the visual scene are encoded at the level of the outer retina, through photoreceptor cells and second order neurones. Retinal photoreceptors contain visual pigment molecules that are composed of an opsin protein coupled with a light sensitive chromophore. The capture of photons initiates a cascade of events leading to a decrease of the sodium conductance in the cytoplasmic membrane. The photoreceptor cell hyperpolarises and this potential change causes the release of neurotransmitters at the synapses with second order neurones. As a consequence the membrane potential changes also in those neurones. Neurotransmitters are released and visual information in transferred to postsynaptic neurones.

Communication occurs between adjacent cells, thus neurones do not need real axons to transfer and regenerate their responses. Thus, in outer retina, cell responses to visual stimuli are graded with the light intensity. The polarity of the response depends upon the wavelength, and, in the case of bipolar cells, also on the position of the visual stimulus on the retinal field[1, 3, 8, 9].

2.1 Photoreceptors

Two types of photoreceptors are active in different photic conditions. Rods mediate dim-light vision, cones function in bright light and are responsible for colour vision. The sensitivity of rods is so high that the absorption of a single photon can elicit a photoresponse. In bright light rod responses saturate. In these conditions, photopic vision, are the cones to be active. Cones were the first photoreceptors to develop during phylogeny [60]. Rods were produced by the transmutation of cones and this process is occasionally repeated whenever needed by habitat changes [21, 63].

Photoreceptors are differentiated morphologically on the basis of the shape of their outer segments. They can be conical, with comb-like foldings of the photosensitive membrane, or cylindrical, containing stacks of photosensitive membrane discs. The two types of photoreceptors differ also for the shape and location of their synaptic terminals. Rod spherules end in the sclerad side of the outer plexiform layer (OPL). Cone pedicles terminate on the vitread side of the OPL. Thus postsynaptic cells can contact specifically and separately rods and cones extending their processes at different levels [50].

Most species of vertebrates have multiple types of cones, usually three, that absorb maximally in the red, green and blue region of the spectrum, providing the structural basis for colour vision. Some species, as birds and reptiles, have a more complex situation with coloured oil droplets filtering the light before it reaches the visual pigments. Thus, by combining different visual pigments with different oil droplets, such retinae have up to six spectral classes of cone receptors [5].

Photoreceptor cells respond only to light impinging on an area (receptive field) slightly larger than their physical size [3]. The interactions between photoreceptors are limited, thus, in general the grain, or acuity, of a retina is higher for smaller photoreceptor size, as is the case in mammals.

2.2 Horizontal cells

Horizontal cells are connected to photoreceptors and bipolar cells. They are electrically coupled to one another to form a syncitium through which they mediate the lateral processing of information. In several species the input from cones and rods is segregated to specific types of horizontal cells. They feedback to cones for chromatic signalling and temporal sensitivity and provide their postsynaptic neurones, bipolar cells, with surround information [9, 27, 31, 44].

Most vertebrates have at least two subtypes of horizontal cell. A cell with a short axon-like process, running for several hundreds microns in the inner nuclear layer (INL) and axonless cells. In the cat the dendrites of the short axon cell contact cone pedicles, whilst the axon terminal connects to rod spherules[21]. In species with a single cell type, as the salamander, cell body and axon terminals are joined with a thin fibre. In dim-light the two parts of the cell are isolated and they act as separate functional subunits, in bright light they are isopotential and behave like a single cell [31].

Horizontal cells in teleost fish retina are particularly large, over 20μm. They form usually three layers along the distal margin of INL. Each layer contain a specific morphological and functional subtype contacting specifically different photoreceptor families[10, 16, 17].

Horizontal cells are divided into two main physiological types. Luminosity-cells respond to the intensity of the light stimulus with hyperpolarisations, irrespective of the wavelength. Chromaticity units produce multiphasic responses according to the colour of the light stimulus [21].

Recently, horizontal cells have been proven to be an interesting model to study functional plasticity in the retina. It has been found that external horizontal cells, which are intimately involved in colour processing, go through major structural changes, extending and retracting finger-like processes called 'spinules', according to the state of light adaptation. The formation of spinules occurs within minutes and is apparently determined by the gaseous transmitter nitric oxide [16, 62].

2.3 Bipolar cells

Bipolar cells distinguish between signals impinging on the centre or on the periphery of their receptive field, producing responses of opposite polarity. This antagonist organisation of their receptive field is the basis for shape discrimination [26]. Bipolar cells receive their input from photoreceptors and horizontal cells. They signal directly to ganglion cells and, through feedback circuits, to amacrine cells. In the primate retina midget bipolar cells contribute to the high acuity of the central region mediating the contact between a single cone terminal and a ganglion cell[4, 29].

There are different morphological subtypes of bipolar cells, that contact preferentially or exclusively different types of photoreceptors. The dendritic field of rod contacting bipolar cells is considerably larger that the dendritic field of cone bipolar cells. As a consequence the rod-pathway active in dim light, is less acute, but more sensitive, than the cone pathway active in bright light.

Bipolar cells establish also separate inhibitory and excitatory pathways for visual information within the inner plexiform layer (IPL), where their terminals contact amacrine and ganglion cells processes. Famiglietti and co-workers [22, 23] demonstrated originally in the carp retina that the IPL could be considered as made up of two sublaminae, with the distal half (sublamina *a*) subserving centre-hyperpolarising responses and the proximal half (sublamina *b*) the centre-depolarising responses. The interaction between these two pathways provide the basis for direction selectivity of third order neurones [41, 45, 53, 54].

3. Inner retina

Functional processing as well as morphological variety for a given cell types increases from the outer to the inner retina. The codification of the visual signal is no longer only in amplitude, but also in latency and frequency. Spikes and wavelets are part of the responses of third order neurones [11, 13, 22, 25]. A structural dominance of the inner retina over the outer retina indicates a high degree of processing with contrast discrimination dominating over colour vision.

3.1 Amacrine cells

The amacrine cells, whose name means axonless[10], are the first neurone in the retinal network to exhibit spiking responses. The input to amacrine cells come from bipolar cells, ganglion cells and from the centrifugal fibres, that feedback to the retina information coming from higher centres. Amacrine cells respond to intensity, colour and movement of the visual stimulus [11, 13, 26, 53, 54].

These cells contain most, if not all, the neurotransmitter/modulator substances found in the rest of the central nervous system[7, 38, 46, 51], thereby implying that they play a widely important role in synaptic integration of visual information. In fact, the greater the proportion, in relation to bipolar cells, of amacrine cells input to ganglion cells, the more sophisticated is thought to be the visual processing capability of the latter[17, 21].

Amacrine cells have a large morphological heterogeneity [29, 43, 52, 56, 57]. Their dendritic field can be small (50 μm) and diffuse in the whole IPL or large (over 800 μm) and narrowly stratified in a single sublayer. Large monostratified cells may receive different inputs on the periphery and on the centre of their dendritic field. Thus, a single cell may act as a set of separate subunits that process locally the visual information. The function of the cell body is to provide the metabolic support for the processing machinery [12, 55].

A novel type of dendritic arrangement is found in some amacrine cells subtypes, in addition to the usual dendritic field. A fine fibre departs from a dendritic tip, branches several times covering a distance of several millimeters from the cell body. The resulting dendritic field is asymmetric, like a fan. These 'distant dendrites' are very fine (< 1 μm), with constant calibre. The function of distant dendritic field is still unknown[58].

Amacrine cells responses are classified into three main groups: sustained depolarising and sustained hyperpolarising responses, recorded in cells with dendrites monostratified in the IPL, and transient responses recorded in multistratified or diffuse cells[10, 23, 25, 26, 43, 52]. The synaptic bases of the different light-evoked amacrine cell response patterns have been elucidated in the carp retina [30, 53, 54]. Sustained responses are thought to be generated by excitatory inputs from ON-centre and OFF-centre bipolar cells, respectively [34, 47, 53]. Transient amacrine cell receive inputs from both types of bipolar cells [30].

3.2 Interplexiform cells

Interplexiform cells, because of their morphology, have been considered for some time as a subtype of amacrine cells. Now, they have been established as an important constituent neurone of the vertebrate retina [14, 24]. Unlike amacrine cells they establish processes in both plexiform layers, providing a further direct pathway between photoreceptors and ganglion cells. Interplexiform cells are mainly a centrifugal type of neurone carrying information from the inner toward the outer plexiform layer [13, 18, 37].

They appear to have a diverse role in retinal functioning, including regulation of electrotonic coupling of horizontal cells [52], modulation of the cone to horizontal cell transmission [20, 27], as well as modulation of the horizontal cell to cone negative feed-back [11]. In spite of the existing impressive body of information on neurochemistry, synaptic organisation and functional role of interplexiform cells, their light-evoked activity is hardly known [14]. The main reason for this is probably the low population of these cells amongst units of the inner nuclear layer (<<1% of the amacrine cell population in carp [46]).

3.3 Ganglion cells

Ganglion cells are the output neurone of the retina. They communicate with bipolar cells, amacrine cells and other ganglion cells. In the primate retina it has also been described a type of ganglion cell, called biplexiform, contacting directly photoreceptors [39]. Ganglion cells code luminosity, colour, edges and movement of objects in the visual scene [2, 32, 40]. The information, collected by their dendrites, is integrated in their cell body and transferred to brain through their long (true) axons. Ganglion cells, as amacrine cells, show a large morphological diversity, especially in the case of cold-blooded vertebrates where they perform much of the visual information processing [17, 23, 28, 33, 40, 43, 47, 59].

The responses of ganglion cells to light stimuli are the most varied in the retina [61]. They can be depolarisation or hyperpolarisation sustained with the light stimulus, transient or bursts of spikes at the onset and offset of light. As for bipolar cells, they have an opponent organisation of their receptive field. Centre-hyperpolarising units have dendrites layered in sclerad half of the IPL, whilst dendrites of centre-depolarising cells are normally found in the vitread part. Transient and spiking cells are either multistratified or diffuse in the IPL. However, these patterns are not mutually exclusive [29, 59]. This could be due to difficulty in generally defining 'sustained' and 'transient' responses of ganglion cells. The character of the response depends on the properties of the centre of the receptive field. Cells with different dendritic architectures might have receptive field centres of different sizes. If so, according to the size of the light stimulus, in different morphological cell types might be elicited a central response or a central response mixed with contribution of the periphery, resulting in different waveforms.

4. Adaptation and plasticity in the retina.

The vertebrate visual systems, retinae included, are the result of a dynamic process of adaptation to the visual tasks to be performed. This process has led to a number of diverse systems that respond optimally to different visual tasks. Adaptation is an in built mechanisms and can be observed, not only during phylogeny and ontogeny, but also during the life course of individuals as they are exposed to environmental changes [35, 36, 42].

Plasticity is a further, much faster, adaptive mechanism, that occurs in the retina following prolonged exposure to dark or light. This mechanism has been studied with considerable intensity in neural systems, as hippocampus and cerebellum, but little is known about the role of plasticity in retinal functional organisation.

There are sound indications that retinal synaptic connectivity is plastic, with subcircuits modifying their functionality during normal course of light-induced activity. Dopamine has been shown to be involved as a modulator in this process [20, 27]. However, recent findings suggest that light adaptive effects in the retina can occur independently of dopamine and that nitric oxide could be a novel modulator controlling synaptic plasticity in the outer retina [15, 16]. The relationship between the two signalling pathways is still unknown.

Nitric oxide is thought to induce in outer horizontal cells the formation of spinules, dendritic structures normally seen during light adaptation. Formation of horizontal cell spinules is one of the fastest, reversible, morphological change known to occur at central synapses, the time scale of formation plus degradation being only a few tens of minutes [62]. It has been shown that adaptational signals spread in the retina. Thus, an adapting light falling at a given point on the retina can affect signalling at a distant point thus raising the possibility of a diffusible messenger of adaptation, nitric oxide is the putative candidate [15].

5. Conclusion.

Studies on the vertebrate retina have potential importance in several different areas of science. They are of course very important fr vision since phototransduction and initial visual processing occur in the retina. For other sensory systems, since we are now learning that the molecular bases of the process of sensory transduction are similar in vision and olfaction. Neurobiology, generally, since the retina is part of the central nervous system. Artificial vision, since the study of retinal organisation could provide indications for the design of new architectures. Neural computers, since the retina has well-defined input and output, relationships, and utilises 'parallel' and 'serial' pathways for information processing. Furthermore, the interest use-dependent plasticity in the horizontal cells is a process analogous to 'learning'.

References

[1] S .Archer, M.B.A Djamgoz and S. Vallerga, Eds. (1995) *The Neurobiology of Vertebrate Outer Retina.* Chapman & Hall, London.

[2] H.B. Barlow (1953) Summation and inhibition in the frog's retina. *J. Physiol.* 119, 69-88.

[3] D.A. Baylor, M.G.F. Fuortes and P. M. O'Bryan (1971) Receptive field of cones in the retina of the turtles. *J. Physiol.* 214, 265-294.

[4] D.Beauchamp and N. W. Daw (1972) Rod and cone input to single goldfish optic nerve fibres. *Vision Res.* 12, 1201-1212.

[5] J.K. Bowmaker and A. Knowles.(1977) The visual pigments and oil droplets of the chicken retina, *Vision Res.* 17, 755-764.

[6] B.B Boycott and H. Wassle (1974) The morphological types of ganglion cells of the domestic cat's retina. *J. Physiol.* 240, 379-419.

[7] N. Brecha (1983) Retinal neurotransmitters: Histochemical and biochemical studies. In : *Chemical Neuroanatomy.* P.C. Emson, Ed. New York, Raven Press, pp. 85-129

[8] D.A. Burkhardt (1993) Synaptic feedback, depolarization, and color opponency in cone photoreceptors. *Visual Neurosci.* 10, 981-989.

[9] A.L. Byzov, Golubtzov, K.V. and Trifonov, Y.A. (1977) The model of mechanism of feedback between horizontal cells and photoreceptors in the vertebrate retina. In: *Vertebrate Photoreception.* H.B. Barlow and P. Fatt, Eds. New York, Academic Press, pp. 265-274.

[10] S.R. y Cajal (1972). *The Structure of the Retina.* Translation by S. A. Thorpe and M. Glickstein, Thomas Springfield, Ill.

[11] R.Y. Chan and K.-I. Naka (1976) The amacrine cell. *Vision Res.* 16 1119-1129.

[12] M.B.A. Djamgoz and Vallerga, S. (1989) Structure-function correlation: amacrine cells of fish and amphibian retinae. In: *Neurobiology of the inner retina*, NATO ASI Series, Vol H31 R. Weiler and N.N. Osborne, Eds., Springer-Verlag, Berlin, pp 195-208.

[13] M.B.A Djamgoz, L. Spadavecchia, C. Usai and S. Vallerga (1990) Variability of light-evoked response pattern and morphological characterisation of amacrine cells in goldfish retina. *J. Comp. Neurol.* 301,171-190

[14] M.B.A Djamgoz and S. Vallerga (1991) An interplexiform cell in the goldfish retina: light-evoked response pattern and intracellular staining with horseradish peroxidase. *Cell and Tissue Res.* 2641,111-116

[15] M.B.A Djamgoz, J.R. Cunningham, P.H Hutson, F. Murray and M.J. Neal, (1995) Nitric oxide donors inhibit release of endogenous dopamine in the rabbit retina. *Br. J. Pharmacol.* In the press

[16] M.B.A Djamgoz, H.-J. Wagner and P. Witkovsky, (1995) Photoreceptor-horizontal cell connectivity, synaptic transmission and neuromodulation. In *Neurobiology and Clinical Aspects of the Vertebrate Outer Retina.* Eds. M.B.A. Djamgoz, S.N. Archer and S. Vallerga, Chapman & Hall, London, 155-193.

[17] R.H. Douglas and M. B. A. Djamgoz, Eds. (1990) *The Visual System of Fish.* Chapman & Hall; London.

[18] J.E. Dowling and B. Ehinger (1975) Synaptic organization of the amine-containing interplexiform cells of the goldfish and Cebus monkey retinas. *Science* 188, 270-273.

[19] J.E. Dowling (1979) Information processing by local circuits: The vertebrate retina as a model system. In: *The Neuroscience Fourth Study Program,* F.O. Schmitt and F.G. Worden, Eds, Cambridge, MIT Press, pp. 163-181.

[20] J.E. Dowling (1986) Dopamine: a retinal neuromodulator? *Trends Neurosci.* 9, 236-240.

[21] J.E. Dowling (1987) *The Retina. An Approachable Part of the Brain.* Cambridge, The Belknap Press.

[22] E.V. Famiglietti and H. Kolb (1976) Structural basis for "ON" and "OFF"-center responses in retinal ganglion cells. *Science* 194, 193-195.
[23] E.V. Famiglietti, A. Kaneko and M. Tachibana (1977). Neuronal architecture of 'on' and 'off' pathways to ganglion cells in carp retina. *Science* 198, 1267-1269.
[24] M. Kallionatis and R.E Marc (1990). Interplexiform cells in the goldfish retina: a combined Golgi and autoradiographic study. *J. Comp. Neurol.* 297, 340-358.
[25] A. Kaneko and H. Hashimoto (1969) Electrophysiological study of single neurones in the inner nuclear layer of the carp retina. *Vision Res.* 9, 37-55.
[26] A. Kaneko (1973) Receptive field organization of bipolar and amacrine cells in the goldfish retina. *J. Physiol.* 235, 133-153.
[27] A.G. Knapp and J.E. Dowling (1987) Dopamine enhances excitatory amino acid-gated conductances in retinal horizontal cells. *Nature* 325, 437-439.
[28] J.-H. Kock, E. Mecke, O.Yu. Orlov, T. Reuter, R. Vaisanen and J.E.C. Wallgren, (1989) Ganglion cells in the frog retina, discriminant analysis of histological classes. *Vision Res.* 29, 1-18.
[29] H. Kolb (1982) The morphology of bipolar cells, amacrine cells and ganglion cells in the retina of the turtle, Pseudemys scripta elegans. *Phil. Trans. R. Soc* B 298, 355-393.
[30] T. Kujiraoka, T. Saito and J.-I. Toyoda (1988) Analysis of synaptic inputs to ON-OFF amacrine cells of the carp retina. *J. Gen. Physiol.* 92, 475-487
[31] A. Lasanky and S. Vallerga (1975) Horizontal cell response in the retina of the larval tiger salamander. *J. Physiol.* 251, 145-165
[32] J.Y. Lettvin, W.S. Maturana, W.S. McCulloch and W.H. Pitts (1959). What the frog's eye tells the frog's brain. *Brain. Proc. Inst. Radio Eng.* 47, 1940-1951
[33] M.W. Levine and J. M.Shefner (1979) X-like and not X-like cells in the goldfish retina. *Vision Res.* 19, 95-98.
[34] J.C. Low, M Yamada and M.B.A. Djamgoz (1989) Voltage clamp study of amacrine cells in the carp retina. *In: Neurobiology of the Inner Retina,* NATO ASI Series, Vol. H31, R. Weiler and N.N. Osborne, Eds., Springer-Verlag Berlin Heidelberg.
[35] J.N. Lythgoe (1977) *The Ecology of Vision,* Academic Press, London.
[36] J.N. Lythgoe (1979) The adaptation of visual pigments to the photic environment. In: *Handbook of Sensory Physiology. Vol. VII/1.* Springer-Verlag, Berlin, Heidelberg, New York.
[37] R.E. Marc and W.-L.S. Liu (1984) Horizontal cell synapses onto glycine-accumulating interplexiform cells. *Nature* 312, 266-269.
[38] R.E. Marc (1986) Neurochemical stratification in the inner plexiform layer of the vertebrate retina. *Vision Res.* 26, 223-238.
[39] A.P. Mariani, (1982) Biplexiform cells: Ganglion cells of the primate retina that contact photoreceptors. *Science* 21 , 1134-1136.
[40] H.R. Maturana, J. Y. Lettvin, W. S. McCulloch. and W. H. Pitts (1960) Anatomy and physiology of vision in the frog (Rana pipiens). *J Gen. Physiol.* 43, 129-175.
[41] J.S. McReynolds and P. D.Lukasievicz (1989). Integration of synaptic input from ON and OFF pathways in mudpuppy retinal ganglion cells. In: *Neurobiology of the Inner Retina.* NATO ASI Series, Eds. R. Weiler and N. N. Osborne, Springer-Verlag (Berlin). 208-220.
[42] W.R.A. Munz and W.N. McFarland. (1977) Evolutionary adaptation of fishes to photic environment. In: *The Visual System of Vertebrates.* Handbook of Sensory Physiology, VII/5, Ed. F.Crescitelli, Springer-Verlag, Berlin, Heidelberg, New York, 194-274.
[43] M. Murakami and Y. Shimoda (1977) Identification of amacrine and ganglion cells in the carp retina. *J. Physiol.* 26, 801-818.
[44] K.-I. Naka and W.A.H. Rushton (1966) S-potentials from colour units in the retina of fish (Cyprinidae). *J. Physiol.* 185, 587-599.

[45] K.-I. Naka (1980) A class of catfish amacrine cells respond preferentially to objects which move vertically. *Vision Res.* 20, 961-965.

[46] K. Negishi, T. Teranishi and S. Kato (1981) Density of retinal dopaminergic cells and indoleamine-accumulating cells in different-sized carp. *Acta Histochem. Cytochem,* 596-606.

[47] H.M. Sakai and K.-I. Naka (1987) Signal transmission in the catfish retina. IV. Transmission to ganglion cells. *J. Neurophysiol.* 58, 1307-1328.

[48] S. Sinclair (1985) *How Animals See. Other Visions of Our World* (Crom Helm, London & Sidney.

[49] H. Spekrejise, H.G. Wagner and M.L. Wolbarsht (1972) Spectral and spatial coding of ganglion cell responses in goldfish retina. *J. Neurophysiol.* 35, 73-86

[50] W.K. Stell (1972) The morphological organization of the vertebrate retina. In: *Handbook of Sensory Physiology VII/2.* Ed. M.G.F. Fuortes, Berlin, Springer, pp. 111-214.

[51] W.K. Stell (1985) Putative peptide transmitters, amacrine cell diversity and function in the inner plexiform layer. In: *Neurocircuitry of the Retina, A Cajal Memorial.* Eds. A.Gallego and P.Gouras, New York, Elsevier, pp. 171-187.

[52] T. Teranishi, K. Negishi and S. Kato (1987) Functional and morphological correlates of amacrine cells in carp retina. *Neurosci.* 20, 935-950.

[53] J.-I. Toyoda, H. Hashimoto and K. Ohtsu (1973) Bipolar-amacrine transmission in the carp retina. *Vision Res.* 13, 295-307.

[54] J.-I. Toyoda and M. Fujimoto (1984) Application of transretinal current stimulation for the study of bipolar-amacrine transmission. *J. Gen. Physiol.* 84, 915-925

[55] S. Vallerga. (1981) Physiological and morphological identification of amacrine cells in the retina of the larval tiger salamander. *Vision Res.* 21,1307-1317.

[56] S. Vallerga and S. Deplano (1984) Differential extent and layering of amacrine cell dendrites in the retina of a sparid fish. *Proc. R. Soc. (Lond.)* B. 221, 465-477.

[57] S. Vallerga (1986) Functional properties of amacrine cells architecture. *Proc. British Photobiology Soc.* 40-44.

[58] S. Vallerga, M.B.A. Djamgoz and C. Usai (1989) Distal (d-type) dendrites of retinal amacrine cells. *Invest. Ophthal. Visual Sci.* 30, 3, 67.

[59] S. Vallerga and M.B.A. Djamgoz (1991) Ganglion cells in the goldfish retina: correlation of light-evoked response and morphology. *Vision Res.* 31, 3,487-498.

[60] S. Vallerga (1994) The phylogenetic evolution of the visual system. In *Human and Machine Vision,* Ed V. Cantoni, Plenum Press, New York & London, 1-12.

[61] H.G. Wagner, E.F. MacNichol, Jr and M.L. Wolbarsht (1960) The response properties of single ganglion cells in the goldfish retina. *J. Gen. Physiol.* 43, 45-62.

[62] H.J. Wagner and M.B.A. Djamgoz (1993) Spinules - a case for retinal synaptic plasticity. *Trends Neurosci.* 16,201-206.

[63] G.L. Walls (1942). *The Vertebrate Eye and Its Adaptive Radiation.* Cranbrook Inst. of Sci. Bull. 19. Bloomfield Hills, Mich.

STRUCTURAL AND FUNCTIONAL ORGANISATION OF THE CEPHALOPOD RETINA

John Patterson
Department of Physiology
Faculty of Basic Medical Sciences
Queen Mary and Westfield College
University of London
LONDON E1 4NS

Cephalopod molluscs, such as octopus and squid, are highly evolved, visual predators possessing large, camera-type eyes which show striking evolutionary convergence with the eyes of vertebrates. Nevertheless, the two types of eye differ greatly in their detailed organisation. Unlike vertebrates, cephalopods have a non-inverted retina, so that incoming light falls directly onto that part of the photoreceptor which is the equivalent of the vertebrate rod or cone outer segment, without passing through overlaying retinal structures, as occurs in vertebrates. Also, in common with most invertebrate eyes, cephalopod photoreceptors depolarise on illumination, while those of vertebrates hyperpolarise. In comparing cephalopod and vertebrate eyes, the most significant differences are found in the arrangement and properties of the photoreceptors and the locations of those areas of neuropil where signal processing of visual information takes place. Each cephalopod eye is connected to the adjacent optic lobe of the brain by a large number of optic nerves composed of glial elements, afferent photoreceptor fibres and the fibres of neurons which are centrifugal to the retina. Some, at least, of these centrifugal fibres transmit efferent information. In terms of signal processing, the cephalopod retina lacks the complex arrangement of inner and outer plexiform layers, so evident in the vertebrate retina, and the initial stages of synaptic integration are divided between a rather simple synaptic plexus in the retina and a complex, layered zone which constitutes a 'deep retina' in the cortex of the optic lobe. For a comparative review of molluscan visual systems, see [1].

The cephalopod visual system

Figure 1 shows the general structural organisation of the retina in octopus, cuttlefish and squids and shows the origins and terminations of centrifugal neurons and photoreceptor axons within the optic lobe of the brain. This diagram summarises the many studies of the cephalopod retina which have used Golgi impregnation, electron microscopy (EM), neural degeneration and cobalt impregnation techniques. The retina can be sub-divided into several, more or less discrete, layers. The distal portions of the photoreceptors, containing photopigment-bearing microvilli, together with very thin extensions of retinal supporting cells, form the most distal layer of the retina, closest to the lens of the eye. The distal tips of the photoreceptors are separated from the fluid of the posterior chamber of the eye by a well defined but acellular limiting membrane. At the proximal end of the photoreceptor outer segments is a layer of screening pigment. The pigment granules undergo migrations with light and dark adaptation. The pigment is largely confined to the layer in the dark-adapted eye, but with light adaptation much of the pigment migrates to the distal tips of the photoreceptors [2, 3].

Proximal to the pigmented layer lies a thin zone formed from a single layer of retinal supporting cell bodies. Beneath this is a much thicker zone formed from the inner segments of the photoreceptors. This contains photoreceptor nuclei arranged as several layers. At the proximal edge of this layer an axon arises from each photoreceptor. These axons pass in the retina for a short distance, giving rise to short collaterals, before the axons turn into the tangential plane, converge into groups and pierce the sclera of the eye to become one of numerous, optic nerves.

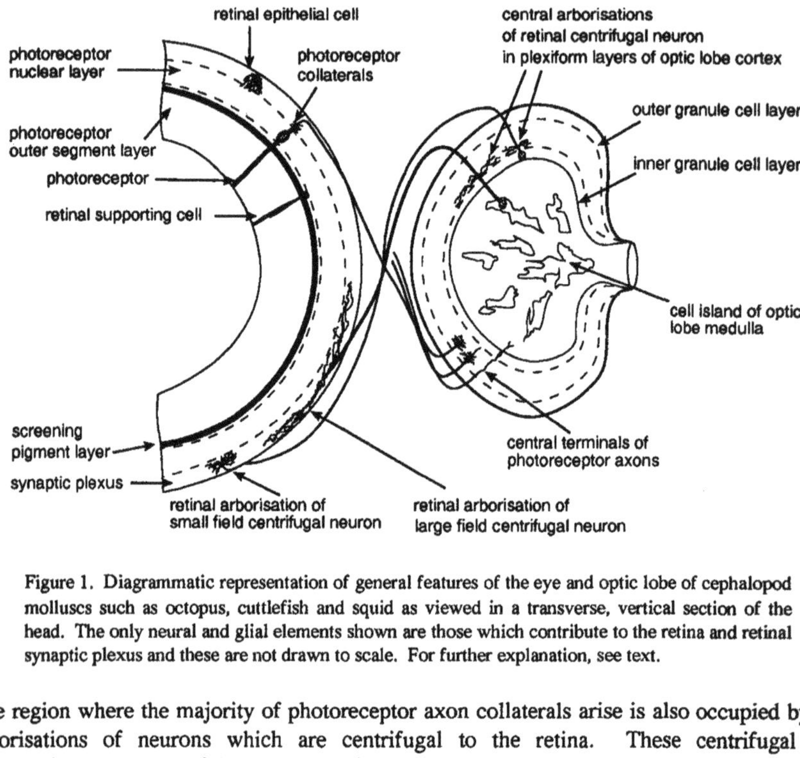

Figure 1. Diagrammatic representation of general features of the eye and optic lobe of cephalopod molluscs such as octopus, cuttlefish and squid as viewed in a transverse, vertical section of the head. The only neural and glial elements shown are those which contribute to the retina and retinal synaptic plexus and these are not drawn to scale. For further explanation, see text.

The region where the majority of photoreceptor axon collaterals arise is also occupied by the arborisations of neurons which are centrifugal to the retina. These centrifugal cell arborisations seem to fall into two size classes, large and small, and together with photoreceptor collaterals they form a complex area of neuropil known as the plexiform zone or retinal synaptic plexus. Small epithelial cells, with cell bodies at the proximal side of the synaptic plexus, send many fine branches into the plexus and in between the photoreceptor inner segments. These have the appearance of glial cells [4].

The bundles of photoreceptor and centrifugal cell axons which form the optic nerves show complex branching, described in some detail by Young [4]. Perhaps the most obvious feature of the arrangement of the optic nerves is the dorso-ventral chiasma between retina and optic lobe. The optic lobe is subdivided into an outer cortex and inner medulla. The cortex, where optic nerves enter the lobe, has outer and inner granule cell layers. These cell layers are formed by the cell bodies of amacrine, centripetal and centrifugal neurons which contribute arborisations to a multi-layered plexiform zone of neuropil lying between the two granule cell layers. The plexiform zone forms a deep retina where further visual processing occurs. The medulla contains neuropil and tracts of fibres together with islands of neuron somata [4]. Photoreceptor axons enter the optic lobe and each forms one of three different forms of termination when studied in Golgi impregnated material. One type terminates as a swelling in the outer radial layer of the cortical plexiform zone, a second type has dendritic extensions from the swelling which extend deeper into the plexiform zone and the third, less common, form lacks a terminal swelling but gives rise to branches in the plexiform zone and continues into the inner granule cell layer, with further branching [4]. Cobalt impregnation of optic nerves [5] has revealed two separate locations for the cell bodies of neurons with axons centrifugal to the retina. Centrifugal neuron cell bodies are found in both the inner granule cell layer and in medullary islands. These cells give rise to axons which pass back through the plexiform layer, where they arborise in a tangential direction, and into the optic nerves to terminate in the synaptic plexus region of the retina. Although it appears that two size classes of centrifugal arborisation are present in the retina of the octopus [6] and in cuttlefish (see Figure 3), it is not possible to say whether these classes correspond to the two locations for centrifugal cell somata in the optic lobe.

Photoreceptor organisation

Photoreceptor outer segments are rectangular in cross section and EM studies show that each outer segment contains stacks of microvilli, each about 1 µm in length, which form rhabdomeres on two opposing sides of each cell [7]. The two areas of microvilli in each cell are separated by a thin region of cytoplasm containing vacuoles and screening pigment granules. The cross sectional width of the outer segment, between the microvillar tips on each side, is relatively constant from cell to cell, but the length between avillous borders of the cell varies greatly in different cells. Figure 2A shows the arrangement of photoreceptors in a resin embedded tangential section of the retina of the small squid, *Alloteuthis subulata*. The general arrangement of the photoreceptors is shown diagrammatically in Figure 2B, where arrows indicate the distance referred to as the cross-sectional length. Since the width between the villous borders of the cell is relatively constant, the other cross-sectional length is the main determinant of the area of the photoreceptor as seen in cross section. If a large number of these cross-sectional lengths are measured from a small region of a single section of retina, the data plotted as a frequency distribution histogram produce results such as those seen in Figure 2C (see also [8]). These show a predominance of smaller area cells over larger ones and there is frequently the suggestion of three or four overlapping sub-groups of cells with normal distributions of cell cross-sectional areas. Cohen [9] describes a similar distribution of photoreceptor cross sectional areas in the squid, *Loligo pealei*, although the sub-grouping is less evident. This sub-grouping is difficult to confirm by pooling of data from different regions of retina because size varies across the retina [10] and may be expected to vary from animal to animal. This results in smearing of the sub-groups, but still shows the predominance of smaller cells.

This heterogeneity of cross sectional-areas provokes the questions, what is the functional significance of this arrangement and how may it arise? In terms of function, the variation in cross sectional area of photoreceptors presumably affects both the capture of photons and the acuity of the photoreceptor. Larger cells might be expected to have lower absolute thresholds, since photons are more likely to interact with photopigment. Such cells may also have wider receptive fields than smaller cells. It would be interesting to know whether different sub-populations of receptor exist in the retina, with differing dynamic ranges and acuities, which perhaps parallel the balance between acuity and sensitivity found in the vertebrate retina. As to how the size variations arise, it is tempting to speculate that the largest cells are the oldest and the smallest are those which arise later in development and growth. Yamamoto [11] has shown that towards the end of embryonic development of *Sepiella japonica,* some variation in the cross-sectional area of photoreceptors is already present. Finally, a repeat pattern for the arrangement of different sized photoreceptors is not evident in the *Alloteuthis* retina.

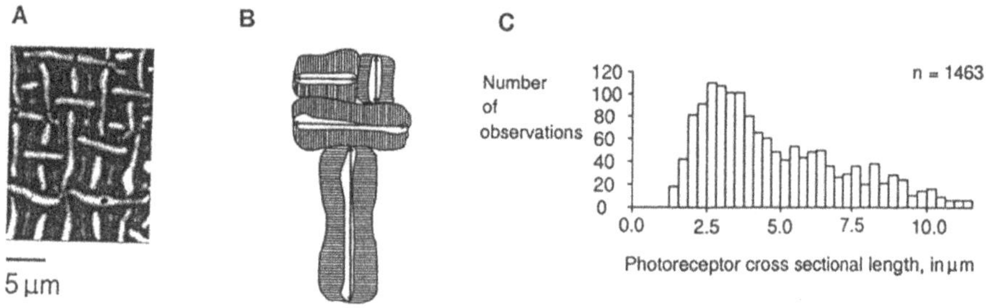

Figure 2. Photoreceptors in the retina of *Alloteuthis subulata.*
A. A light micrograph of a 1.0 µm tangential section of the retina from resin embedded material fixed in 2% paraformaldehyde + 2% glutaraldehyde, buffered to pH 7.4 with sodium cacodylate and with sodium chloride, magnesium chloride and dextran added to the fixative. (Full details of the method are given in [8]).
B. Diagram to show arrangement of photoreceptor outer segements as seen in tangential section. Note that microvilli of outer segments are orientated in one of two orthogonal planes. Double-headed arrows show the cross-sectional distance between avillous borders of the cell used to estimate cross-sectional area.
C. Frequency distribution histogram of the cross sectional distances described in B. For discussion, see text and [8].

Another significant feature of the cephalopod retina is the arrangement of the microvilli in different cells into one of two orthagonal planes in the horizontal (tangential) plane (Figure 2B). The microvillar arrays forming the rhabdomeres of arthropod compound eye photoreceptors have long been known to produce sensitivity to the plane of polarisation of incoming light, with the photoreceptor response varying in magnitude as the plane of polarisation is altered. Similar sensitivity is found in the frequency of action potentials recorded from the optic nerves of squid [12, 13]. The presence of two orthagonal planes of orientation of the microvilli presumably provides the nervous system with difference signals which assist further analysis of polarised light distribution in the visual environment. Saidel et al [13] suggest that analysis of polarised light by the cephalopod visual system may serve to enhance object contrast underwater and it has also been suggested that such a system may serve to defeat the cryptic countershading of many marine prey organisms, in which the dorsal surface is darker than the ventral. The countershading balances the relative intensities of downwelling and upwelling light in seawater. Since daylight descending in seawater retains a greater degree of polarisation than the weaker, upwelling and back-scattered illumination, the analysis of polarised light may serve to unmask countershaded animals.

Beneath the basal membrane which lies proximal to the layer of supporting cell nuclei, lie the inner or proximal segments of the photoreceptors, where the photoreceptor nuclei are located. The most distinctive features of this layer when viewed with the EM are the presence of large lamellated, myeloid bodies [14, 15] and the presence of tight junctions between receptors and of desmosome-like junctions between receptor and supporting cell bodies at the transition from outer to inner segment regions [14]. The layer of tight junctions may produce an electrical or ionic barrier in the retina and may be responsible for the reversal of the polarity of the electroretinogram observed by Tasaki *et al*[16]. It is also worth noting that the ionic composition of the watery, "vitreous" fluid in cephalopods differs from that of cephalopod plasma [17]. Proximally, the cell bodies of the photoreceptors produce axons and give rise to short collateral fibres from the cell body and initial region of the axon. In Golgi impregnated *Octopus* retina these collaterals extend for 2-10 μm in the tangential plane, passing laterally for about 1 to 2 cell diameters and contributing to the retinal synaptic plexus [6]. Similarly impregnated photoreceptors of the cuttlefish, *Sepia officinalis*, are shown in Figure 3A (see also [18]. At the EM level, squid [15] and cuttlefish [19] photoreceptor collaterals form gap junction connections with each other, while in octopus, Tonosaki [20] describes these as containing small, clear vesicles. Obviously these collaterals are well suited to mediating lateral interactions between photoreceptors.

Centrifugal cells in the retina

In addition to the axons and collaterals of photoreceptors, the synaptic plexus region in Golgi impregnated *Octopus* retina, is also seen to contain the retinal arborisations of centrifugal neurons with cell bodies in the optic lobe [4, 6, 21]. Figure 3B shows the appearance of these arborisations in similar preparations of *Sepia* retina as seen in radial sections of the retina and Figure 3C their appearance in tangential sections.

Viewed in the tangential plane, these centrifugal cell arborisations seem to fall into two size classes in both *Sepia* (Figure 3C) and *Octopus* [6]. The small-field arborisations, seen in the top part of Figure 3C, have diameters of 50 -150 μm, while those of large field arborisations shown lower in the figure range from 400 to over 900 μm. Similar values are found in *Octopus* and partially impregnated, large field arborisations can be traced for many hundreds of micrometers in some sections. There is a considerable body of evidence, dating from Lund's [22] neural degeneration studies, which suggests that two classes of centrifugal cell arborisation do indeed exist in the retinal synaptic plexus, although the arbitrary nature of Golgi staining and the difficulty of obtaining sufficient numbers of well stained cells make a statistical treatment difficult. Seen in Golgi preparations, the small field arborisations seem to have a more delicate branching pattern and often show beading of the finer processes, while the larger arborisations have a coarser branching pattern. At the EM level, Tonosaki [20] described, in the octopus retina, two forms of synaptic vesicle, small clear vesicles (typical of vertebrate cholinergic neurons) and larger vesicles with an electron dense core (typical of aminergic neurons). Because Gray [23] observed small, clear vesicles in the readily identifiable optic lobe terminals of photoreceptor axons, it seems likely that such vesicles would be found in the receptor collaterals of the retina. Nevertheless, for the squid retina, Cohen [15] commented on the presence of small, clear vesicles in large transverse process of the synaptic plexus, readily distinguishable from photoreceptor collaterals. Cohen [15] also observed transverse elements containing dense-cored vesicles and those containing a mixture of many small, clear vesicles with a few dense-cored vesicles. It seems that within the synaptic plexus region, two or perhaps three different forms of centrifugal neuron may co-exist. As a final point, the two locations for centrifugal cell bodies, in the inner granule cell layer and the medullary islands of the optic lobe [5] could be the source of the small and the large field arborisations, respectively. Some simple calculations for *Octopus* retina [6] have suggested that this could be the case.

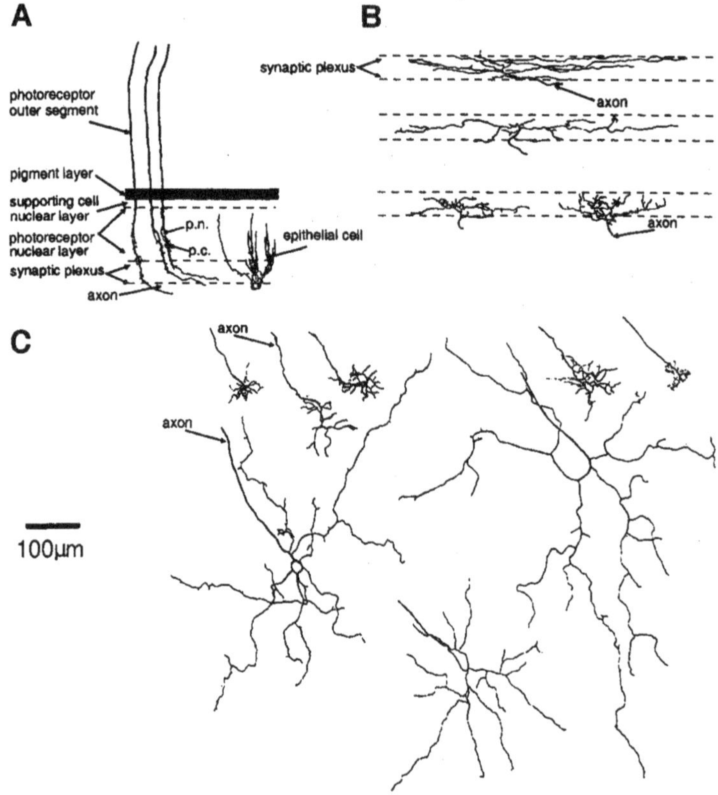

Figure 3. Camera lucida drawings of Golgi impregnated cells in the retina of *Sepia officinalis*. For the Golgi-rapid method used, see [18]

A. A montage of photoreceptors and a retinal epithelial cell as seen in radial sections of retina. p.n., photoreceptor nucleus; p.c., photoreceptor collateral.

B. Retinal arborisations of centrifugal cell arborisations as seen radial sections. Larger and smaller field arborisations are shown, but those with smaller fields may be partially impregnated large cells, rather than the "small field" cells referred to in the text. Such small-field cells are only rarely encountered in radial section. Dotted lines show the approximate boundaries of the retinal synaptic plexus.

C. Montage of retinal arborisations of centrifugal neurons as seen in tangential sections of retina which pass between the sclera and pigment granule layer. Four small-field arborisations are shown at the top of the figure and below these are three large-field arborisations.

Electrophysiology of cephalopod optic nerves

The most interesting features of the cephalopod visual system are the physical separation of the eye and optic lobe and the presence of numerous optic nerves containing both photoreceptor axons (which transmit afferent signals to optic lobe neuropil) and centrifugal axons (some of which may be expected to transmit efferent signals to the retina). The separation of early visual information processing between the retinal synaptic plexus and the plexiform layers of the optic lobe, with both areas connected by fibres contained in easily accessible optic nerves, makes this system particularly favourable for the study of the role played by efferent feedback mechanisms. Each optic nerve contains many bundles of fibres, each bundle with 20 or more axons wrapped in a mesaxon system [24].

Several authors [6, 12, 13, 23, 25, 26, 27, 28, 29, 30] have recorded activity from optic nerves in octopus or squid, mainly using suction electrodes to recorded either massed action potential activity, or in some cases, single units. Patterson and Silver [6] and Patterson [30] used suction electrodes to record activity from optic nerves in isolated eye plus optic lobe preparations, which were super-perfused with an appropriate, cooled, physiological saline. Figure 4 shows responses typical of those recorded from these preparations. Figure 4A shows activity in an *Alloteuthis subulata* optic nerve in response to illumination of the retina with various intensities of white light. The dynamic range of the responses extends for between 3 and 4 logarithmic units of stimulus intensity. High intensities produce a profound inhibition of the response to illumination, after an initial burst of light-evoked activity. This inhibition outlasts the period of illumination and some time after the return to darkness a prolonged afterdischarge, suggesting post-inhibitory rebound, is seen. Comparing the responses at 0.0 and -1.0 Log_{10} light levels shows that the degree of inhibition is intensity dependent. At lower intensities the light-evoked response is more monotonic and graded with intensity. Section of the optic nerve produces little if any alteration to these responses and in fact there is relatively little evidence of efferent activity in these whole nerve preparations of the squid, *Alloteuthis*. Therefore, the responses observed are those of photoreceptor axons and the inhibition and prolonged depolarising afterdischarge seen at high intensities are due to interactions in the retina, possibly mediated by the photoreceptor collaterals. The interactions between photoreceptors are not limited to simple, inhibitory mechanisms however, since Hartline and Lange [27, 28] demonstrated a property of 'background facilitation' in which photoreceptor responses to illumination are enhanced by weak background illumination. Even more complex and time dependent alterations in photoreceptor response dynamics have also been described by these authors, which give an indication of the complexity of the interactions possible within the retinal synaptic plexus.

Figure 4B shows recordings obtained with similar techniques from the optic nerves of *Octopus vulgaris*. These differ from those presented for *Alloteuthis* in two obvious ways. There is no sign of inhibitory and post-inhibitory rebound effects in the responses to higher light intensities and the dynamic range of responses seems to be shifted to higher light intensities, perhaps reflecting the higher ambient light levels experienced in natural life by these animals. Secondly, there is a clearly defined afterdischarge which follows a brief silent period at the end of a period of illumination. When the optic nerve is cut on the optic lobe side of the recording electrode, this afterdischarge is invariably abolished, showing it to be the result of efferent action potentials passing to the retina from the optic lobe. Experiments involving the section of octopus optic nerves have been performed in some detail [6], and recordings from the central stumps of severed optic nerves show efferent action potential discharges both during the period of illumination and, transiently, on the return to darkness. Such efferent activity must, in this case, be driven by photoreceptor input to the optic lobe via intact, adjacent optic nerves, with lateral interactions occurring through centrifugal cell and other cell arborisations within the plexiform zone of the optic lobe.

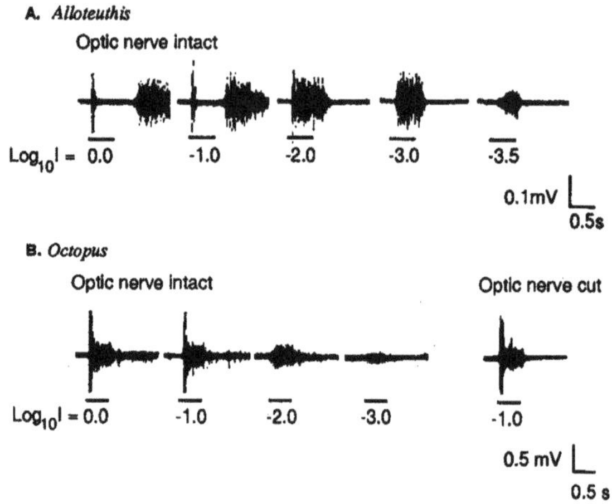

Figure 4. Suction electrode recordings of activity in an optic nerve in isolated, superperfused, eye-optic lobe preparations of: A, the squid, *Alloteuthus subulata*; B, *Octopus vulgaris*. Preparations were stimulated with white light from a quartz halogen light source of maximum intensity (Log_{10} I = 0.0) of 310 mW/cm^2 for the period shown by the black bar beneath each record. The stimuli were presented to local regions of the retina by means of a fibre optic and a small visual perimeter device. The details of the method are given in [30] and details of the superperfusion saline used in octopus experiments are given in [6]. 4A shows an intensity series derived from an intact optic nerve in *Alloteuthis*. The four panels at left of 4B show a similar series obtained from an intact optic nerve in *Octopus,* while the trace at right shows the reponses recorded after the optic nerve was severed between the recording electrode and the optic lobe.

To date the most comprehensive studies of the response characteristics of fibres present in the cephalopod optic lobe, remain those of Hartline and Lange [27, 28]. These authors demonstrated that at least three distinct categories of unit could be characterised, on the basis of their responses and receptive field dimensions, in multi-unit and single unit recordings from squid and octopus optic nerves. These three types are described as, fast-afferent (photoreceptor axon), slow afferent and efferent. Patterson and Silver [6] attempted to relate these classes of unit to what is known of the anatomy of photoreceptors and centrifugal cell arborisations in the octopus retina. These possible relationships are summarised in Table I.

Table I

Retina: Golgi impregnation From Patterson and Silver [8]		**Optic nerve: Physiology** From Lange and Hartline [28]	
Cell type	**Arborisation spread**	**Unit type**	**Receptive field**
photoreceptors	circa 0.25°	fast afferents	11° - 4°
small field centrifugals	0.6° - 1.7°	efferents	circa 30°
large field centrifugals	3.5° - >12°	slow afferents	15° - 180°

The angular dimensions do not compare well, since anatomical measurements are based on collateral or arborisation spread in the retina, while the physiological measurements of receptive fields will also include the acceptance angles of the photoreceptors and any interactions between cells in the retinal synaptic plexus, and also, in the case of efferent fibres, lateral interactions in the optic lobe. At least on the basis of a rank-order of size, the data suggest that efferent fibres may correspond with the small-field centrifugal arborisations in the retina and the slow afferents with some, or all, of the large field centrifugal cells.

Histochemistry of the retina

Biochemical studies of cephalopod retina [31] have shown that this tissue can synthesise acetylcholine and dopamine. Fluorescence studies using a modified glyoxylic acid method [32] clearly show the presence of dopamine-like material in the proximal part of the retinal synaptic plexus of *Octopus, Sepia* and *Alloteuthis*, in the region where centrifugal cells arborise. In *Sepia* and *Alloteuthis*, the dopamine-like fluorescence in concentrated into two layers in the tangential plane, while in *Octopus*, the fluorescence is more uniformly distributed throughout the synaptic plexus. Tansey [33] has shown both 5-hydroxytryptamine- (5-HT) and dopamine-like fluorescence in the optic lobe. Using immunocytochemical localisation techniques, Kito-Yamashita *et al* [34] have demonstrated 5-HT immunoreactivity in the *Octopus* retina, where fluorescence is associated with the cell bodies and processes of retinal epithelial cells. Taken together, these observations suggest that dopamine is the putative neurotransmitter in some, at least, of the centrifugal neurons which pass to the retina, while acetylcholine is likely to be present in the photoreceptors. The application of exogenous dopamine to the octopus retina [35] produced enhancement of the early receptor potential and proximal migration of retinal screening pigment, both characteristic of dark adaptation. These authors also showed that section of the optic nerves markedly reduced the degree of migration of screening pigment from the light to the dark adapted position, implicating efferent feedback to the retina in the production of the dark adapted state. Earlier studies [36], had already shown that the microinjection of dopamine into the synaptic plexus region of the retina enhanced the electroretinogram (ERG) in a dose-dependent manner, but that similar injections into the outer segment layer of the retina had no effect. Suzuki and Tasaki [36] also studied the effects of electrical stimulation of numerous optic nerves while recording a single optic nerve's discharges. Periods of electrical stimulation produced enhancement of the ERG, in a similar manner to the injection of dopamine, suggesting that this effect is produced by efferent feedback to the retina. Electrical stimulation also produced an inhibition of afferent, photoreceptor action potentials. This latter effect could result from either lateral inhibition, produced by antidromic action potentials in photoreceptor axons, or from efferent cells, stimulated to produce orthodromic action potentials, causing inhibition of the photoreceptors.

Figure 5 shows, in a series of recordings from an intact, octopus optic nerve, the effects on efferent feedback to the retina of the previous light history of the preparation. A one minute period of dark adaptation followed by a 0.5 second test flash of illumination produces little visible afterdischarge. Following 25 seconds of continuous illumination, to light adapt the area of retina with which the optic nerve is connected, a 5 second period in darkness is followed by a test flash which now results in a significant afterdischarge. (These afterdischarges are efferent and invariably abolished when the optic nerve is cut, see Figure 4B). After a further 10 seconds darkness, the test flash now elicits only a weak afterdischarge and after 30 seconds in darkness, little or no afterdischarge is elicited. Thus, even brief

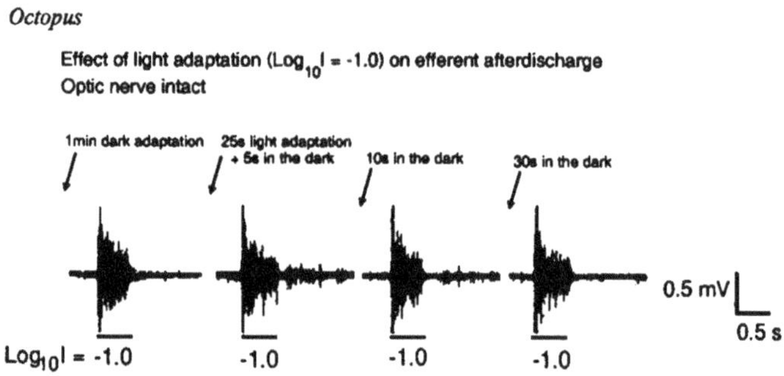

Figure 5. Suction electrode recordings of activity in an *Octopus* optic nerve, in response to periods of illumination with white light of intensity approximately 30 mW/cm^2 (Log_{10} I = -1.0) in an isolated eye-optic lobe preparation. The retina was illuminated for the period shown by the black bar beneath each record. Between periods of illumination, the preparation was subjected to dark adaptation or light adaptation (by opening the shutter on the photostimulator and illuminating the retina locally with light of 30mW/cm^2) for the periods indicated above each record. The details of the preparation are the same as for Figure 4.

periods of light and darkness can profoundly affect the degree of efferent feedback to the retina.

Perhaps the most simplistic hypothesis, which can unify many of these observations, is that efferent feedback to the retina causes the release of dopamine in the retinal synaptic plexus, and that this results in a hyperpolarisation of the photoreceptor inner segments. Such hyperpolarisation would decrease the likelihood of photoreceptor axons generating action potentials in the region of the synaptic plexus, where receptor potential amplitude becomes encoded as action potential frequency. Also, since light-induced depolarisation of photoreceptors produces distal migration of the retinal screening pigment, even in isolated eyes with severed optic nerves [2], it seems likely that hyperpolarisation might be sufficient stimulus to restore the pigment to the proximal, dark adapted position. However, Gleadall *et al*'s [35] observation that pigment migration to the dark position is dramatically impaired by optic nerve section, or by depletion of dopamine by reserpine treatment, are serious complications to this simplistic view. Perhaps, in the absence of efferent feedback, the repolarisation of light adapted photoreceptors to normal dark adapted membrane voltages, is impaired. This might result if there are excitatory connections between photoreceptors in the synaptic plexus, which become unstable in the dark if efferent feedback is abolished. The recordings from *Alloteuthis* preparations, where there is little evidence of ongoing efferent feedback (Figure A), are suggestive of disturbed interactions between photoreceptors at high light intensities, where profound inhibition during illumination and a prolonged depolarising afterdischarge are seen in photoreceptor axons. If hyperpolarising input from efferents maintains the stability of the system and ensures normal membrane voltage and therefore pigment migration in the dark, then the enhancement of the ERG and ERP produced by exogenous dopamine could, as Gleadall *et al* [35] point out, be a simple consequence of pigment migration to the dark adapted position allowing greater amounts of photopigment to interact with incoming light. Efferent induced hyperpolarisation of photoreceptors also fits quite well with the behaviour of the efferent activity shown in Figure 5. When screening pigment is in the dark adapted position a brief period of illumination produces little or no efferent afterdischarge, but following light adaptation the efferent afterdischarge produced by illumination could simply act to hyperpolarise the photoreceptors, hastening the process of dark adaptation and restoring photoreceptor interactions to the dark adapted state.

Clearly, there is an increasingly well-established role for dopaminergic efferent feedback to the cephalopod retina. Only further experiment will determine whether the observed effects result from relatively simple alterations in membrane conductance and voltage or are produced by more complex, second messenger systems. Also awaiting further investigation are the roles of other elements of the synaptic plexus, such a epithelial cells, and the whole question, long unanswered, of whether more than one type of centrifugal cell arborisation exists in the synaptic plexus. Such subdivision of centrifugal cell arborisations could provide the anatomical substrate for large field, slow afferent input to the optic lobe.

References

1 J.B. Messenger (1981). Comparative physiology of vision in molluscs. In *Handbook of Sensory Physiology* Vol VII/6C Ed. by Autrum,H. Springer Berlin-Heidelberg-New York pp 93-200.

2 J.Z. Young (1963b). Light and dark adaptation in the eyes of some cephalopods. *Proc. Zool. Soc. Lond.* 140 (2): 255-272.

3 N.W. Daw and A.L. Pearlman (1974). Pigment migration and adaptation in the eye of the squid, *Loligo pealei*. *J. gen. Physiol.* 63: 22-36.

4 J.Z. Young (1971). *The Anatomy of the Nervous System of Octopus vulgaris* Oxford University Press.

5 W.M. Saidel (1979a). Relationship between photoreceptor terminations and centrifugal neurons in the optic lobe of octopus. *Cell. Tiss. Res.* 204: 463-472.

6 J.A. Patterson and S.C. Silver (1983). Afferent and efferent components of *Octopus* retina. *J. Comp. Physiol.* 151: 381-387.

7 H. Saibil and E. Hewat (1987). Ordered transmembrane and extracellular structure in squid photoreceptor microvilli. *J. Cell Biol.* 105: 19-28.

8 E.J. Fisher and J.A. Patterson (1988). Photoreceptor size in the retina of the squid, *Alloteuthis subulata*. *J.Physiol. Lond.* 398: 97P.

9 A.I. Cohen (1973a). An ultrastructural analysis of the photoreceptors of the squid and their synaptic connections. I. Photoreceptive and non-synaptic regions of the retina. *J. Comp. Neurol.* 147: 351-378.

10 J.Z. Young (1963a). The number and sizes of of nerve cells in Octopus. *Proc. Zool. Soc. Lond.* *140* (2): 229-254.

11 M. Yamamoto (1985). Ontogeny of the visual system in the cuttlefish, *Sepiella japonica*. I. Morphological differentiation of the visual cells. *J. Comp. Neurol.* 232: 347-361.

12 J.Y. Lettvin and W.H. Pitts (1962) Neapolitan studies. *MIT Quart. Prog. Rep.* 64: 288-292.

13 W.M. Saidel, J.Y. Lettvin and E.F. MacNichol (1983) Processing of polarised light by squid photoreceptors. *Nature* 304: 534-536.

14 T. Yamamoto, K. Tasaki, T. Sugawara and A. Tonosaki (1965). Fine structure of the octopus retina. *J. Cell Biol.* 25: 345-359.

15 A.I. Cohen (1973b). An ultrastructural analysis of the photoreceptors of the squid and their synaptic connections. II. Intraretinal synapses and plexus. *J. Comp. Neurol.* 147:

379-398.

16 K. Tasaki. A.C.Norton, Y. Fukada and K. Motokawa (1963). Further studies on the dual nature of the Octopus ERG. *Tohoku J. Exper. Med.* 80: 75-88.

17 J.D. Robertson (1953). Further studies on ionic regulation in marine invertebrates. *J.exp. Biol.* 30: 277-296.

18 J.A. Patterson (1985). A modified, prefixed, Golgi-rapid technique for the cephalopod retina. *J. Neurosci. Methods.* 12: 219-225.

19 M. Yamamoto (1984). Photoreceptor collaterals in cuttlefish retina. *Zool. Sci.* 1: 501-503.

20 A. Tonosaki (1965). The fine structure of the retinal plexus in *Octopus vulgaris*. *Z. Zellforsch. Mikrosk. Anat.* 67: 521-532.

21 J.Z. Young (1962). The retina of cephalopods and its degeneration after optic nerve section. *Phil. Trans. Roy. Soc. Lond.* B 245: 1-18.

22 R.D. Lund (1966). Centrifugal fibres to the retina of *Octopus vulgaris*. *Exp. Neurol.* 15: 100-112.

23 E.G Gray (1970). A note on the synaptic structure of the retina of *Octopus vulgaris*. *J.Cell Sci.* 7: 203-215.

24 P.N. Dilly, E.G. Gray and J.Z Young (1963). Electron microscopy of the optic nerves and optic lobes of *Octopus* and *Eledone*. *Proc. Roy. Soc. Lond.* B. 158: 446-456.

25 J.Y. Lettvin, and H.R.Maturana (1961). *Octopus* vision. *MIT Quart. Prog. Rep.* 61:194-209.

26 B.B. Boycott, J.Y Lettvin, H.R Maturana and P.D Wall (1965). Octopus optic responses. *Exp. Neurol.* 12: 247-256.

27 P.H. Hartline and G.D. Lange (1974). Optic nerve responses to visual stimuli in squid. *J.Comp. Physiol.* 93: 37-54.

28 G.D. Lange and P.H. Hartline (1974). Retinal responses in squid and octopus *J.Comp. Physiol.* 93: 19-36.

29 W.M Saidel (1979b). Efferent neurons in the retina of *Octopus*: anatomy and physiology *Soc Neurosci(Abstr)* 5: 260.

30 J.A. Patterson (1984). An in vitro preparation for the study of visual physiology in the squid, *Alloteuthis subulata*. *J. Physiol. Lond.* 360: 18P.

31 D.M.K. Lam, T.M. Wiesel and A. Kaneko (1974). Neurotransmitter synthesis in cephalopod retina. *Brain Research* 82: 365-368.

32 S.C. Silver, J.A. Patterson and P.G. Mobbs (1983). Biogenic amines in cephalopod retina. *Brain Research* 273: 366-368.

33 E.M. Tansey (1980). Aminergic fluorescence in the cephalopod brain. *Phil. Trans. Roy Soc. Lond.* B. 291: 127-145.

34 T. Kito-Yamashita, C. Haga, K. Hirai, T. Uemura, H. Kondo and K. Kosaka (1990). Localisation of serotonin immunoreactivity in cephalopod visual system. *Brain Res.* 521: 81-88.

35 I.G Gleadall, K. Ohtsu, E. Gleadall and T. Tsukahara (1993). Screening-pigment migration in the octopus retina includes control by dopaminergic efferents. *J. exp. Biol.* 185: 1-16.

36 H. Suzuki and K. Tasaki (1983). Inhibitory retinal efferents from dopaminergic cells in the optic lobe of the octopus. *Vision Res.* 4: 451-457.

The Organization of Gaze Control in the Blowfly *Calliphora erythrocephala*

Roland Hengstenberg

Max-Planck-Institut für biologische Kybernetik, D-72076 Tübingen, Germany

Abstract - Flies have a distinct body posture during walking and flight. Hence head and eyes have mostly a preferred alignment with the surroundings. Voluntary and involuntary changes of body orientation elicit compensatory head turns which stabilize the original, correct orientation of the eyes relative to the surroundings. Compensatory head turns are elicited by several sensory cues which simultaneously influence the fly's walking posture or flight attitude, depending upon the locomotor state. Head/trunk-coordination is achieved by directionally specific neck sense organs. Visual and mechanosensory response components have complementary kinetic and dynamic characteristics. This enables the fly to correct unwanted misalignments of its eyes rapidly and precisely over a large range of angular velocities. The fly thus optimizes its visual abilities despite its formidable maneuverability.

Introduction

Flies have, like other animals, a preferred body posture. When walking, it is determined by the contact with the ground: the body axis is roughly horizontal, and the back is directed upwards. The flight attitude is similar but the body axis may be more elevated, especially in slow flight (Wagner 1985, David 1985). Very probably, the fly's sense organs and sensory nervous system have been adapted, during evolution, to work best in this preferred orientation.

Flying insects may turn around the vertical (yaw) without affecting flight equilibrium. Any rotation about a horizontal axis (transverse axis: pitch turns; longitudinal axis: roll turns) bears, however, the risk to crash unless flight balance is quickly stabilized by equilibrium reflexes (Schöne 1980, Hengstenberg et al. 1986).

The tasks of good vision and equilibrium control may require different alignments with the surroundings: during fast and narrow turns, for example, the fly has to bank sideways to account for the centrifugal acceleration (Hengstenberg et al. 1986). The optical surround, on the other hand, retains its usual orientation. Therefore it may be advisable to keep the eyes aligned with the vertical i.e. to counter-roll the head during banked turns.

Insects have two kinds of eyes: (1) the three dorsal ocelli are small camera eyes, with a single underfocussed wide-angle lens, which are best suited to sense global brightness in the dorsal hemisphere. (2) the two large compound eyes, consisting of several thousand mosaic eyes (ommatidia) which resolve spatial details in the lateral hemispheres of visual space. Both kinds of eyes are part of the head capsule and can therefore only be moved by turning the head.

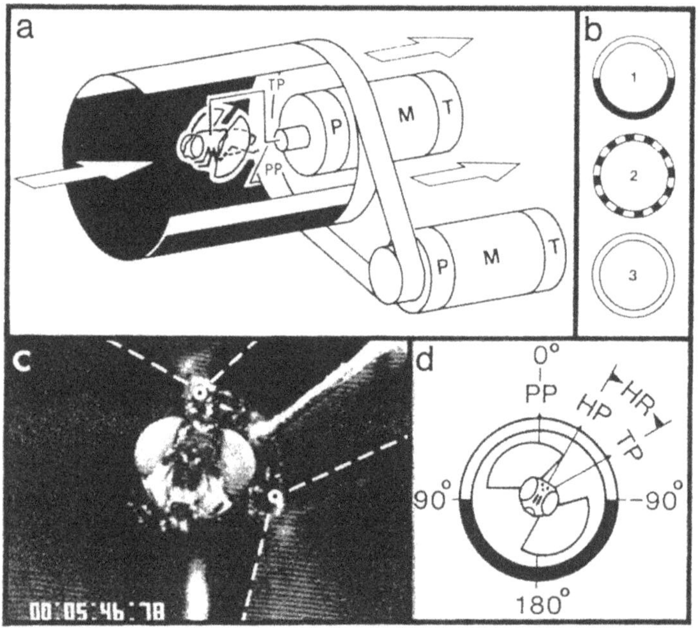

Figure 1:
(**a**) Wind-tunnel to elicit and observe head movements in the blowfly during stationary flight. Servomotors (M), fitted with position (P) and speed-sensors (T) allow to roll independently the fly and its optical surroundings. Angular positions are indicated in the video records by pointers (TP, PP). Wind flowing through the tunnel (arrows, 2 m/s) increases the flight mood of flies. (**b**) Cross sections of patterns lining the tunnel. (**c**) Videoframe as recorded through the entrance nozzle. Note that the fly generates simultaneously a compensatory head turn and a corrective difference in wingbeat amplitudes when tilted to its left. (**d**) Measures and sign conventions; HR [°] = head roll angle, referred to the fly's trunk.

The blowfly *Calliphora erythrocephala* can turn its head about all three principal axes (yaw and pitch ±20°, roll ±90°; Hengstenberg et al. 1986). The mechanics of the neck joint, the structure and insertions of the neck muscles, and their motoneurons have been studied in detail (Strausfeld et al. 1987). Physiological properties of motoneurons are partly known by electrophysiological investigations (Milde et al. 1987). Head movements of insects are very difficult to observe in free flight (Wagner 1985, Zeil 1983). Voluntary, stabilizing movements of the head could, however, be demonstrated in tethered flight (Land 1975, Hengstenberg 1992). Head movements in response to unexpected perturbations of the correct flight attitude can be studied during stationary flight in a wind-tunnel (Hengstenberg et al. 1986).

This article demonstrates gaze-stabilizing head movements made by the fly *Calliphora* in response to roll stimuli i.e. apparent rotations about the fly's body axis.

Methods

Flies were fixed coaxially to the shaft of a servomotor that allowed to tilt or turn them during stationary flight about the body axis (Fig 1a). The animals were surrounded by a translucent cylinder that could be lined with various patterns (Fig 1b), illuminated diffusely from the outside, and turned by a second servomotor. Air was flowing past the fly, due to a suction ventilator, at a comfortable travelling speed of 2 m/s. Head movements of the fly were observed by a small video telescope through the entrance nozzle of the windtunnel (Fig 1c), and measured by single frame analysis. The angle between the sagittal planes of the head and that of the fly's trunk is the head roll angle (HR [°] = HP-TP; Fig 1d).

If a fly is rolled sinusoidally during flight (Fig 2a; 1 Hz, ±90°) in an environment simulating "sky" and "ground" in their normal orientation, it generates phase-locked head movements of opposite sign (Fig 2b). They reduce the overall rotation of the head, relative to the surroundings by about 2/3, and thus tend to stabilize the orientation of the eyes. At the same time the fly generates compensatory steering maneuvers by changing wing pitch and wingbeat amplitude differentially on the left and right (Fig 1c). Hence, in free flight the actual stabilization of the eyes would be even better than suggested by the head movements alone.

The experiment above does not indicate whether the fly senses its own motion, irrespective of the optical surround, or whether it makes use of the relative motion between its eyes and the surroundings. Two complementary experiments show that flies use both sources of information: For the result of Fig 2c the fly was held stationary and the surround was rolled sinusoidally; the synphasic head movements prove that the fly uses visual cues to control head turns. For the reciprocal experiment (Fig 2d) the sky-and-ground pattern was removed, and the fly had to fly in optically featureless surroundings. Nevertheless, when the fly was rolled sinusoidally, it generated antiphasic head movements, proving that it also senses rotations of its body independent of visual cues i.e. through a mechanosensory organ.

The visual and mechanosensory roll responses can be further decomposed by specific modification of the stimuli, combined with surgical manipulations of various sense organs (Hengstenberg 1992, 1988).

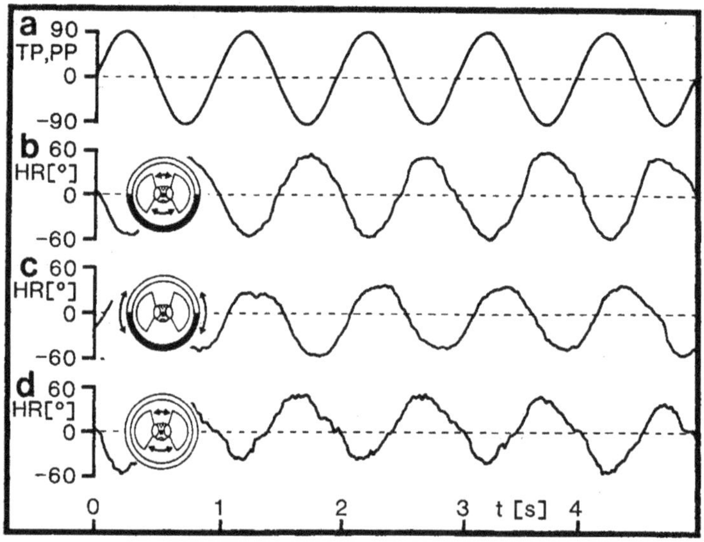

Figure 2:
Compensatory head roll of *Calliphora* in tethered flight. (**a**) Sinusoidal roll stimulus (1 Hz, ±90°). (**b**) Rolling the fly inside the stationary sky-and-ground pattern elicits an antiphasic head roll compensating about 2/3 of the imposed misalignment. (**c**) Rolling, instead, the pattern while holding the fly stationary elicits a synphasic head turn, due to visual roll perception, which reduces the apparent misalignment. (**d**) Rolling the fly in optically homogeneous surroundings elicits stabilizing head turns due to body motion perception.

Figure 3 gives an overview of the response components that could be isolated so far, and are now amenable to detailed anatomical and physiological investigations. Each component is specified by its name, the mediating sense organ, and a pictogram illustrating the stimulus used for its characterization.

The *pattern motion response* (Fig 3a) is elicited by coherent roll motion of extended patterns, irrespective of the kind of pattern, as long as it can be resolved by the compound eyes. Visual motion is detected by thousands of elementary motion detectors simultaneously, where each has a very small receptive field, at a distinct location in visual space, and a specific preferred direction (Reichardt 1987, Götz et al. 1979). Widefield motion sensitive neurons in the third visual neuropil of the fly integrate spatially a selected fraction of local motion signals (Hausen 1984). Thereby signals are generated that can be used for specific control of head posture and locomotor behaviour (Hengstenberg et al., this volume). The *contour orientation response* (Fig 3b) is due to horizontal contrast edges in the frontal part of the visual field and mediated through the compound eyes.

The dorsal light response (Fig 3c, d) has two components of unequal strength: the *tonic dorsal light response* (Fig 3c) is elicited by a difference in mean brightness in the visual fields of the two compound eyes. This response is especially important for the fly because it is the only cue indicating the absolute orientation in space. The *phasic dorsal light response* (Fig 3d) is mediated by transient illumination differences of the two lateral ocelli. Compared to other insects (Stange et al. 1979, Taylor 1981) this response is very small in flies.

Among the mechanosensory roll responses, the *self-motion response* (Fig 3e) is most prominent. It is mediated by the "halteres". They are homologous with the hindwings of other insects and have been transformed, in Diptera, into "gyroscopic" sense organs for rotations (Pringle 1948, Nalbach 1993). A haltere looks like a tiny pendulum of 1 mm length: It consists of a distal knob that comprises most of haltere mass, a thin stiff stalk, and a basal swelling that is attached to the side of the fly's trunk by a hinge joint. In flight, the halteres oscillate at wingbeat frequency, but in antiphase with the wing, in a vertical plane, tilted 30° backwards from the fly's transverse plane (Pringle 1948, Hengstenberg 1988). Rotations of the fly generate periodic coriolis forces that depend upon the direction and speed of rotation, and act upon the haltere's center of mass, i.e. the haltere knob (Nalbach 1993). The strain, caused by these periodic forces, is sensed by ca. 450 highly specialized cuticular mechanoreceptors that are located in the basal swelling and organized in seven distinct fields (Pflugstaedt 1912, Gnatzy et al. 1987, Hengstenberg 1988). The projections of haltere receptors in the central nervous system are at present only coarsely known (Strausfeld and Seyan 1985, Hengstenberg et al. 1988).

Properties and processing of the receptor signals are now studied electrophysiologically (Nalbach pers. comm.). A *difference in wing-load* (Fig 3f) that may occur during righting of the body after banking, is sensed probably by cuticular strain receptors at the wing base and elicits a transient righting response of the head.

The *direction of gravity* (Fig 3g) is sensed by leg proprioceptors through the distribution of weight among the legs when the fly is walking. Depending upon substrate

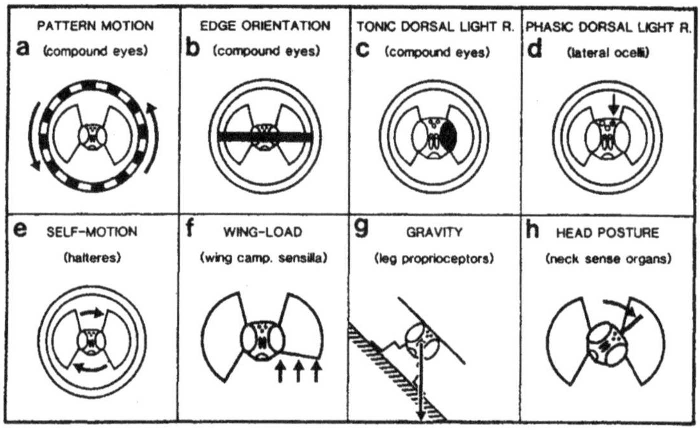

Figure 3:
Cues and sense organs for roll perception in *Calliphora*. Four separate visual (**a-d**) and four mechanosensory (**e-h**) response components can be distinguished by application of specific stimuli and elimination of particular sense organs. Gravity (**g**) is effective only in the walking fly. Most cues also influence the flight torque and walking posture.

inclination flies turn their head about the transverse and longitudinal axis towards an upright position (Horn et al. 1978). Since flies have no specific graviceptors, their sense of gravity is switched off automatically when they start flying. Consequently no responses to direction of gravity can be observed in flight (Hengstenberg 1988). Paradoxically, as it may seem, this is a patent means to prevent possible confusions that any kind of graviceptor would suffer from the oblique and varying linear accelerations generated along the aerobatic flight trajectories that are typical for flies.

Finally, *head posture* (Fig 3h) depends also on the orientation of the head. Direction-specific neck sense organs control head posture in a feedback configuration (Horn et al. 1978, Preuss and Hengstenberg 1992) and flight steering in a feed forward configuration (Liske 1977). There are two kinds of sense organs: the prosternal organs are external mechanoreceptive hair fields which sense pitch- and roll-turns of the head (Preuss and Hengstenberg 1992), and internal chordotonal organs whose role is not yet clear.

Most of the signals that affect head posture are also used by the fly to control walking posture (Hengstenberg unpublished) and flight steering (Gaffron 1934, Götz et al. 1979, Nalbach et al. 1994, Srinivasan et al. 1977). In each of these locomotor states head- and body-turns are performed simultaneously (see Fig 1c). Therefore the respective angles and angular velocities have to be added if the efficiency of gaze realignment is to be judged in an unrestrained fly. Figure 4 shows a signal flow diagram summarizing the results reported above: there are two optical and three mechanical cues, transduced by six kinds of sense organs that feed into neural circuits which extract roll-specific signals. These are, in turn, distributed to motor control circuits which coordinate the activity of neck-, leg-, and steering muscles. Compensatory movements reduce the roll-error-signals physically. Switches are opened by different locomotor states (F = flight, W = walking, S = standing) and disable inputs from mechanoreceptive sense organs and inappropriate outputs.

From these findings it emerges that the orientation control system of *Calliphora* is a highly developed network for simultaneous processing of many sensory signals. Its configuration changes with locomotor states, but little is known of how these changes are brought about.

Why should *Calliphora* use that many visual and mechanosensory mechanisms to stabilize its head? Signals from a single sensory system may depend upon the instantaneous orientation and motion of the fly and, in case of the visual system, also on the particular structure and lighting of the surroundings. Signals may, therefore, be not precise enough for optimal correction of attitude and course. Simultaneous use of independent signal sources is likely to increase the axis-fidelity and accuracy of the estimated self-motion.

The use of multiple sensory channels may also be advisable if the dynamic range of any one channel is smaller than required for locomotor control. In Crustacea, Molluses, and Vertebrates, including man, eye movements are controlled bimodally. In many instances the mechanosensory mechanisms tend to be fast but not very accurate, whereas visual mechanisms tend to be slower but more precise (Budelmann 1990, Dichgans 1983, Dieringer 1986, Sandeman 1983). Does this notion also apply for *Calliphora*?

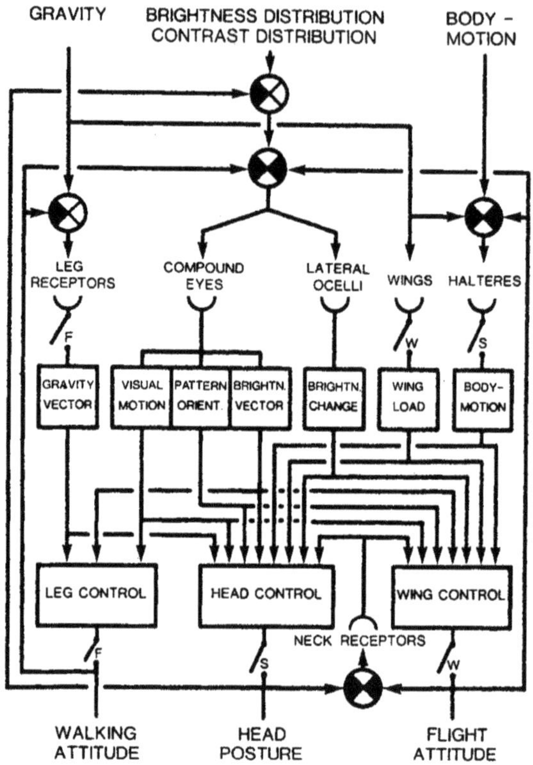

Figure 4:
Signal flow diagram of roll control in *Calliphora.* It illustrates the effective cues, the sense organs involved, the signal processing stages that can be logically separated, the distribution of signals on motor control networks, and the physical reduction of error signals by motor actions (black sectors of summation points). Switches are opened by different locomotor states (F = flying, W = walking, S = standing) to ensure appropriate signal routing.

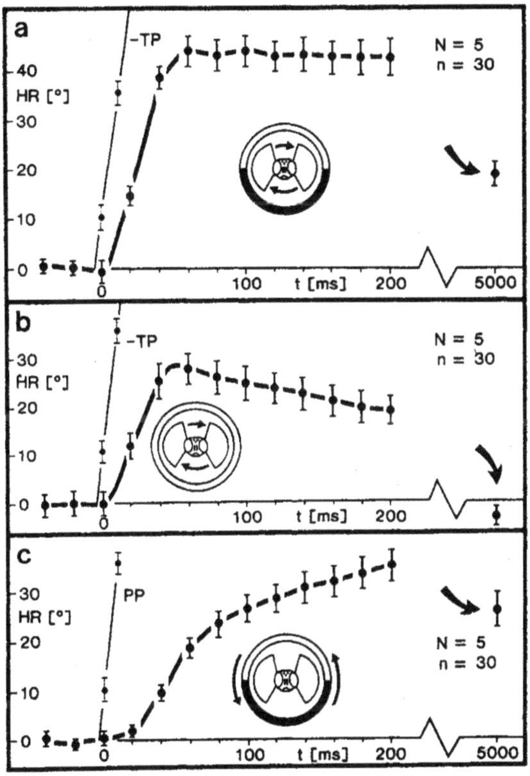

Figure 5:
Time course of head turns in response to step stimuli (90°, 2000°/s). (**a**) A fly, tilted by -90° within the sky-and-ground pattern generates, after a short delay (ca. 10 ms) a fast head roll (1200°/s) which is maintained for many seconds (see arrow and time scale). (**b**) Tilting the fly in visually homogeneous surrounding: elicits, after a short delay (ca. 10 ms), a fairly fast head roll (700°/s) that disappears within 5 seconds (arrow). (**c**) Stepwise displacement of the pattern elicits a head roll after a much longer delay (ca. 30 ms) which rises slowly (450°/s) to a maintained level. Apparently, the combination of visual and mechanosensory mechanisms (**b**, **c**) yields the superior performance shown in (**a**).

To test this hypothesis the time course of head turns was recorded in response to angular step displacements (Fig 5a-c; 90° at 2000°/s; note that the sign of stimuli is chosen to facilitate the comparison of response curves). If the fly is rolled in optically structured surroundings (Fig 5a) it turns its head after a short latency (10 ms) with high speed (1200°/s) against the imposed misalignment and maintains, after some adaptation, a compensatory posture for a long time (Fig 5a; arrow). If the sky-and-ground pattern is removed and the fly is turned through 90° (fig 5b), it responds equally after a short delay (ca. 12ms) and with fairly high speed (700°/s) but the response fades completely within 5 seconds (Fig 5b, arrow). This proves that the mechanosensory channel is fast and that the steady state response is caused visually. If, conversely, the fly is held stationary and the sky-and-ground pattern is rolled through 90° (fig 5c), the fly starts to turn its head only after a much longer delay (30 ms) and at a lower speed (450°/s). The corrective head turn is, however, maintained for a long time (Fig 5c, arrow). The combination of the mechanosensory response component (Fig 5b) and the visual component (Fig 5c) provides the fly with the quick, fast, and maintained head realignment shown in Fig 5a. For this „dynamic complementation“ of responses it is necessary, and has been demonstrated, that the component responses can be elicited independently. As a consequence of this, *Calliphora* can stabilize its head effectively at all speeds that a fly may encounter in real life.

Conclusions

Flies have a sophisticated and powerful gaze control system. It allows them to stabilize, during walking and flight, the orientation of their eyes relative to the surroundings. Hence they can see better during locomotion than their visual system would allow if the eyes were fixed to the trunk. This is especially true while flies proceed along their aerobatic flight trajectories.

The striking similarities of functional organization in gaze control systems of different animals and the accessibility of neural constituents in the fly makes this system particularly attractive for physiological studies.

Acknowledgments - I wish to thank Norbert Bayer for his excellent assistance with the experiments, Ute Bolze and Karin Bierig for the careful production of figures, Ursula Steiner for typing the manuscript, and Prof. K.G. Götz for numerous discussions on the perceptual abilities of insects.

References

1. Budelmann BU: (1990) The statocysts of squid. In: DL Gilbert, WJ Adelman and IM Arnold (eds.), Squids as Experimental Animals. Plenum Press, New York, 421-439.
2. David CT: (1985) Visual control of the partition of flight force between lift and thrust in free-flying *Drosophila*. Nature **313**: 48-50.
3. Dichgans J: (1983) Psychophysics of visual vestibular interaction. In: E Horn (ed.), Multimodal Convergences in Sensory Systems. Fortschr. Zool. **28**: 253-260.
4. Dieringer N: (1986) Vergleichende Neurobiologie von blickstabilisiernden Reflexsystemen bei Wirbeltieren. Naturwissenschaften **73**: 299-304.

5. Gaffron M: (1934) Untersuchungen über das Bewegungssehen bei Libellenlarven, Fliegen und Fischen. Z. Vergl. Physiol. **20**: 299-337.
6. Gnatzy W, Grünert U, Bender M: (1987) Campaniform sensilla of *Calliphora vicina* (Insecta, Diptera) I. Topography. Zoomorphology **106**: 312-319.
7. Götz KG, Hengstenberg B, Biesinger R: (1979) Optomotor control of wing beat and body posture in *Drosophila*. Biol. Cybern. **35**: 101-112.
8. Hengstenberg R, Sandeman DC and Hengstenberg B: (1986) Compensatory head roll in the blowfly *Calliphora* during flight. Proc. R. Soc. Lond. **B 227**: 455-482.
9. Hengstenberg R: (1988) Mechanosensory control of compensatory head roll during flight in the blowfly *Calliphora erythrocephala*, Meig. J. Comp. Physiol. **A 163**: 151-165.
10. Hengstenberg R, Hausen K, Hengstenberg B: (1988) Cobalt pathways from haltere mechanoreceptors to inter- and motoneurons controlling head posture and flight steering in the blowfly Calliphora. In: N Elsner, FG Barth (eds) Sense organs. Interfaces between environment and behavior. G. Thieme Verlag, Stuttgart, 215.
11. Hengstenberg R and Bayer N: (1988) Die Bedeutung der Schwerkraft für die Roll-Regelung im Flug bei der Schmeißfliege *Calliphora*.. Verh. Dtsch. Zool. Ges. **81**: 203.
12. Hengstenberg R: (1992) Stabilizing head/eye movements in the blowfly *Calliphora erythrocephala*. In: A Berthoz, W Graf, and PP Vidal (eds.), The Head-Neck Senosory-Motor System. Oxford University Press, New York, 49-55.
13. Horn E and Lang HG: (1978) Positional head reflexes and the role of the prosternal organ in the walking fly, *Calliphora erythrocephala*. J. Comp. Physiol. **126**: 137-146.
14. Land MF.: (1975) Head movements and fly vision. In: GA Horridge (ed.), The Compound Eye and Vision of Insects. Clarendon Press, Oxford, 469-489.
15. Liske E: (1977) The influence of head position on the flight behavior of the fly *Calliphora erythrocephala*. J. Insect Physiol. **23**: 375-379.
16. Milde JJ, Seyan HS and Strausfeld NJ: (1987) The neck motor system of the fly *Calliphora erythrocephala*. II. Sensory organization. J. Comp. Physiol. **A 160**: 225-238.
17. Nalbach G: (1993) The halteres of the blowfly *Calliphora*. I. Kinematics and dynamics. J. Comp. Physiol. **A 173**: 293-300.
18. Nalbach G and Hengstenberg R: (1994) The halteres of the blowfly *Calliphora* II. Three-dimensional organization of compensatory reactions to real and simulated rotations. J. Comp. Physiol A (in press).
19. Preuss T and Hengstenberg R: (1992) Structure and kinematics of the prosternal organs and their influence on head position in the blowfly *Calliphora erythrocephala* Meig., J. Comp. Physiol. **A 171**: 483-493.
20. Pringle JWS: (1948) The gyroscopic mechanism of halteres of Diptera. Phil. Trans. R. Soc. Lond. **B 233:** 347-385.
21. Reichardt W: (1987) Evaluation of optical motion information by movement detectors. J. Comp. Physiol. **A 161**: 533-547.

22. Sandeman DC: (1983) The balance and visual systems of the swimming crab: their morphology and interaction. In: E Horn (ed.), Multimodel Convergences in Sensory Systems. Fortschr. Zool. **28**: 213-230.
23. Schöne H: (1980) Orientierung im Raum. Wissenschaftliche Verlagsgesellschaft, Stuttgart.
24. Srinivasan MV: (1977) A visually evoked roll response in the housefly. J. Comp. Physiol. **119**: 1-4.
25. Stange G and Howard J: (1979) An ocellar dorsal light response in a dragonfly. J. Exp. Biol. **83**: 351-355.
26. Strausfeld NJ, Seyan HS and Milde JJ: (1987) The neck motor system of the fly *Calliphora erythrocephala*. I. Muscles and motor neurons. J. Comp. Physiol. **A 160**: 205-224.
27. Strausfeld NJ, Seyan HS: (1985) Convergence of visual, haltere, and prosternal inputs at neck motor neurons of *Calliphora erythrocephala*. Cell Tissue Res. **240**: 601-615.
28. Taylor CP: (1981) Contribution of compound eyes and ocelli to steering of locusts in flight. I. Behavioural Analysis. J. exp. Biol. **93**: 1-18.
29. Wagner H: (1985) Aspects of the free flight behaviour of houseflies (*Musca domestica*). In: M Gewecke and G Wendler (eds.), Insect Locomotion. Paul Parey Verlag, Berlin, 223-232.
30. Zeil J: (1983) Sexual dimorphism in the visual system of flies: The free flight behaviour of male Bibionidae (Diptera). J. Comp. Physiol. **150**: 395-412.

Visual Sensation of Self-Motions in the Blowfly *Calliphora*

Roland Hengstenberg, Holger Krapp, Bärbel Hengstenberg
Max-Planck-Institut für biologische Kybernetik, D-72076 Tübingen, Germany

Abstract - Insects flying across their habitat generate in their eyes complex motion patterns. They contain information about the animal's instantaneous direction and speed of locomotion as well as about the three-dimensional layout of the surroundings. In flies, visual motion is analyzed by large arrays of small field motion detectors. For each location in visual space, motion is detected in different directions. Local motion signals of distinct preferred directions are projected retinotopically into different layers of the lobula plate (third visual neuropil). There, neurons with extended dendritic arborizations across the retinotopic input array select from different depths of the neuropil many local signals, each corresponding to a particular direction of motion at a particular location in visual space. Spatial integration of thousands of well-selected local inputs over tailor-suited areas of visual space would allow to create specific filters for distinct optic flow patterns. By intracellular recordings from wide-field neurons, receptive field mapping with local motion stimuli and cell identification by dye injection and 3-D-reconstruction we demonstrate the ways how flies overcome the multiple ambiguities of local motion signals and generate useful representations of different self-motions.

1. Habitats and lifestyles of blowflies:

Sheep blowflies (*Lucilia spec.*) live in open country, e.g. sheep pasture; hence their visual surroundings are simple and coarsely structured. Blue bottles (*Calliphora spec.*) live in shrubland at forest edges or even within not too dense forests (Colyer and Hammond 1968). Their habitat is characterized by many obstacles to be circumvented.

The visual surroundings are richly by structured and the global features of sky and ground are often masked by vegetation. Blowflies (Fam. Calliphoridae), in general, have thus to be able to manoever artistically through unknown and very complex surroundings, i.e. to perceive quickly and reliably their motions in space in order to control their flight attitude and flight trajectory as required. To facilitate this perceptual task, flies stabilize the orientation of their eyes, relative to the surroundings, against voluntary and unexpected rotations of their body (Hengstenberg 1991 and this volume).

2. Motion in space and optical flow patterns:

Flies have, like other animals, a preferred body posture when walking on the ground and a similar attitude when flying through still air: the body axis is more or less horizontal and the back is directed upwards (Fig. 1a). Flying insects have, in principle, all six degrees of freedom to move in space: they can translate along their body axes (lift, side-slip, thrust) and rotate about these axes (yaw, pitch, roll; Fig. 1a) or perform certain combinations of these motions.

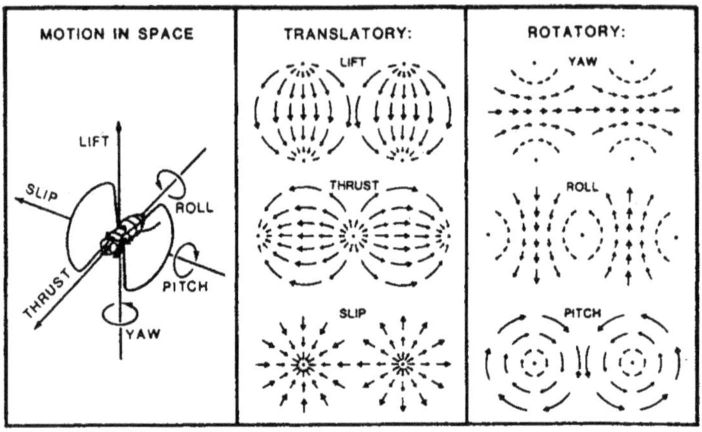

Fig. 1:
Motion in space and retinal motion patterns. (**a**) Degrees of freedom of a flying insect. (**b**) Retinal motion patterns caused by translations along the fly's principal axes. (**c**) Retinal motion patterns caused by rotations about the fly's principal axes.

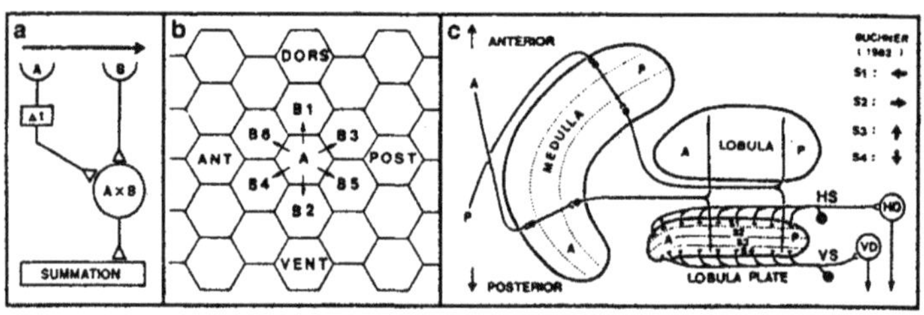

Fig. 2:
Elementary motion detection in insects (**a**) Functional structure of a unidirectional motion detector. (**b**) Motion is sensed along all directions of the facet array. (**c**) Local motion signals of different preferred directions are transferred to different layers of the lobula plate (see inset).

The visual field of the fly's two compound eyes covers almost the whole sphere around the animal. Each eye surveys roughly its respective hemisphere with a narrow zone of binocular overlap along the mid-sagittal plane and a small blind area at the back of the visual field where the fly's body obstructs vision straight backwards.

The principal motions (Fig. 1a) create, in the eyes, characteristic motion patterns that are illustrated, in a cartoon fashion, in Fig. 1b, c. The hemispherical visual fields of the two eyes are shown as circular areas, as if peeled off the fly's eyes, and flattened into a transverse plane in front of the animal. Fig. 1b illustrates flow patterns caused by translatory motions of the fly: upward lift (top), forward thrust (middle) and slip to the right (bottom). Fig. 1c illustrates flow fields of rotatory motions: yaw turn to the left (top), roll to the right (middle) and upward pitch (bottom). Each of these motion patterns is characterized by the arrangement of its motion axis (Fig. 1b: dots = focus of expansion/contraction; Fig. 1c: dots = center of rotation) and a typical flow pattern, where arrows of different size are meant to indicate motions of different speed. Translatory motion patterns (Fig. 1b) depend upon the distance of visual objects and vanish if things are infinitely far away. Hence for Fig. 1b a finite distance of the surroundings was assumed. Rotatory motion patterns (Fig. 1c) are independent of distance but the apparent speed across sampling stations of finite separation on the eye (see below) depends upon the latitude of the region considered from the equator.

Optic flow thus depends upon the three-dimensional layout of the surroundings, on the instantaneous motion and on the direction of gaze (Koenderink and van Doorn 1987). Conversely, extraction of translatory components from the complex flow pattern provides information about the three-dimensional layout of the surroundings. Motion discontinuities reveal steps in depth, as occurring at the edges of objects. Rotatory flow components, on the other hand, reveal unambiguously body-rotations of the fly because they are independent of object distance. Fast and reliable sensation of body rotations would allow the fly in the first place to control its body attitude in order to keep balance and additionally to stabilize the orientation of its eyes in order to reveal and probably make use of the translatory flow components (Hengstenberg et al. 1986, Srinivasan 1993, Colett et al. 1993).

In this paper we demonstrate how certain neurons in the visual system of the blowfly *Calliphora* extract specific rotatory components from arbitrary optic flow patterns. Before dealing with this issue, however, it is necessary to introduce briefly the anatomy of the fly's eyes and visual nervous system, as well as some of our present knowledge about the process of visual motion detection.

3. Anatomy of the fly's visual system

Flies have two kinds of eyes: (1) the three dorsal ocelli are small camera eyes with a single, underfocussed wide-angle lens and a few hundred photoreceptors; they are best suited to sense global brightness in the dorsal hemisphere (Schuppe and Hengstenberg 1993). (2) the two large compound eyes consist, in *Calliphora*, of about 2 x 6000 ommatidia, overlooking the two lateral hemispheres. Each ommatidium has its own little lens which provides for a small receptive field of about 2° diameter and thus allows to resolve spatial details. An ommatidium contains 8 photoreceptors which fall in two classes; the peripheral receptors R1-R6 are mainly used for motion detection and two central photoreceptors R7, R8 which serve other, less well defined purposes (Heisenberg and Buchner 1977, Heisenberg and Wolf 1984, Hardie 1985).

The optic lobes of the central nervous system consist of three successive areas (neuropils): the outermost lamina ganglionaris, the medulla and the lobula complex which, in flies, consists of two parts, the anterior lobula and the posterior lobula plate.

All three neuropils are retinotopically organized i.e. they consist of a repetitive multitude of similar small field neurons that preserve the two-dimensional organization of the retina. The restricted lateral extent of their arborizations (Strausfeld 1976) and the small receptive fields found in electrophysiological recordings (Devoe and Ockleford 1976, Gilbert and Strausfeld 1992, Gilbert, Penisten, Devoe 1991) suggest that these neurons perform local interactions which are, however, only partly understood due to the technical difficulties associated with recordings from these very small fibres.

At several depths of the visual neuropils tangential neurons contact extended areas of the retinotopic array of smallfield neurons (Strausfeld 1976, Fischbach and Dittrich 1989) and seem to collect layer-specific information over large parts of the visual field. Some of these have big enough fibres to be thoroughly studied by intracellular recordings. This is especially true for motion sensitive tangential neurons of the lobula plate (Hausen 1993).

4. Elementary motion detection by small-field columnar neurons

Studies of the optomotor turning response of the beetle *Chlorophanus* (Hassenstein and Reichardt 1956) have revealed that motion is detected in insects by a second order nonlinear interaction between input signals arising in two adjacent facets of the compound eye. Directional specificity is achieved by delaying one of the two input signals.

The simplest, unidirectional structure fulfilling these requirements is illustrated in Fig. 2a. Since the beetle's response was found to be bidirectional, an antisymmetrical scheme was proposed to account for the observed behaviour (Reichardt 1961). Recently it was shown that the subtraction stage of this scheme increases the directional specificity by elimination of non-directional signals of the imperfect detector stages (Egelhaaf and Borst 1993).

Different physiological mechanisms for the essential step of motion detection have been proposed (rev. Franceschini et al. 1989). At present, however, physiological evidence is still too sparse to finally decide which mechanism is used by the fly. Similarly it is not quite clear which small field neurons perform the essential operation of motion detection. (Egelhaaf and Borst 1993, Hausen 1993). The application of modern methods of optical microstimulation (Franceschini et al. 1989), whole cell patch recording, optical recording of calcium fluxes (Borst and Egelhaaf 1992), genetic microsurgery (Pflugfelder and Heisenberg 1994 in press), immune cytology (Buchner et al. 1988) and computer-assisted neuroanatomy (Bausenwein et al. 1992) will probably allow to settle these questions in the near future.

Studies of different optomotor responses in walking and flying flies have revealed that pattern motion is sensed along all principal directions of the facet array and probably everywhere in the fly's visual field (Fig. 2b; Götz 1964, Götz et al. 1979).

Local motion signals of any one direction class (Fig. 2b) are directed to the lobula complex in a retinotopic manner (Strausfeld 1976, Fischbach and Dittrich 1989) and are represented in distinct layers of the lobula plate (Fig. 2c; Buchner und Buchner 1984). Local motion signals, however, cannot be directly used for motor control because they are ambiguous in several respects:

(1) The direction of motion is uncertain within ± 90° of the preferred direction. The detector output depends on the cosine of the angle between the direction of motion and the preferred direction of the detector. Therefore each output value corresponds to two possible input directions of stimulus motion. This problem can be solved by comparison of detector signals with preferred directions that share the same receptive field (Reichardt et al. 1988).

(2) The speed of motion is uncertain for similar reasons: the output of a directionally specific motion detector is zero in the absence of motion and at infinite speed, and has a working range inbetween. Therefore, again, each output value corresponds to two possible input speeds. Comparison of signals from detectors overlapping receptive fields the same preferred direction but different sampling bases or different delays would allow to solve this problem (Buchner 1984).

(3) Finally, local motion signals are ambiguous with respect to the type of self-motion by which they may have been generated. A vertical downward motion at the equator in the left lateral visual field, for example, may either be caused by an increase in lift (translatory Fig. 1b top) or by a roll turn to the right (rotatory; Fig. 1c, middle). Similarly, a horizontal front-to-back motion in the right lateral field may be caused by forward progression in confined surroundings (translatory; Fig. 1b, middle) or by a yaw turn to the left (rotatory; Fig. 1c, top). These ambiguities can be resolved by comparison of motion signals arising from diametrically opposed parts of the visual field (Götz 1972, Kern et al. 1993).

5. Wide-field integration of motion in the third visual neuropil

The lobula plate (Fig. 2a) contains numerous tangential neurons. They have flat, often fan-shaped dendritic arborizations which spread over large areas of the neuropil and correspond to large receptive fields (Hausen 1993). The dendritic arborizations are located in distinct depths of the neuropil and this correlates with distinct preferred directions: neurons lying more anteriorly respond preferentially to horizontal pattern motion and more posteriorly situated cells prefer vertical motions in the visual field (Hausen 1993, Hengstenberg 1982). Axons of tangential neurons may project to different brain areas, to the contralateral optic lobe or terminate in small "optic foci" on the postero-lateral side of the brain. Here are also the input areas of descending neurons which transfer visual and other signals through the neck to motor coordination areas in the thoracic compound ganglion (Gronenberg and Strausfeld 1990, Gronenberg and Strausfeld 1992). Two groups of tangential neurons of the lobula plate are particularly conspicuous because of their comparatively thick fibres: The "Horizontal System" (HS;

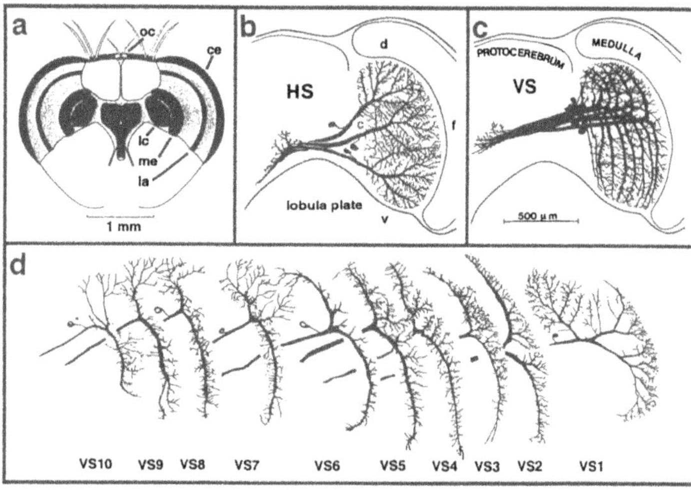

Fig. 3:
Neurons of the horizontal (HS) and vertical system (VS). (**a**) Rear aspect of the fly's brain, showing the location of the ocelli (oc) and compound eyes (ce), the three visual neuropils: lamina (la), medulla (me), lobula complex (lc) and central brain areas with the neck connectives. (**b**) The three HS-neurons in the anterior layer of the lobula plate; letters indicate visual field coordinates: d = dorsal, v = ventral, f = frontal, c = caudal. (**c**) VS-neurons in the posterior layer of the lobula plate. Cobalt stainings courtesy K. Hausen. (**d**) Individual structures of dendritic arborizations of the VS-neurons revealed by intracellular recording and staining.

Fig. 3b) consists of three neurons and the "Vertical System" (VS; Fig. 3c, d) of usually ten cells. Fig. 3a shows a fly's head in frontal aspect from behind with the brain dissected free. Eyes and neuropils are shown in black and fibre tracts are shaded grey.

The bean shaped area in Fig. 3b shows the anterior layer of the right lobula plate with indications of the visual field coordinates (d = dorsal, v = ventral, f = frontal, c = caudal).

Each of the three HS-neurons spans completely the fronto-caudal extent of the lobula plate. The dorso-ventral extent of the lobula plate is, however, divided among the three HS-neurons: HSN occupies the dorsal third, HSE the equatorial third and HSS the ventral third with some overlap of dendritic fields at the borders. The inconspicuous cell bodies lie at the medial border of the lobula plate, detached from the main fibres as usual in insects. The HS-axons terminate with a few stubby collaterals in the ventrolateral protocerebrum (Fig. 3b; Hausen 1993).

The 10 VS-neurons (Fig. 3c,d) occupy the posterior layers of the lobula plate. Their dendritic fields span the whole dorso-ventral extent of the lobula plate but in stripes of limited fronto-caudal extent. The dendritic arborizations show characteristic features that can be recognized in different individuals and justify a specific labeling of each VS-neuron (Fig. 3d, bottom). A particuliar feature of VS1 and VS7-VS10 is that their broader dorsal dendrites invade the anterior layers of the lobula plate. The cell bodies of VS-neurons form a cluster at the medial border of the neuropil and the axons project as a bundle to a target area a little more dorsal than that of the HS-neurons (Hengstenberg et al. 1982, Strausfeld and Seyan 1985).

Intracellular recordings from HS- and VS-cells reveal that these neurons respond mainly with graded changes of membrane potential although they are capable of generating action potentials (Dvorak et al. 1975, Hengstenberg 1977, Hausen 1984).

Both, HS- and VS-neurons respond bidirectionally to visual motions in the ipsilateral hemisphere: VS-cells are excited by vertical motion (but see below). HS-cells are excited by vertical motion and inhibited by the reverse motion; they are additionally excited by contralateral back-to-front motion. The responses increase with pattern size are invariant against contrast polarity and depend upon the speed of motion much like different optomotor responses (Hausen 1993, Hengstenberg 1982, Egelhaaf and Borst 1993). These findings suggest that HS- and VS-neurons monitor self-motions of the fly and may play a crucial role in the control of posture and locomotion. This conjecture was corroborated by different experiments made to eliminate HS- and VS-neurons:

(1) Transsection of the axons on one side and subsequent study of turning responses showed that pattern motion towards the operated side did no more evoke an optomotor following response whereas motion towards the control side released a normal response as in intact animals (Hausen and Wehrhahn 1983).

(2) Laser ablation of the precursor cells of HS- and VS-neurons in the larval stage of *Musca* produced adult flies lacking these neurons on the operated side and showing severe deficits in turning responses with wide-field motion towards the operated side (Geiger and Nässel 1982).

(3) *Drosophila* mutants (omb^{H31}, rb^{5}) which fail to make HS- and VS-neurons during metamorphosis show none of the characteristic responses to wide-field pattern motion

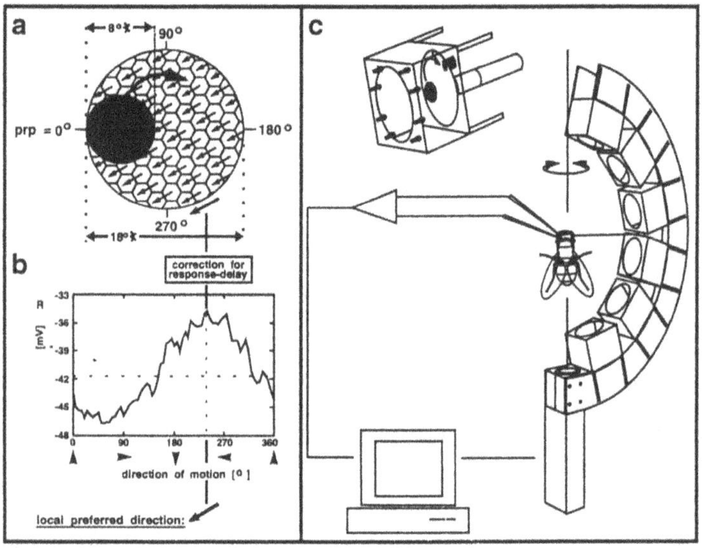

Fig. 4:
Mapping of local motion tuning. (**a**) Circulating dot stimulus passing quickly through all directions of motion. (**b**) Local motion tuning curve, specifying directional preference (center of gravity) and local motion sensitivity (modulation) of the response. (**c**) Pivoted stimulus meridian, carrying six "dot-circulators" allows to map local motion responses almost everywhere in visual space.

(Heisenberg and Wolf 1984, Pflugfelder and Heisenberg 1994, Hengstenberg 1991). HS- and VS-neurons are obviously important for the sensation and control of self-motions but the specific role of single neurons could not be revealed by these experiments.

6. Functional specification of single tangential neurons

The characteristic abilities of single neurons are revealed by mapping their motion sensitivity and directional preference throughout the receptive field (Hengstenberg 1981). Local motion stimuli are presented at many places in the visual surround and cell responses are recorded intracellularly. Neurons are unambiguously identified by fluorescent dye injection and 3D-reconstructions from serial sections (Hengstenberg et al. 1983).

Local preferred direction and motion sensitivity are measured quickly and accurately by a specifically developed stimulus (Fig. 4a): a single black dot (8° •) is moved on a circular path (10° •) at constant speed (2cps) somewhere in the visual field. Whenever the direction of dot motion coincides with the preferred direction of motion detectors, looking at that spot and feeding into the recorded neuron, the response becomes maximal. At the opposite position of the dot path, the response becomes minimal. Reversing the direction of dot motion allows, by comparison of responses, to eliminate the effects of dot location and of the response delay. One such local motion tuning curve (Fig. 4b) can be measured within 5 seconds; its center of gravity gives the local preferred direction and its modulation is a measure of local motion sensitivity (Menzel and Hengstenberg 1991, Krapp and Hengstenberg 1992).

Figure 4c shows a stimulator with six "dot-circulators" (Fig. 4c, inset) that can be activated singly, in sequence or simultaneously in defined phase relationships. The devices are mounted on a meridian that can be rotated about the vertical. Thereby local motion stimuli can be almost applied everywhere in the fly's visual field (Krapp and Hengstenberg 1993).

Figure 5 shows motion response maps and camera lucida drawings of three representative VS-neurons. The dendritic arborizations in the right lobula plate are shown in frontal aspect to facilitate comparisons with the response maps. The retinotopy of the lobula plate is illustrated by letters, outside of the neuropil, designating locations in the visual field and in the motion response maps (d = dorsal, v = ventral, f = frontal, c = candal).

The response maps are Mercator-projections of the ipsilateral hemisphere. Measurements were taken at the positions marked by circles. The local motion responses are represented as arrows originating at the site of measurement. The direction of the arrow indicates the local preferred direction and its length indicates the motion sensitivity, normalized to the maximum response. At positions where no measurements had been made, interpolated arrows were filled in. The Mercator projection is veridical for angles relative to the circles of longitude and latitude. For surface area, however, it is non-veridical: it overrepresents area with increasing elevation. Therefore the dorsal and ventral areas of response maps are overemphasized and must be interpreted with consideration.

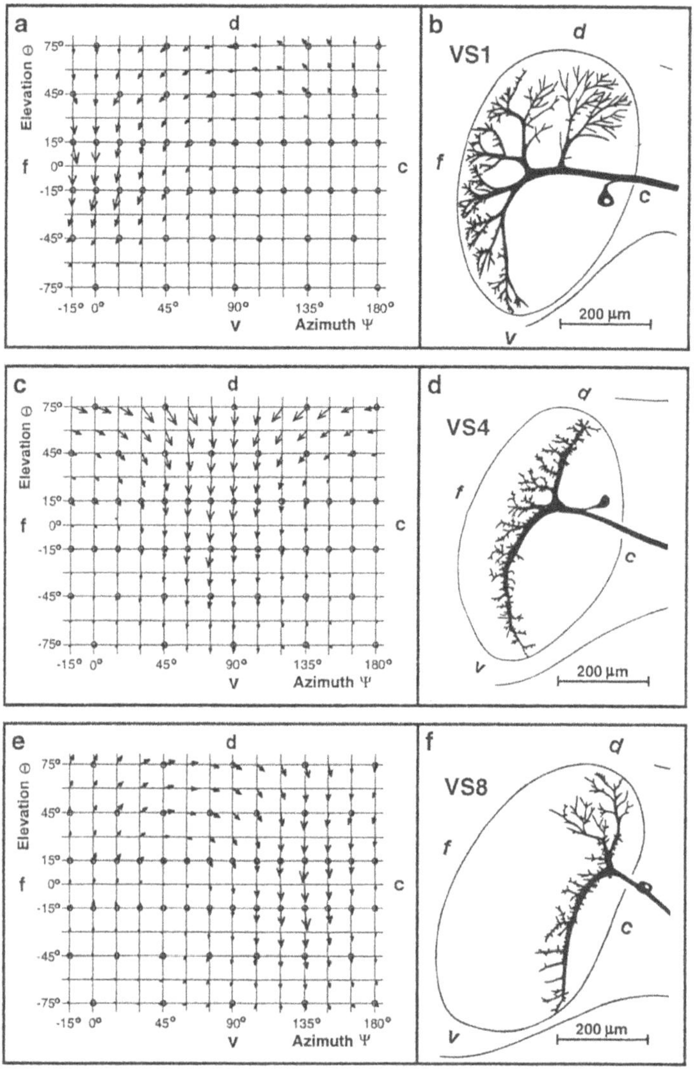

Fig. 5:
Dendritic structures and motion response maps of different VS-neurons. Mercator projection of the ipsilateral hemisphere. Stimulus positions are marked by small circles. Arrows indicate the local preferred direction. The length of the arrows represents local sensitivity, normalized for the maximum response. Arrows without circles are interpolated. Note that each of the three VS-neurons is maximally excited by a distinct rotation about a horizontal axis.

The VS1-neuron (Fig. 5b) occupies frontal and dorsal areas of the lobula plate; frontal branches are situated in the posterior layer associated with vertical motions and the medio-dorsal branch lies in the anterior layer associated with horizontal motions (cf section 5; Hengstenberg and Hengstenberg 1980). The response map of VS1 (Fig. 5a) shows a high sensitivity to downward motion in frontal parts of the visual field, responses to back-to-front motion in dorso-lateral parts and weak but significant responses to upward motion in the dorso-caudal region. This distribution of motion sensitivity and directional preference agrees very well with the notions about the general architecture of the lobula plate (section 5; Hausen 1993) and the detailed structure of this neuron. Apparently, VS1 is strongly excited by counterclockwise pattern rotation about the transverse axis ($\Psi = 90°$, $\theta = 0°$) i.e. during upward pitch turns of the fly.

The VS4-neuron (Fig. 5d) occupies a strip-like area of the lobula plate, extending from the dorsal to the ventral margin of the lobula plate, centered near the middle of the fronto-caudal extent, and the dendrites are located in the posterior layer, associated with vertical motions.

The response map of VS4 (Fig. 5c) shows a belt of high sensitivity to downward motion in the lateral part of the visual field ($60° < \Psi < 105°$ at $\theta = 0°$). The consistency of directional tuning along the meridian of maximal sensitivity ($75° < \Psi < 90°$) is quite amazing. In the dorsal half of the field the overall motion sensitivity is larger than in the ventral half. The response pattern widens in angular extent with increasing elevation and the local preferred directions turn increasingly towards the horizontal: a front-to-back component is added in the fronto-dorsal quadrant ($\Psi < 90°$; $\theta > 0°$) and a back to front component in the caudo-dorsal quadrant ($\Psi > 90°$; $\theta > 0°$). This motion response pattern suggests that VS4 is maximally excited when the fly rolls to the left (compare Fig. 1c, middle).

VS8 (Fig. 5f) occupies more caudal areas of the lobula plate with a strip-like ventral dendrite situated in the posterior layer and a dorsal fan-shaped arborization situated in a more anterior layer. The response map of VS8 (Fig. 5e) reveals a belt of downward motion sensitivity along a caudo-lateral meridian ($\Psi = 135°$). Front-to-back motions in dorso-lateral areas and even upward motions in dorso-frontal parts of the visual field contribute to the response of the cell. It seems that VS8 is maximally excited if the fly turns downwards about an oblique horizontal axis originating at $\Psi = 45°$, $\theta = -15°$.

The response maps shown in Fig. 5 and those of the other VS-cells demonstrate that these neurons are vigorously excited by rotations of the fly about different horizontal axes. It should be stressed, however, that the motion response patterns of VS-neurons are not exact neural representations of optical flow patterns. In Fig. 6 are compared the measured motion response pattern of the neuron VS6 (Fig. 6a) and the calculated optic flow pattern for a roll turn to the left (Fig. 6b). It is obvious that optic flow as shown in Fig. 6b would maximally excite VS6, but although the directions of roll flow are also well represented in the ventral part of the receptive field, the motion sensitivity in this area is much reduced. It is not clear at present whether VS-neurons are meant to extract exclusively rotatory components of optic flow and are therefore ventrally insensitive, where translatory flow prevails during locomotion (Collett 1980, Preiss 1991).

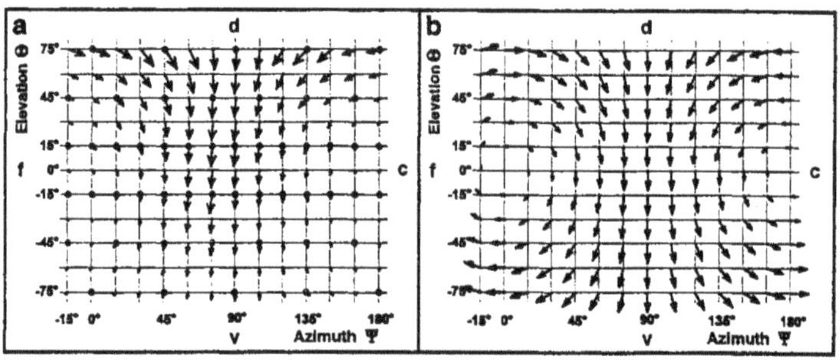

Fig. 6:
Motion response pattern compared to optic flow pattern. (**a**) ipsilateral map of local motion responses of VS6 (**b**) computed optic flow pattern for a roll turn to the left. Note the striking similarity in the direction of flow but the reduced sensitivity in the ventral part of the response field of VS6.

Alternatively the response profiles of VS-neurons could be exquisitely tuned to a mixture of rotatory and translatory components that may occur during flight or may be best suited for motor control.

7. Discrimination of rotations from translations

A local motion signal does not indicate whether it has been caused by a translation or a rotation of the fly (see section 4). Therefore such signals cannot be used directly for motor control. Signals from diametrically opposite areas of the visual field have the same sign in case of translations and opposite signs in case of rotations. Hence best discrimination is achieved by convergence of the appropriate signals from the corresponding areas of the visual field (Götz 1972, Kern et al. 1993).

HS-neurons are known to receive input from the contralateral eye which makes them most sensitive to yaw rotations (Hausen 1984, 1993). Figure 7 shows a receptive field map of the equatorial horizontal neuron (HSE) in the right lobula plate. On the ipsilateral side

(Fig. 7; $\Psi < 0°$) the cell responds to horizontal front-to-back motions within an equatorial zone ($-45° < \theta < 45°$) and over the whole azimuth range of the eye ($0° < \Psi < 180°$). This corresponds nicely with the extent of the dendritic arborization of HSE in the lobula plate (Fig. 3b). Stimuli in the contralateral hemisphere (Fig. 7; $\Psi < 0°$) are effective in the mirror-symmetrical part of the visual field. The preferred directions are tilted fronto-ventrally but there is a considerable sensitivity to horizontal back-to-front motions.

Selectivity for rotations can be further enhanced when rotatory motion is sensed in a region of the visual field where rotation components are large and translation components are small (Collett 1980, Junger and Dahmen 1992). For a flying insect, and yaw rotations, there are two such exceptional areas: one straight ahead and the other straight behind the fly, but there it cannot see very well.

Do HS-neurons make use of this topological asset? In Fig. 7a the length of the arrows indicating the local motion sensitivities is scaled logarithmically to show the homogeneous orientation of preferred directions even in areas with low sensitivity. When this accentuation is not made and the responses are plotted linearly (Fig. 7b) it becomes obvious that HSE is maximally motion-sensitive in straight ahead direction ($\Psi = 0°$, $\theta = -15°$) and that a comparatively small area of 30° x 30°, centered around the gaze axis ($\Psi = 0°$, $\theta = 0°$), dominates the overall response of HSE.

The huge size of the receptive field of the HSE-neuron, the prevailing sensitivity to horizontal motion, the appropriate convergence of signals from the ipsi- and contralateral eye and the sensitivity maximum straight ahead make this neuron most sensitive to yaw turns of the fly to the left. None of these processes acts as a logical discriminator. All add up to increase gradually the neuron's specificity for one kind of self-motion.

Similar mappings are presently being made for the other two HS-neurons (HSN, HSS) for the ten VS-neurons (Krapp in prep.) and other tangential neurons that can be recorded long enough. The neurons VS4-VS6 which respond to downward motion in the lateral part of the ipsilateral hemisphere (Figs. 5, 6) could be complemented by contralateral downward motion resulting in a preference for lift translations or by

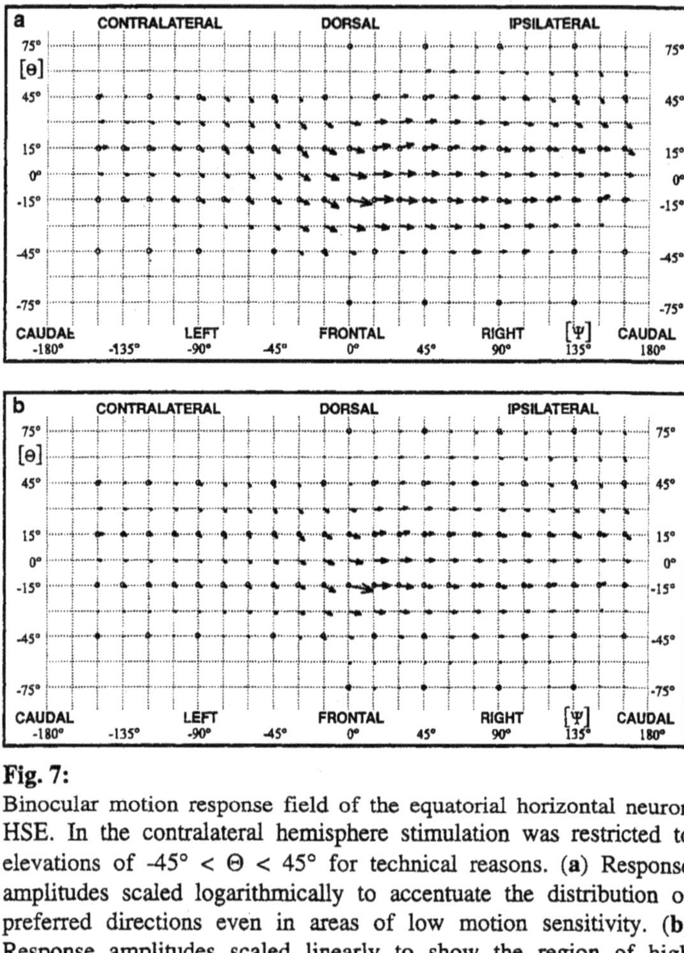

Fig. 7:

Binocular motion response field of the equatorial horizontal neuron HSE. In the contralateral hemisphere stimulation was restricted to elevations of $-45° < \Theta < 45°$ for technical reasons. (**a**) Response amplitudes scaled logarithmically to accentuate the distribution of preferred directions even in areas of low motion sensitivity. (**b**) Response amplitudes scaled linearly to show the region of high sensitivity around the straight ahead direction ($\psi = 0°$, $\theta = 0°$). Note the gigantic size of the receptive field, its mirror-symmetrical extent in the two hemispheres, the essentially horizontal directions (but see text) and the sensitivity maximum straight ahead. This combination of properties makes the HSE-neuron in the right lobula plate most sensitive to yaw turns to the left.

contralateral upward motion to yield a specificity for roll rotations. VS1-VS3 which respond to downward motion along a frontal meridian (Krapp and Hengstenberg 1994) could be supplemented by upward motion along the posterior meridian to create a preference for upward pitch rotations. None of these interactions has to be made necessarily on HS- or VS-neurons. That could equally take place on dendrites of descending neurons. If convergence does take place on tangential neurons of the lobula plate, as in case of HSE (Fig. 7), the specific role of that neuron for the sensation of self-motions and its possible role in locomotor control can be stated more clearly.

Conclusions

The functional organization of wide-field motion-sensitive neurons in the fly can be revealed by mapping sequentially the responses to local motion stimuli at many places in the receptive field. The motion response map of a certain neuron shows the particular self motion of the fly by which that neuron would be maximally excited.

The experiments show further that the high specificity is achieved gradually by a combination of different mechanisms: (1) selection of small-field motion signals from specific areas of the visual field, (2) selection of a signal with an appropriate preferred direction from the ensemble of local signals, (3) weighting of the local signals according to the desired sensitivity distribution, (4) wide-field spatial integration and (5) convergence of wide-field signals from ipsi- and contralateral hemispheres.

The functional representations of different self-motions in single tangential neurons can be conveniently used by the fly for the control of posture and locomotion.

Acknowledgments - We thank Mrs. K. Bierig for the careful production of the figures, Mrs. U. Steiner for typing the manuscript and Prof. K.G. Götz for many inspiring discussions.

References

1. Bausenwein B, Dittrich APM and Fischbach K: (1992) The optic lobe of *Drosophila melanogaster*. II. Sorting of retinoptic pathways in the medulla. Cell Tissue Res. **267**: 17-28.
2. Borst A, Egelhaaf M: (1992) In vivo imaging of calcium accumulation in fly interneurons as elicited by visual stimulation. Proc. Natl. Acad. Sci. USA **89**: 4139-4143.
3. Buchner E: (1984) Behavioral analysis of spatial vision in insects. In: MA Ali (ed). Photoreception and Vision in Invertebrates. Plenum Press, New York, 561-621.
4. Buchner E and Buchner S: (1984) Neuroanatomical mapping of visually induced nervous activity in insects by H^3-Deoxyglucose. In: MA Ali (ed). Photoreception and Vision in Invertebrates. Plenum Press, New York, 623-634.
5. Buchner E, Bader R, Buchner S, Cox J, Emson PC, Flory E, Heizmann CW, Hemm S, Hofbauer A and Oertel WH: (1988) Cell-specific immuno-probes for the brain of normal and mutant *Drosophila melanogaster*. I. Wildtype visual system. Cell Tissue Res. **253**: 357-370.

6. Collett TS: (1980) Some operating rules for the optomotor system of a hoverfly during voluntary flight. J. Comp. Physiol. **138**: 271-282.
7. Collett TS, Nalbach HO and Wagner H: (1993) Visual stabilization in arthropods. In: Miles FA, Wallman J (eds). Visual Motion and its Role in the Stabilization of Gaze. Elsevier, Amsterdam. Rev. Oculomotor Res. **5**: 239-263.
8. Colyer CN and Hammond CO: (1968) Flies of the British Isles. Frederick Warne Ltd, London.
9. Devoe RD and Ockleford EM: (1976) Intracellular responses from cells of the medulla of the fly *Calliphora erythrocephala*. Biol. Cybern. **23**: 13-24.
10. Dvorak DR, Bishop LG, Eckert HE: (1975) Intracellular recording and staining of directionally selective motion detecting neurons in fly optic lobe. Vision Res. **15**: 451-453.
11. Egelhaaf M and Borst A: (1993) Movement detection in arthropods. In: FA Miles, J Wallman (eds.). Visual Motion and its Role in the Stabilization of Gaze. Reviews of Oculomotor Research **5**: 53-77.
12. Fischbach KF, Dittrich APM: (1989) The optic lobe of *Drosophila melanogaster*. I. A Golgi analysis of wild-type structure. Cell Tiss. Res. **258**: 441-475.
13. Franceschini N, Riehle A and le Nestour A: (1989) Directionally selective motion detection by insect neurons. In: DG Stavenga, RC Hardie (eds). Facets of Vision. Springer Verlag, Berlin 360-390.
14. Geiger G and Nässel DR: (1982) Visual processing of moving single objects and wide-field patterns in flies: behavioural analysis after laser-surgical removal of interneurons. Biol. Cybern. **44**: 141-149.
15. Gilbert C, Penisten DK, Devoe RD: (1991) Discrimination of visual motion from flicker by identified neurons in the medulla of the fleshfly *Sarcophaga bullata*. J. Comp. Physiol. **A 168**: 653-673.
16. Gilbert C and Strausfeld NJ: (1992) Small-field neurons associated with oculomotor and optomotor control in muscoid flies - functional organization. J. Comp. Neurol. **316**: 72-86.
17. Götz KG: (1964) Optomotorische Untersuchungen des visuellen Systems einiger Augenmutanten der Fruchtfliege *Drosophila*. Kybernetik **2**: 77-92.
18. Götz KG: (1972) Principles of optomotor reactions in insects. Bibl. ophthalm **82**: 251-259.
19. Götz KG, Hengstenberg B and Biesinger R: (1979) Optomotor control of wing beat and body posture in *Drosophila*. Biol. Cybern. **35**: 101-112.
20. Gronenberg W and Strausfeld NJ: (1992) Premotor descending neurons responding selectively to local visual stimuli in flies. J. Comp. Neurol. **316**: 87-103.
21. Hardie R: (1985) Functional organization of the fly retina. In: Ottoson D (ed). Springer Verlag, Berlin. Progress in Sensory Physiol. **5**: 4-79.
22. Hassenstein B and Reichardt W: (1956) Systemtheoretische Analyse der Zeit- und Reihenfolgen- und Vorzeichenauswertung bei der Bewegungsperzeption des Rüsselkäfers *Chlorophanus*. Z. Naturforsch. **11b**: 513-524.

23. Hausen K: (1984) The lobula complex of the fly: Structure, function and significance in visual behavior. In: MA Ali (ed). Photoreception and Vision in Invertebrates. Plenum Press, New York, 523-559.
24. Hausen K: (1993) The decoding of retinal image flow in insects. In: FA Miles, J Wallman (eds). Visual Motion and its Role in the Stabilization of Gaze. Reviews of Oculomotor Research, Elsevier, Amsterdam, **5**: 203-235.
25. Hausen K and Wehrhahn C: (1983) Microsurgical lesion of horizontal cells changes optomotor yaw responses in the blowfly *Calliphora erythrocephala*. Proc. R. Soc. Lond **B 129**: 211-216.
26. Heisenberg M and Wolf R: (1984) Vision in *Drosophila*. Springer Verlag, Berlin.
27. Heisenberg M and Buchner E: (1977) The role of retinula cell types in visual behavior of *Drosophila melanogaster*. J. Comp. Physiol. **117**: 127-162.
28. Hengstenberg R: (1977) Spike responses of "non-spiking" visual interneurons. Nature **270**: 338-340.
29. Hengstenberg R: (1981) Rotatory visual responses of vertical cells in the lobula plate of *Calliphora*. Verh. Dtsch. Zool. Ges. **74**: 180.
30. Hengstenberg R: (1982) Common visual response properties of giant vertical cells in the lobula plate of the blowfly *Calliphora erythrocephala*. J. Comp. Physiol. 149: **179**-193.
31. Hengstenberg R: (1991) Spontaneous and stabilizing head movements in wildtype and optomotor blind *Drosophila*. 12th EDRC, Mainz, p. 100.
32. Hengstenberg R: (1991) Gaze control in the blowfly *Calliphora*: A multisensory two-stage integration process. The Neurosciences **3**: 19-29.
33. Hengstenberg R: (1994) The organization of gaze control in the blowfly *Calliphora erythrocephala*. In: Taddei-Ferretti (ed). Biocybernetics of Vision: Integrative Mechanisms and Cognitive Processes (in press).
34. Hengstenberg R, Bülthoff H and Hengstenberg B: (1983) Three-dimensional reconstruction and stereoscopic display of neurons in the fly visual system. In: NJ Strausfeld (ed). Functional Neuroanatomy. Springer Verlag, Berlin, 183-185.
35. Hengstenberg R, Hausen K and Hengstenberg B: (1982) The number and structure of giant vertical cells (VS) in the lobula plate of the blowfly *Calliphora erythrocephala*. J. Comp. Physiol. **149**: 163-177.
36. Hengstenberg R, Sandeman DC and Hengstenberg B: (1986) Compensatory head roll in the blowfly *Calliphora* during flight. Proc. R. Soc. Lond. **B 227**: 455-482.
37. Junger W and Dahmen HJ: (1991) Response to self-motion in waterstriders: visual discrimination between rotation and translation. J. Comp. Physiol **A 169**: 641-646.
38. Kern R, Nalbach HO and Varjú: (1993) Interactions of local movement detectors enhance the detection of rotation. Optokinetic experiments with the rock crab, *Pachygrapsus marmoratus*. Vis. Neurosci. **10**: 643-65.
39. Koenderink JJ and van Doorn AJ: (1987) Facts on optic flow. Biol. Cybernetics **56**: 247-254.
40. Krapp H, Hengstenberg B and Hengstenberg R: (1994) Correspondence of dendritic field structure, receptive field organization and specific flow patterns in interneurons

of the blowfly *Calliphora*. In: N Elsner, H Breer (eds). Sensory Transduction. Thieme, Stuttgart, p. 453.

41. Krapp H and Hengstenberg R: (1992) Reliability of a fast method to determine locally the preferred direction of motion sensitive neurons. In: N Elsner, DW Richter (eds). Rhythmogenesis in Neurons and Networks. Thieme, Stuttgart, p. 306.
42. Krapp H and Hengstenberg R: (1993) Representation of specific optical flow fields in lobula plate neurons of the blowfly *Calliphora*. In: N Elsner, M Heisenberg (eds). Gene - Brain - Behavior. Thieme, Stuttgart, p. 357.
43. Menzel J and Hengstenberg R: (1991) A fast method to determine the distribution of local preferred directions within the receptive field of motion sensitive neurons. In: N Elsner, H Penzlin (eds). Synapse, Transmission, Modulation. Thieme, Stuttgart, p. 274.
44. Pflugfelder GO and Heisenberg M: (in prep.) Optomotor blind of *Drosophila melanogaster*: A neurogenetic approach to optic lobe development and optomotor behavior.
45. Preiss R: (1991) Separation of translation and rotation by means of eye-region specialization in flying gypsi moths (Lepidoptera. Lymantriidae). J. Insect Behavior **4**: 209-219.
46. Reichardt W: (1961) Autocorrelation, a principle for the evaluation of sensory information by the central nervous system. In: WA Rosenblith (ed). Sensory Communication. MIT Press + J. Wiley & Sons, New York, 303-317.
47. Schuppe H and Hengstenberg R: (1993) Optical properties of the ocelli of *Calliphora erythrocephala* and their role in the dorsal light response. J. Comp. Physiol. **A 173**: 143-149.
48. Schwarz U and Miles FA: (1989) Ocular responses to linear motion are inversely proportional to viewing distance. Science **45**: 1394-1396.
49. Srinivasan MV: (1993) How insects infer range from visual motion. In: FA Miles, J Wallman (eds). Visual Motion and its Role in the Stabilization of Gaze. Elsevier Publ. Co., Amsterdam. Reviews of Oculomotor Research **5**: 139-156.
50. Strausfeld NJ: (1976) Atlas of an Insect Brain. Springer Verlag, Berlin.
51. Strausfeld NJ and W Gronenberg: (1990) Descending neurons supplying the neck and flight motor of Diptera-organization and anatomical characteristics. J. Comp. Neurol. **302**: 954-972.
52. Strausfeld NJ and HS Seyan: (1985) Convergence of visual, haltere and prosternal inputs at neck motor neurons of *Calliphora erythrocephala*. Cell Tissue Res. **240**: 601-615.

INFORMATION PROCESSING IN THE INSECT OCELLAR PATHWAY

John Patterson
Department of Physiology
Faculty of Basic Medical Sciences
Queen Mary and Westfield College
University of London
LONDON E1 4NS

In addition to compound eyes, capable of analysing contrast, movement and colour in the visual environment, many insects also possess simple eyes, known as ocelli. Lateral ocelli or stemmata are found on the head of the larvae of holometabolous insects. These larvae lack compound eyes. Dorsal ocelli, which form the subject of this account, are located on the frontal region of the head in many adult insects and their nymphal stages, where they supplement the visual input provided by the compound eyes. Lateral and dorsal ocelli have been reviewed recently by Toh and Tateda [1], and the dorsal ocellar system has been the subject of a comprehensive review by Goodman [2]. Although absent or rudimentary in some insect groups, most insect species have from two to four (usually three) dorsal ocelli and an associated system of interneurons, providing additional visual information. Although the basic features of the ocelli have been described in many species, the structure and function of these eyes has been most fully studied in locusts, dragonflies, flies and honeybees. In these animals there are three ocelli, two lateral and one median, although the anatomical arrangement varies, as does the field of view. In locusts and dragonflies each lateral ocellus lies close to the medial margin of the compound eye, while the median ocellus lies in the midline and in a more rostral position. In flies and bees the three ocelli are grouped closely together and positioned on the dorsal surface of the head. The cockroaches possess two lateral ocelli which have also been studied in some detail. These cockroach studies (reviewed in [1]) have paralleled those on locust and dragonfly ocelli which form the basis for this account.

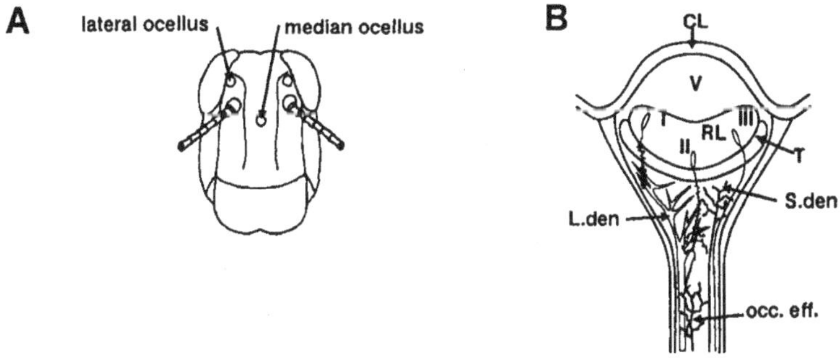

Figure 1 A. The general arrangement of the lateral and median, dorsal ocelli of the locust (*Schistocerca gregaria*). B. Diagrammatic representation of a section through the locust median ocellus and distal ocellar nerve. Both are redrawn from Patterson [3]. CL, corneal lens; V, vitreous cell layer; RL, receptor cell layer T, tapetum; L.den, L-neuron dendrite; S.den, S-neuron dendrite; occ.eff., terminal arborisation of putative ocellar efferent fibre, RI, RII, RIII refer to three classes of photoreceptor termination described by L.J. Goodman [4].

A feature of almost all insect ocelli is the presence of a single cuticular lens which overlies some tens to several hundred photoreceptor cells to form a cup-shaped photoreceptive organ. This lens is underfocussed [5, 6], and it is widely stated that ocelli are only capable of, at best, very limited form or directionally selective vision. There is no evidence of the multifaceted dioptrics or ommatidial structures which typify the highly organised compound eyes, but in other respects photoreceptors appear similar in both types of eye. Many insect ocelli also have a reflecting tapetal layer beneath the photoreceptors and/or screening pigment, which serves to enhance or regulate photon capture. Most photoreceptors have short axons which terminate in the ocellar cup or a short distance along the ocellar nerve which connects each ocellus to the insect brain. Beneath the receptor layer is a synaptic plexus where photoreceptors and first order interneurons form synaptic connections. The general anatomical arrangement of the ocelli of the locust and the anatomical arrangement of the ocellar cup and distal part of the ocellar nerve in the locust are summarised in Figure 1.

Projections of ocellar L-neurons

The first-order interneurons are of basically two types, a small number of large and uniquely identifiable L-neurons [1, 2, 3, 7, 8, 9, 10, 11] and a larger number of smaller S-neurons [1, 2, 12]. L-neurons, particularly in locust and dragonfly, have been studied extensively because their large size suits them to microelectrode studies of their physiology, often linked to the injection of intracellular markers [1, 2, 3, 5, 10, 13, 14, 15, 16, 17, 18, 19, 20, 21]. L-neurons fall into three distinct classes, although not all classes are present in every insect species [2]. The first type of L-neuron passes from the ocellus to the posterior slope of the brain where it terminates. In locusts, such L-neurons from the lateral ocelli terminate in the ipsilateral side of the brain. In many other species both ipsi- and contra-lateral sides receive such L-neurons. Similar L-neurons from the median ocellus typically terminate in either or both sides of the brain. L-neurons make synaptic contact with, among others, second-order ocellar interneurons which descend to arborise in sub-oesophageal and thoracic ganglia. The second class of L-neuron, found in flies and honeybees, passes from the ocellus directly to the thoracic ganglia, with arborisations at various locations en route. In many other species this role is fulfilled by third-order neurons (see below). The third type of L-neuron exists in the locust, where two L-cell axons pass between each lateral ocellus and the median ocellus. Within any given species, the L-neurons have been shown to show a remarkable degree of constancy and uniqueness of anatomical arrangement [8].

Figure 1 compares, in highly diagrammatic form, the distribution of L-neurons within the brains of the locust [7,9] and the dragonfly [10, 11]. Both locusts and dragonflies lack L-neurons of the second type, which pass directly to the ventral nerve cord. The main differences between the L-neuron maps in these two insect groups is the presence in locusts of L-neurons (of the third type) forming direct connections between lateral and median ocellar nerves and their absence in dragonflies. and the presence in dragonflies of lateral ocellar L-neurons which terminate in the contralateral half of the brain, while in the locust lateral ocellar L-neurons are wholly confined to the ipsilateral side of the brain. In both locusts and dragonflies, L-neurons (of the second type) passing from the ocellus directly to the thoracic neuropile have not been described, although the trans-synaptic migration of cobalt ions has allowed description of some, at least, of the third-order neurons of the system, with which the L-neurons connect [9, 10]. It is interesting to note that in locust, where lateral ocellar L-neurons are confined to the ipsilateral side of the brain, third-order neurons pass to both contralateral circumoesophageal connectives from each side of the brain, while in the dragonfly comparable third-order neurons only appear to pass to the ipsilateral connective. The projection of lateral ocellar information to the contralateral side of the ventral nerve cord seems to be achieved by the arrangement of the third-order neurons in the locust system and by the decussation of lateral ocellar L-neurons in the dragonfly system.

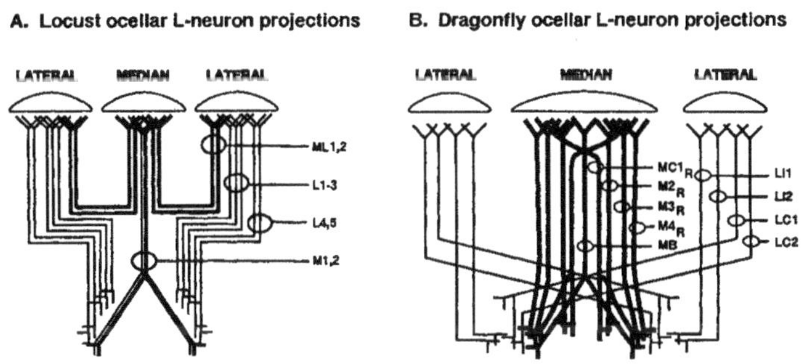

Figure 2. Highly diagrammatic representations of the central projections of median and lateral ocellar L-neurons in A, the locust, *Schistocerca gregaria*, based on [7, 8, 9] and B, in dragonflies (*Anax* and *Aeschna* species), based on [10, 11]. For discussion, see text. Note that the arrangement of L-neuron dendrites in ocelli is only figurative except for those of dragonfly MC1 cells, which are known to pass from one half of the median ocellus to the contralateral side of the brain. Lateral ocellar L-neurons are drawn with thin lines for clarity only.

Information about the arrangement of L-neuron arborisations within the ocellus itself is scant, but in the dragonfly median ocellus, at least one pair of L-neurons ($MC1_L$ and $MC1_R$) have ocellar arborisations confined to one half of the bi-lobed ocellus[10]. In the locust Simmons[19] describes examples of neurons L4-5 and L1-3 which have restricted projections within the ocellar cup, and an example of L1-3 with a more widespread projection. These arrangements suggest that some L-neurons may view restricted regions of the ocellar visual field, despite the underfocussed nature of ocellar dioptrics. Where the receptive fields of L-neurons have been measured [5], this has not been found to be the case.

The responses of ocellar photoreceptors and L-neurons

Figure 3 shows some examples of the responses to wide-field illumination, of photoreceptors and L-neurons of the dragonfly median ocellus. In general terms the responses of these cells are very similar in both locust [14] and dragonfly [13, 20], although there are some differences of detail. The depolarising responses of dragonfly ocellar photoreceptors are graded with light intensity, in common with those of many insect photoreceptors. The relationship between the logarithm of the light intensity and the amplitude of the response shows a typical tail at low light intensities and a more or less linear dynamic range where depolarisation increases linearly as light intensity increases logarithmically (see Figure 4). At high light intensities the response amplitude saturates, although sufficiently high intensities to do this were not used in the experiments which form the basis for Figures 3 and 4. In terms of detail, the responses of dragonfly ocellar photoreceptors are rather unusual in frequently showing a spike-like transient at the onset of illumination and one or more distinctive oscillations of membrane voltage on a return to darkness [13].

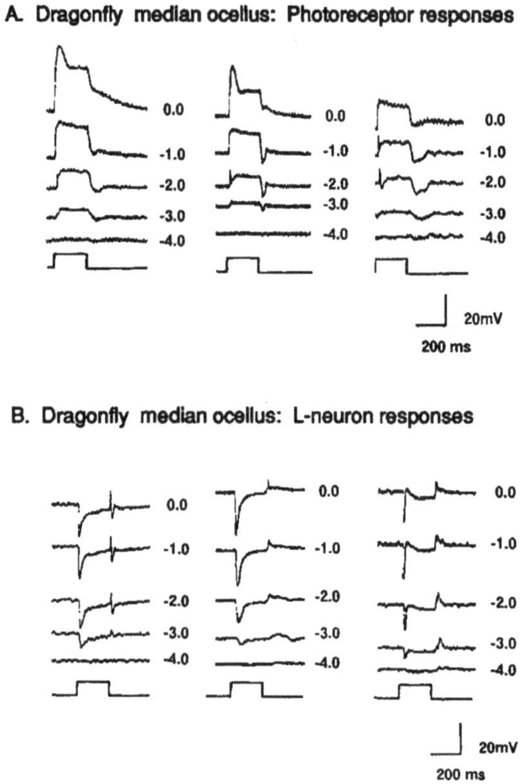

Figure 3. Intracellular recordings of the responses of: A dragonfly median ocellar photoreceptors and; B, L-neurons. These are previously unpublished recordings obtained in a study of the morphology and responses of median ocellar neuron neurons in *Anax junius* and *Aeschna tuberculifera* by Patterson and Chappell [10], which provides full details of the experimental method. The responses were obtained with square wave pulses of white light of Log_{10} relative intensity as indicated at left of each record. The maximum intensity (0.0) was 635 footcandles.

Ocellar L-neurons receive input from large numbers of photoreceptors and intracellular recordings from identified (as well as presumed) L-neurons show that these second-order neurons hyperpolarise on illumination of the ocellus and generate one to a few action potentials on the return to darkness (see Figure3). L-neurons have dark adapted dynamic ranges located at lower intensities than the photoreceptors which provide their synaptic input and the dynamic range can sometimes occupy a narrower range of intensities. The saturation of the initial transient hyperpolarisation begins at lower intensities than that of photoreceptor initial depolarisation, so that L-neurons respond more strongly than photoreceptors at lower light intensities and may even produce sizeable hyperpolarisations at intensities producing scarcely detectable photoreceptor responses (see Figure 4). The responses of dragonfly ocellar L-neurons are broadly similar in locusts and dragonflies, but those in the dragonfly often show a greater cut back of membrane voltage after the initial, rapid hyperpolarisation at the onset of illumination. The rapid spiking potential at the end of a period of illumination, which typifies locust L-neuron responses, is less evident in dragonfly responses. Instead, rapid, but not spike-like oscillations of the membrane voltage are frequently seen at the end of a period of illumination [15].

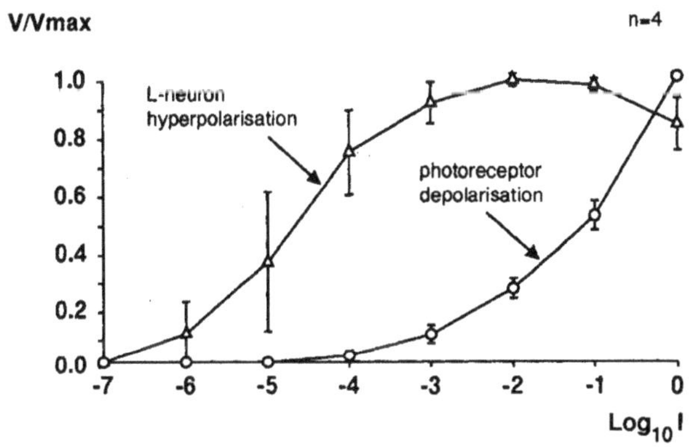

Figure 4. Response versus Intensity (VLogI) curves for photoreceptors (o) and L-neurons (Δ) of the dragonfly median ocellus. The vertical axis represents the peak depolarisation of photoreceptors and the peak hyperpolarisation of L-neuron responses. Both have been normalised by expressing the voltage change at any intensity (V) as a fraction of the maximum response (Vmax) produced for each cell (at Log_{10} I = 0.0 for photoreceptors and either -1.0 or -2.0 for L-neurons). Each point is the average for 4 cells. Error bars are ± 1.0 SD. Horizontal axis is Log_{10} relative light intensity. The maximum intensity (0.0) was 635 footcandles. The data include the responses shown in Figure 3. Figure 3 also provides information as to the species and methods used.

Since the physiology of ocellar and and compound eye cells has been well studied in both locusts and dragonflies, it is possible to compare the properties of equivalent cells in both types of eye. In general terms the properties of ocellar photoreceptors and L-neurons and compound eye photoreceptor and lamina monopolar cells (LMCs) are similar, although the optical properties of the two systems are very different. In both eyes the dark adapted responses of photoreceptors show a dynamic range extending over several decades of light intensity. The VLogI curves for L-neurons and for LMCs of the compound eye system, are not only shifted to lower intensities with respect to those for the photoreceptors which provide their input, but the dynamic range of the responses may also have a steeper slope than that of photoreceptors. Clearly, synaptic gain is occurring at the first visual synapse. This gain can be estimated in a variety of ways, including comparison of response amplitudes or slope of initial response to illumination. Early estimates of the gain at the first ocellar synapse in the locust [14] yielded values of 1.8 to 7.8 depending on the stimulus intensity, while similar measurements for the dragonfly compound eye [22] give values of up to 14. In technically elegant experiments, Simmons [20] has measured the synaptic gain between a single dragonfly ocellar photoreceptor and an L-neuron and found the gain to be about nine and shows that this in comparable with a value of four obtained from similar experiments on the fly compound eye [23]. Therefore, there is little to suggest that great differences exist between the synaptic gain of the first visual synapse in ocellar and compound eye pathways. This is perhaps somewhat surprising given the much greater convergence of photoreceptor input onto the large second-order neurons of the ocellar as compared with the compound eye system. In a careful and comparative study, Wilson [5] has shown that locust ocellar L-neurons are some five times more sensitive to a point source than comparable cells of the lamina of the compound eye, and that L-neurons could be up to 5000 times more sensitive to an extended source.

Extensive study of the behaviour of the photoreceptor-LMC synapse both in the dark adapted state and at various degrees of light adaptation in flies and dragonflies has led Laughlin and Hardie [24] to propose a 'log transfer-subtraction-amplification' model for the operation of the first visual synapse. Such a mechanism is said to 'match the bandwidth of peripheral neurons to the expected intensity fluctuation about any one mean' [21]. Essentially, this means that contrast features will be reliably detected in the face of wide variations in the mean level of illumination of the visual field. The general features of ocellar photoreceptor and L-neuron responses suggest that this system has similar properties, and in experiments to evaluate Laughlin and Hardie's model in the locust ocellus, Simmons [21] has shown this to be the case. Simmons has however noted some important differences. Ocellar L-neurons are slow to undergo full light adaptation [5] and Simmons [21] has also shown that slow or incomplete light adaptation of locust L-neurons means that they retain an ability to detect slow changes in wide-field light intensity and that they may differ from dragonfly LMCs in this respect. This ability, of L-neurons to detect slow changes in intensity input, could be one area in which the ocelli provide information to the central nervous system which is not readily obtained from the compound eyes.

The presence of both a cut back or sag of membrane voltage following the initial, transient responses to illumination of dragonfly photoreceptors and L-neurons and the frequent observation of oscillations of membrane voltage at the cessation of illumination are suggestive of feedback mechanisms operating at the first ocellar synapse; electron microscope studies [25] have shown the presence of feedback synaptic structures. On the basis of pharmacological experiments, Klingman and Chappell [26] have proposed a feedback model for synaptic signalling between receptors and L-neurons. It is proposed that GABA is released from L-neurons onto photoreceptors in the dark and that this GABA depolarises the receptor terminals so that with the onset of illumination larger amounts of the photoreceptor transmitter (judged to be acetylcholine) are released leading to large, and frequently saturating, hyperpolarisations of the L-neurons, followed by inhibition of GABA feedback and the development of a new, light dependent equilibrium. At light-off membrane voltage oscillations in receptors and L-neurons occur as the dark adapted equilibrium is restored. Despite the elegance and neural economy of this feedback model, Simmons [20] has been unable to find evidence of connections from dragonfly L-neurons to photoreceptors in experiments involving current injections into L-neurons while recording simultaneously from the L-neuron and a photoreceptor known to synapse with the L-neuron. The feedback to photoreceptors produced by electrical stimulation of the ocellar nerve observed by Stone and Chappell [27] could still be due to the effects of S-neurons. Simultaneous recording from pairs of L-neurons did however show that depolarising current injected into one L-neuron produced an IPSP in the other L-neuron in both dragonflies [20] and locusts [19]. Simmons states that such interactions are unsuited to the cut back in the light induced hyperpolarisation seen in L-neurons and conductance and reversal potential measurements on L-neurons do not support a major role for synaptic input onto L-neurons as the cause for the cut back of hyperpolarisation[20]. Whether the response dynamics of L-neurons are due to direct or indirect synaptic feedback, to the intrinsic properties of the cells or to field potential effects, such as those proposed for the compound eye by Shaw [28] still requires clarification.

<u>Ocellar input to multimodal descending interneurons</u>

Behavioural studies have shown the ocelli to play some role in the maintenance of the dorsal light reaction [29] and phototactic behaviour of insects [30]. An analysis of the effects of ocellar input to higher order interneurons is an important prerequisite for any detailed understanding of the way in which visual information derived from these wide field and underfocussed eyes is integrated into the overall behaviour of the insect. Since the transynaptic migration of cobalt ions from L-neurons into third-order cells reveals interneurons which descend the into the ventral nerve cord (VNC), early studies of higher ocellar driven interneurons focussed on microelectrode studies of VNC units. Ocellar L-neurons seem to be best suited to detecting dimming of the ocelli, producing off-spikes or spike-like potentials, this early work concentrated on follower neurons of the locust VNC which also show brisk, off-spikes on dimming of the ocelli. These off-units responded either to dimming of ipsilateral and median ocelli , or dimming of contralateral and median ocelli or to dimming of the median ocellus alone [3, 31]. Some of these unidentified units showed spontaneous activity which was usually inhibited by illumination of the ocelli which generated off-spikes in the units and, some cases, showed excitation on illumination of an ocellus which did not produce an off-spike in the unit [31]. Simmons [18, 21] has studied the connections between locust L-neurons and third-order ocellar neurons in considerable detail. He has also described the detailed morphology of one identified, third-order ocellar neuron within the brain and made the important observation that the third order neurons respond to air currents across the head, showing that these third-order neurons are multimodal [18].

There is now considerable evidence that input from the ocelli is integrated with that from other sense organs and transmitted to the thoracic ganglia by descending, multimodal interneurons. In the locust, Griss and Rowell [32] have described the arborisations within the brain and the thoracic ganglia of three pairs of such, identified interneurons; DNI; DNM; and DNC. DNI and DNM have cerebral arborisations confined to the side of the brain which is ipsilateral to the axon in the VNC, while DNC has an axon which decussates in the brain so that the cerebral arborisations lie contralateral to the axon in the VNC. Rowell and Reichert [33] have characterised the responses of DNI, DNM and DNC to several different forms of sensory stimulation. Figure 5 is an attempt to summarise some of their complex findings in diagrammatic form. This shows that these neurons respond to compound eye, ocellar and wind hair inputs. Figure 5 does not show the interactions between inputs demonstrated by these authors. Ocellar off stimuli (presented to the appropriate ocellus or ocelli) in the absence of other stimuli reliably generate responses in all three neurons and tonic illumination of these ocelli tends to suppress on-going activity generated by other sensory inputs. Other interactions include a strong enhancement of the wind hair responses, when dimming is detected by the ocelli, and especially, by the compound eyes. There is also an increase in responsiveness to wind during fictive flight. All three neurons are spatially and directionally sensitive and respond to movements of the animal about the three axes of rotation; roll, pitch and yaw. DNI responds best to to a diving banked turn to the ipsilateral side, DNM to downwards pitch and DNC to a diving banked turn to the contralateral side. The visually directional selectivity is not dependent on the modulation of compound eye input by ocellar input, but the directional responses to wind may be derived from ocellar or compound eye inputs.

These findings show how ocellar input is merged with that from other sensory modalities to provide an integrated system for the analysis of the orientation of movement of the animal in

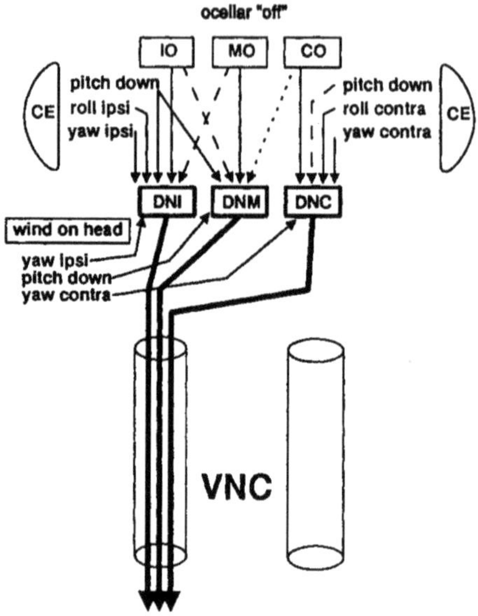

Figure 5. A diagrammatic summary of some of the multimodal inputs to the identified interneurons DNI, DNM and DNC descending from the brain to the ventral nerve cord (VNC) of the locust, *Locusta migratoria*. IO; MO; CO; Ipsilateral, median and contralateral ocellus. CE; compound eye. This diagram is based on Rowell and Reichert [33], but interactions between inputs and the effects of fictive flight, which these authors describe, are omitted. The division of compound eye inputs between ipsi- and contra-lateral eyes is done only for clarity of the diagram. Roll, pitch and yaw stimuli were presented to both compound eyes simultaneously in the actual experiments. Solid lines indicate reliable effects, dotted lines, weaker or less reliable effects.

space. Such a system is, of course, highly suitable for the control of insect flight, and further studies [34, 35] have revealed numerous mesothoracic interneurons which are directly or indirectly postsynaptic to the three, large DN neurons. These interneurons appear to be involved in steering in flight.

Smaller ocellar neurons and efferent feedback to ocelli

While the physiology of L-neurons is now quite well described, much less detail is known about the larger number of ocellar S-neurons. although the action potential activity of visually- and mechanoreceptor-driven cells in extracellular and intracellular recordings from ocellar nerves probably derive from these smaller axons. Some, at least of these smaller cells are efferent to the ocellar synaptic plexus [36, 37]. In the cucumber looper moth, such efferents may alter ocellar spectral sensitivity [38]. Rotzler [39] has shown, that in the locust, units in the lateral ocellar nerve respond to stimulation of the compound eyes, to wind stimuli on the thorax, to sound and to active and passive movements of the wings. Such units can be presumed to be efferent to the ocellar neuropile. Most ocellar interneurons are affected by electrical stimulation of the circumoesophageal connectives and ocellar L-neurons are depolarised by such stimulation and the response to ocellar illumination is altered. As Rotzler [39] points out, presumed efference recorded from the ocellar nerve may simply be activity in multimodal S-neurons with no output to other ocellar neurons, but there is considerable evidence to suggest that some information ascending the ocellar nerves is able to modify L-neuron activity. Should further investigation provide fuller evidence for efferent feedback modulation of L-neuron output, then the interactions within the ocellar system could be very complex, since the stimuli which affect the L-neurons would be very similar to those which make inputs into those DN neurons which are postsynaptic to L-neurons. Another possibility, not widely investigated, is that some of the integration of sensory inputs seen in DN neurons is achieved, in fact, by multimodal S-cells providing input to L-neurons. There is some evidence that this occurs in the cockroach [37].

Clearly, the complexity of anatomical arrangement and the richness of synaptic interactions in the insect ocellar system defies the label of "simple visual system". While a single, well defined role for this system remains elusive, there is clear evidence that the ocelli can provide a rapid conduction pathway which integrates with information descending to centres controlling orientation, locomotion and flight.

References

1 Y. Toh and H. Tateda (1991). Structure and function of the insect ocellus. *Zool. Sci.* 8: 395-413.

2 L.J. Goodman (1981). Organisation and physiology of the insect dorsal ocellar system. In Autrum (Ed) *Comparative Physiology and Evolution of Vision in Invertebrates. Invertebrate Visual Centers and Behavior II. (Handbook of Sensory Physiology: Vol VII/6C)*. Springer, Berlin, Heidelberg, New York. pp 2011-286.

3 J.A. Patterson (1976). Neuronal architecture and function in the ocellar system of the locust. In *Neural Principles in Vision.* Ed by F. Zettler and R Weiler. Springer, Berlin, Heidelberg. pp 334-353.

4 L.J. Goodman (1974). The neural organisation and physiology of the insect dorsal ocellus. *In Compound eye and Vision of Insects*. Ed by G.A. Horridge. Oxford: Clarendon pp 515-548.

5 M. Wilson (1978a). The functional organisation of locust ocelli. *J. Comp. Physiol.* 124: 297-316.

6 H. Schuppe and R. Hengstenberg (1993). Optical properties of the ocelli of *Calliphora erythrocephala* and their role in the dorsal light response. *J. Comp. Physiol.* A. 173: 143-149.

7 L.J. Goodman, J.A. Patterson and P.G. Mobbs (1975). The projection of ocellar neurons within the brain of the locust, Schistocerca gregaria. *Cell. Tiss. Res.* 157: 467-492.

8 C.S. Goodman (1974). Anatomy of locust ocellar interneurons: constancy and variability. *J.Comp. Physiol.* 95: 185-201.

9 C.S. Goodman (1976). Anatomy of the ocellar neurons of acridid grasshoppers. I. The large interneurons. *Cell. Tiss. Res.* 175: 166-183.

10 J.A. Patterson and R.L. Chappell (1980). Intracellular responses of Procion filled cells and whole nerve cobalt impregnation in the dragonfly median ocellus. *J. Comp. Physiol.* 139: 25-39.

11 R.L. Chappell, L.J. Goodman and J.B. Kirkham (1978). Lateral ocellar nerve projections in the dragonfly brain. *Cell Tiss. Res.* 190: 99-114.

12 C.S. Goodman and J.L.D. Williams (1976). Anatomy of the ocellar neurons of acridid grasshoppers. II. The small interneurons. *Cell. Tiss. Res.* 175: 184-203.

13 R.L. Chappell and J.E. Dowling (1972). Neural organisation of the median ocellus of the dragonfly. I. Intracellular electrical activity. *J. Gen. Physiol.* 60: 121-147.

14 J.A. Patterson and L.J Goodman (1974a). Intracellular responses of receptor cells and second-order cells in the ocelli of the desert locust, *Schistocerca gregaria*. *J. Comp. Physiol.* 95: 237-250.

15 M. Wilson (1978b). Generation of graded potential signals in the second order cells of locust ocellus. *J. Comp. Physiol.* 124: 317-331.

16 M. Wilson (1978c). The origin and properties of discrete hyperpolarising potentials in the second order cells of locust ocellus. *J. Comp. Physiol.* 128: 347-358.

17 R.G. Guy, L.J. Goodman and P.G. Mobbs (1979). Visual interneurons in the bee brain: Synaptic organisation and transmission by graded potentials. *J. Comp. Physiol.* 134: 253-264.

18 P.J. Simmons (1981). Synaptic transmission between second- and third-order neurones of a locust ocellus. *J. Comp. Physiol.* 145: 265-276.

19 P.J. Simmons (1982a). Transmission mediated without and without spikes at connexions between large second-order neurones of locust ocelli. *J, Comp. Physiol.* 147: 401-414.

20 P.J. Simmons (1982b). The operation of connexions between photoreceptors and large second-order neurones in dragonfly ocelli. *J. Comp. Physiol.* 149: 389-398.

21 P.J. Simmons (1993). Adaptation and responses to changes in illumination by second- and third-order neurones of locust ocelli. *J. Comp. Physiol.* A. 173: 635-648.

22 S. B. Laughlin (1973). Neural integration in the first optic neuropile of dragonflies. *J. Comp. Physiol.* 84: 335-355.

23 A. Dubs, S.B. Laughlin and M.V. Srinivasan (1981). Single photon signals in fly photoreceptors and first order interneurones at visual threshold. *J. Physiol. (Lond.)* 317: 317-334.

24 S.B. Laughlin and R.C. Hardie (1978). Common strategies for light adaptation in the peripheral visual systems of fly and dragonfly. *J. Comp. Physiol.* 128: 319-340.

25 J.E. Dowling and R.L. Chappell (1972). Neural organisation of the median ocellus of the dragonfly. II. Synaptic structure. *J. Gen. Physiol.* 60: 148-165.

26 A. Klingman and R.L. Chappell (1978). Feedback synaptic interaction in the dragonfly ocellar retina. *J. Gen. Physiol.* 71: 157-175.

27 S. Stone and R.L. Chappell (1981). Synaptic feedback onto photoreceptors in the ocellar retina. *Brain Res.* 221: 374-358.

28 S.R. Shaw (1975). Retinal resistance barriers and electrical lateral inhibition. *Nature (Lond.)* 255:480-483.

29 L.J. Goodman (1965). The role of certain optomotor reactions in regulating stability in the rolling plane during flight in the desert locust, *Schistocerca gregaria*. *J. Exp. Biol.* 42: 385-407

30 R. Jander and C.K. Barry (1968). Die phototaktische Gegenkopplung von Stirnocellen der Hauschrecken und Grillen (Saltatoptera: *Locusta migratoria* und *Gryllus bimaculatus*). *Z. vergl Physiol.* 57: 432-458.

31 J.A. Patterson and L.J Goodman (1974b). Relationships between ocellar units in the ventral nerve cord and ocellar pathways in the brain of *Schistocerca gregaria*. *J. Comp. Physiol.* 95: 251-262.

32 C. Griss and C.H.F. Rowell (1986). Three descending interneurons reporting deviation from course in the locust. I. Anatomy. *J. Comp. Physiol.* 158: 765-774.

33 C.H.F. Rowell and H. Reichert (1986). Three descending interneurons reporting deviation from course in the locust. II. Physiology. *J. Comp. Physiol.* A. 158: 775-

794.
33 C.H.F. Rowell and H. Reichert (1991). Mesothoracic interneurons involved in flight steering in the locust. *Tissue and Cell.* 23: 75-139.
34 C.H.F. Rowell and H. Reichert (1991). Mesothoracic interneurons involved in flight steering in the locust. *Tissue and Cell.* 23: 75-139.
35 C.H.F. Rowell (1991). The organisation of exteroceptive sensory inputs to interneurons of the flight neuropil of locusts. *Tissue and Cell.* 23: 271-276.
36 H. Kondo (1978). Efferent system of the lateral ocellus in the dragonfly: Its relationships with the ocellar afferent units, the compound eyes and the wing sensory system. *J. Comp. Physiol.* 125: 341-349.
37 T. Ohyama and Y. Toh (1986). Multimodality of ocellar interneurons of the american cockroach. *J.Exp. Biol.* 125: 405-409.
38 S. Yamazaki and S. Yamashita (1991). Efferent control in the ocellus of a noctuid moth. *J. Comp. Physiol.* A. 169: 647-652.
39 S. Rotzler (1989). Possible efference to the ocellar system of the locust and its effect on ocellar interneurons. *J. Comp. Physiol.* A. 165: 27-34.

BASIC MECHANISMS OF OCULOMOTOR CONTROL

ANGELO BUIZZA
Dip. di Informatica e Sistemistica, Università di Pavia
via Abbiategrasso 209 - I-27100 Pavia - Italy

INTRODUCTION

In many vertebrates, in particular man and primates, an important part of the information coming from the environment is collected and processed by vision. Then, these animals have developed both sophisticated visual system and effective eye movement, or *oculomotor*, system (OMS). Both in higher and lower animals, the main role of eye movement (EM) is to serve vision and ensure optimal conditions for it. Then EMs may be different in different animals, depending on the particular features and needs of the animal's visual system.

In *afoveate animals*, EM should mainly prevent the image of the environment from slipping on the retina, in order to obtain stable vision. Then OMS in these animals reduces to basic image-holding reflexes (vestibulo-ocular and optokinetic reflex) coupled with a fast position-resetting mechanism (afoveate saccade). Animals with *fovea* (or any other, functionally similar, retinal area, such as cat's *macula*) need both to stabilize images and to foveally fix and pursue specific objects. Then, *foveate animals* have got both image-holding reflexes and other kinds of eye movements aimed at orienting gaze, keeping fixation, and pursuing moving objects. Finally, *frontal-eyed* (usually foveate) animals need vergence to look at different distances and to make the images of one object to fall onto corresponding areas of the right and left retina, in order to be fused.

Then, the oculomotor system is a complex system which evolved from stabilizing reflexes, to either voluntary or involuntary gaze orienting and visual scanning movements, fixation control, pursuit, and vergence.

The link between eye movements and vision is not simply one of solution of stability and control problems. Indeed EMs are relevant to active vision. Like the movement of any other body segment, EM is the result of a complex process starting with information acquisition from the environment, passing through movement goal selection and motor programming, and eventually resulting in movement execution. Eye movements endlessly explore the environment and allow the visual system to collect a continuous flow of data, which, in turn, is used by the central nervous system to build its internal representation of the world and to decide and programme appropriate motor (re)actions, including further EMs aimed at gathering more information. In such a process, binocular vision contributes to depth perception.

The fulfilment of the many diverse duties of the OMS easily may create conflict situations. For instance, foveal pursuit of a moving object is incompatible with scan movements which must stop during pursuit. A hierarchical control structure is then required, in which higher levels are in charge of conflict solution by selecting specific goals and deciding the consequent priorities. Conflicts may also arise from the interaction with head movements. Head is the platform supporting the oculomotor apparatus and, in principle, its movement may disturb oculomotor control. In many situations, stabilizing reflexes help to reduce and possibly eliminate OMS sensitivity to head movements.

Finally OMS is plastic and in proper situations can significantly modify its behaviour in order to adapt to particular conditions and/or stimuli, improve performance by learning, recover from lesions, etc..

Eye movements may appeal to many researchers, from neurophysiologists to people working on machine vision and robotics. This widespread interest has promoted many multidisciplinary studies, with consequent remarkable growth of knowledge. Nevertheless, understanding of many aspects of oculomotor behaviour, in particular plasticity, hierarchical control and relationships with vision, is still unsatisfactory in many respects.

In the following sections a concise account of the overall structure of the oculomotor system, the main features of its components, and possible relationships with vision will be given.

DIFFERENT TYPES OF EYE MOVEMENT

Eye movements are defined in terms of rotation of the eyeball from its primary position (line of sight parallel to the head medial plane). They may be either *vergence* (disconjugate) or *version* (conjugate) movements. The former allow fixation of objects moving away from (*divergence*) or toward (*convergence*) the observer. During vergence the eyes make equal and opposite rotations. During version they rotate by the same angle in the same direction so as to move the line of sight without changing the fixation distance. In natural conditions EM usually consists of both vergence and version components, that easily can be separated in laboratory conditions.

Both vergence and version movements may be either *slow* or *fast*. This distinction relies on different speed but also on different neural substrates and control modes. Fast movements (*saccades*), are ballistic in nature, and behave as if they were controlled in discrete time. Slow movements are smoother and controlled in continuous time.

Like other kinds of movement, EMs may be either *spontaneous*, *reflex*, or *voluntary*.

Spontaneous movements are independent of both subject's will and external stimuli. Most of the saccadic scan, both in the light and in the dark, EMs during sleep, and the so-called *physiologic nystagmus* are typical examples. The latter consists of continuous, high frequency (30-80 Hz), small amplitude (less than 1') vibration of the eyeball around its position (a kind of tremor) with associated drift (amplitude of a few minutes of arc) and *microsaccades* (amplitude of 1' to 20'). Stabilization of the seen object to the eyeball by means of devices preventing the retinal image movement produced by these small EMs (e.g. contact lenses) results in image fade-out. Then the role of physiologic nystagmus seems to be refreshing of the image on the retina in order to prevent fading due to the phasic behaviour of the retinal receptors.

Reflex movements are "automatic" responses to specific stimuli, basically independent of subject will although in many situations voluntary control (e.g. enhancement or reduction) is possible. EMs produced by the vestibulo-ocular and the optokinetic reflexes are the most typical examples. The former is activated by head movements stimulating the vestibular receptors, the latter by visual stimuli.

Finally voluntary movements strictly depend on subject's will. Saccades may be executed voluntarily and independently of external stimuli (e.g. to orientate the gaze to a new fixation point) whereas smooth pursuit is the main example of voluntary movement requiring an external stimulus, i.e. a moving object to be pursued (°).

OCULOMOTOR SYSTEM ANATOMY[1-4]

The *oculomotor plant* consists of the *eye-balls*, the *ocular muscles*, and the relevant *ligaments* and *connective tissue*. The eye rotates in the orbit as a sphere with three degrees of freedom and bounded movements. Translational movements are almost absent and functionally irrelevant. Rotations are produced by coordinated contraction and relaxation of six ocular muscles surrounding the eye-ball, arranged in three agonist-antagonist pairs (*lateral* and *medial rectus*, *superior* and *inferior rectus*, *superior* and *inferior oblique*).

Ocular muscles receive neural command from the *oculomotor nuclei* (OMN) through the *oculomotor nerves*. All neural commands from the different subsystems controlling EMs eventually reach the OMN where a neural signal is built coding the eye movement to be done.

The eye movements elicited by visual stimuli are organized in a loop originating from *retinal receptors* and reaching the OMN through complex neural circuits including, among others, the *optic nerve*, the occipital and pre-occipital *visual areas*, the *striate cortex*, the medial and the superior medial *temporal areas*, the *tectal* and *pretectal nuclei*, and the *superior colliculi*. The *frontal eye fields* and the superior colliculi are two major inputs to saccadic system, the striate cortex and temporal areas are involved in pursuit control.

The *vestibulo-ocular reflex* arc originates from the vestibular receptors (*semicircular canals* and *otolith maculae*), projects onto the *vestibular nuclei* through the *vestibular nerve*, and reaches the oculomotor nuclei both directly through the *medial longitudinal fasciculus* and indirectly through complex pathways involving the *cerebellum*, the *pontine reticular formation*, and possible other structures. The vestibular nuclei are considered as a centre for internal reconstruction of head movement in space. They also receive visual input, through the *optokinetic system*, and input from *neck receptors*. Neck receptors measure the head movement with respect to trunk and drive a further reflex which is referred to as the *cervico-ocular reflex*.

The cerebellum is likely to have a regulatory function in movement coordination[5]. Then it would play a major role in coordinating the different kinds of EM with each other and with the movements of other body segments, in particular the head.

(°) Although smooth pursuit largely depends on will, it is not clear whether it is a voluntary movement or a reflex particularly sensitive to voluntary control and/or requiring voluntary trigger.

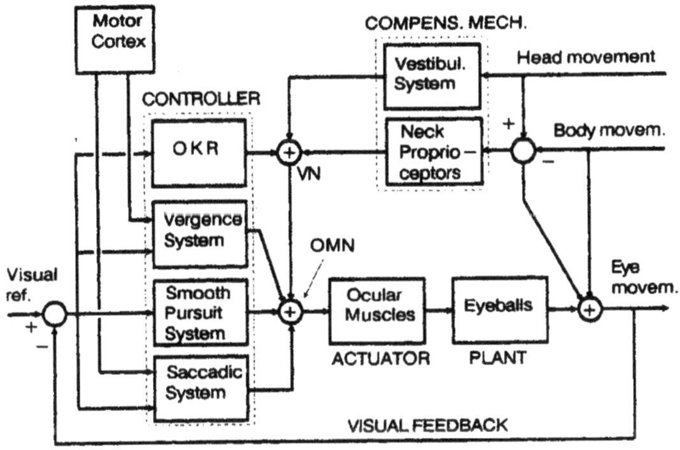

Figure l - A functional scheme of the oculomotor control system.

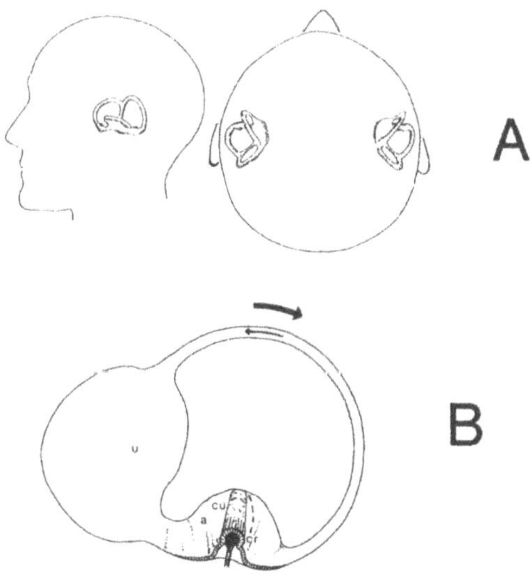

Figure 2 - **A**. Location of the semicircular canals in the skull, lateral (on the left) and top (on the right) views. For the sake of clarity, canal dimensions have been magnified; the actual curvature radius of each canal is of a few millimeters. **B.** Anatomical scheme and working principle of the semicircular canal. **a** = ampulla, **cu** = cupula, **cr** = crista ampullaris; **u** = utricule; the dashed curve shows the shape of the cupula when bent by the pressure produced by the endolymph flow (thin arrow).

The overall structure of the oculomotor system may be synthesized as in the block diagram of Fig. 1. It is possible to identify in this scheme the typical components of a classical control system. The eyeball is the controlled plant with eye position or velocity as controlled variable (output). The ocular muscles are the actuator. The saccade, smooth pursuit, optokinetic, and vergence systems form the controller. The EMs elicited by visual stimuli are controlled by a negative feedback loop with the visual reference movement as input variable. The closed loop structure results from measuring the reference movement by means of the retinal receptors which move with the eye and then measure the reference movement relative to the eye, that is the error signal to be zeroed by the oculomotor control. Finally, the vestibulo-ocular and the cervico-ocular reflexes can be seen as compensatory mechanisms aimed at counterbalancing the destabilizing effect of head and/or body movement.

The description of OMS in terms of control theory[2] provided a sound interpretative framework and helped to understand the properties of its different components. However, a comprehensive picture of the whole system including the interaction among the different subsystems is still lacking.

VESTIBULO-OCULAR REFLEXES

The vestibulo-ocular reflexes[2,4,6,7] represent the oldest oculomotor systems. They are aimed at compensating head movements by equal and opposite eye movements in order to keep the gaze orientation unchanged.

Head movements are sensed by the *vestibular receptors*. The *vestibule* (non-auditory part of the inner ear) contains the *semicircular canals* and the *otoliths*, which basically are inertial accelerometers. The former are sensitive to head angular accelerations and the latter to accelerations with no rotary components (linear accelerations). In man and frontal eyed animals canal-ocular reflexes, compensatory for head rotation, are much more important than otolith-ocular reflexes, which produce rudimentary compensatory EM only.

The semicircular canals form two groups of three mutually orthogonal canals, one group on each side of the skull (Fig. 2-A). When the head is in its normal upright position, the *horizontal canal* lies in an almost horizontal plane, the *anterior* and the *posterior canal* in vertical planes. The canals have a toroidal shape with a wider chamber (*ampulla*) closed in the middle by a transverse, water-tight, gelatinous wall (*cupula*) surrounding the hair processes (*cilia*) of the sensory cells. The canals are filled with a fluid (*endolymph*) and their wall (*membranous duct*) is fixed to the skull. When the head rotates, inertia makes the endolymph to move with respect to the canal's wall and to bend the cupula and, consequently, the cilia (fig. 2-B). Through mechanisms still largely unknown, deformation of the cilia produces electrical signals modulating the discharge rate of primary vestibular neurons. They project onto the vestibular nuclei where a neural signal is made available which basically codes head rotation velocity. It is sent to the OMN both monosynaptically, through the *medial longitudinal fasciculus*, and polysynaptically, through the *pontine reticular formation*. Projections from OMN to ocular muscles are organized so that head

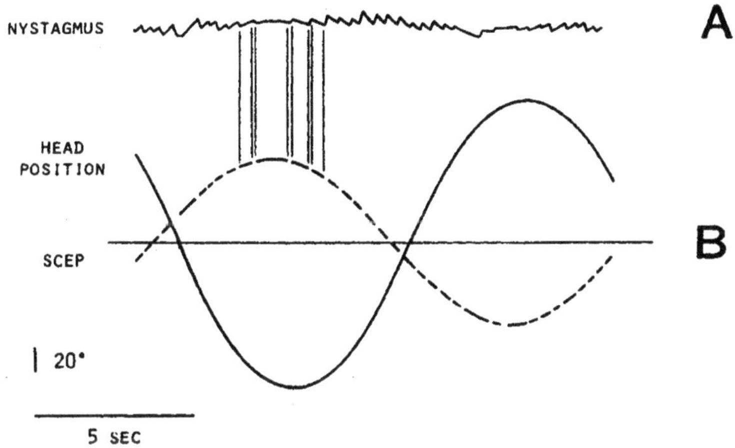

Figure 3 - **A.** Example of horizontal nystagmus induced by harmonic subject oscillation (240 peak-to-peak at 0.07 Hz) around a vertical axis in the dark. **B.** By removing nystagmus fast phases and fitting together the slow ones, the equivalent eye movement produced by the canal ocular reflex can be reconstructed (SCEP), which is almost phase opposite with respect to head rotation. Notice that in this case compensation is uncomplete because SCEP amplitude is less than head rotation amplitude.

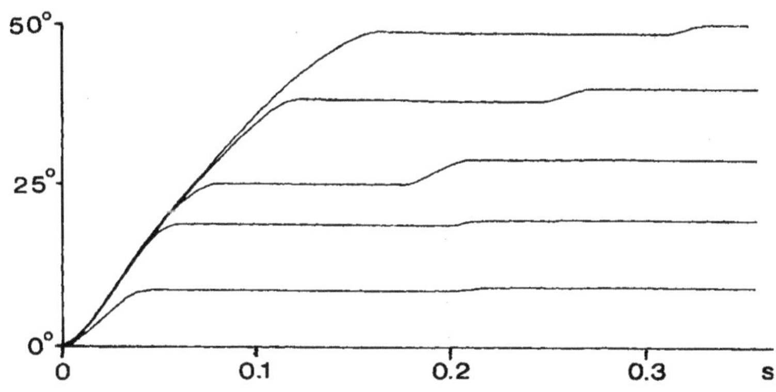

Figure 4 - Time course of eye position during saccades of different amplitudes. Notice the typical sigmoid profile. Eye excursions larger than 20° are usually covered by two saccades (redrawn using data from Becker[18]).

rotation in one direction results in eye rotation in the opposite direction.

The canal-ocular reflex can be tested by submitting subjects to controlled rotations[4,6]. Due to its working principle, each canal is sensitive to rotation in its own plane and the geometric arrangement results in decomposition of head rotation in three mutually orthogonal components. Usually, only the horizontal reflex is tested, which requires rotations around a vertical axis. To this purpose, the subject is seated on a rotating chair, in the dark, to avoid visuo-motor reactions, and with his/her head restrained, to prevent stimulation of the neck receptors and the cervico-ocular reflex. In these conditions a typical pattern of eye movement can be recorded, made of alternating *slow and fast phases* (*vestibular nystagmus*, Fig. 3). Slow phases are *compensatory* in nature and represent the response of the canal-ocular reflex. Fast phases are *anticompensatory*, saccade-like, reset (°°) movements. By removing the nystagmus fast phases and fitting together the slow ones, the overall equivalent response of the canal-ocular reflex can be reconstructed (slow cumulative eye position, Fig. 3-B).

OPTOKINETIC REFLEX

Head movement compensation provided by the canal-ocular reflex may prove inadequate for three main reasons: i) the inertial principle on which its function relies makes the canal scarcely sensitive to slow and/or constant velocity and/or low frequency head rotation; ii) the reflex gain (ratio of the eye to head rotation) may be less than one (see Fig. 3-B); iii) the canal-ocular reflex is open-loop and may not be as accurate as required for full stabilization. When subject is moving in a still visual environment, uncomplete compensation results in residual gaze shift and consequent image slip on the retina. The slip on the retina of the image of a large portion of the visual field is the specific stimulus for a further stabilizing reflex, the *optokinetic reflex*[2,4,6,7]. It produces eye rotations aimed at suppressing the image slip itself, so holding the gaze still. When stimulated by a continuous movement of the scene the optokinetic reflex too produces nystagmus (*optokinetic nystagmus*).

When the subject moves in a stable visual environment, as in every-day life, the eye movements induced by the vestibulo-ocular and the optokinetic reflexes are synergistic and cooperate in better gaze stabilization. The two reflexes are so tightly coupled that the definition of *vestibulo-optokinetic system* has been proposed[4]. Actually, they have fully complementary roles. The vestibulo-ocular reflex is high-pass and open-loop whereas the optokinetic one is low-pass and closed-loop, and provides a feedback also to its companion. Their cut-off frequencies are close to each other so that when they work together head movement can be compensated for on a broad frequency band. They interact at the level of the vestibular nuclei and produce a neural signal representing a likely central estimate of head movement in space. In addition to gaze stabilization, it is probably used for postural model reconstruction, posture control, perceptual stability of the environment, self- and

---

(°°) Probably, anticompensatory phases are not simple reset movements, but do reveal more subtle aspects of vestibulo-saccadic interaction[4].

object-motion sensation, etc.

SACCADIC SYSTEM

Saccades[8] are fast, ballistic eye movements that can rotate the eye with velocities as high as 600°/s in man. Many behaviourally distinct eye movements belong to this class (visual scanning and search movements[9], fast tracking movements, eye movements produced during reading[10], spontaneous movements in the dark and during REM sleep, fast phases of nystagmus) and are probably controlled by the same neural structures. They are used whenever fast gaze shift is needed. It has been suggested that their neural command implements a minimal-time, optimal control law[11].

Saccades may be produced over a large amplitude range, from a few minutes of arc (microsaccades) up to 80°-90°. In laboratory they are elicited by using controlled target jumps. In this case the saccade starts about 200 ms after the stimulus onset and reaches the target with a highly repeatable steplike movement (Fig. 4), very hard to modify after iniziation. Movements larger than about 20° are usually performed with two or more saccades separated by 80-130 ms intervals (Fig. 4). Independently of the number of saccades, the target can always be very accurately foveated.

As in the double-step saccades of Fig. 4, a new saccade cannot be produced before an interval of 80-200 ms from the previous one. Then saccade is commonly thought of as a movement controlled in discrete-time.

SMOOTH-PURSUIT SYSTEM

When tracking continuous target movement, the saccadic system can only provide a staircase approximation of it. To cope with this situation, foveate, frontal-eyed animals have developed a specialized system, the *smooth-pursuit system*[2,4].

Unlike saccades, smooth-pursuit movements may be produced only in the presence of a moving stimulus. A visual target is the specific stimulus, but examples of pursuit of non-visual (e.g. tactile) stimuli have been described. This lead to the hypothesis that pursuit is actually driven by an internal signal coding the movement of an object of interest and reconstructed from information provided by different sensory channels[12]. Of course, visual information, when available, plays a major, dominant role in reconstruction.

Smooth-pursuit control is classicaly thought of as a velocity servomechanism. It has initial response latency of 100-150 ms and can accurately match regular and continuous target movements provided their speed doesn't exceed 90-100 °/s and, in the case of periodic movement, frequency is limited to about 1 Hz. Physical and geometrical stimulus features, stimulation conditions, background, and target movement profile may further reduce the above limits[13], and tracking of fast targets requires corrective saccades whose importance increases with target speed (Fig. 5).

VERGENCE SYSTEM

In spite of their relevance, vergence movements[4,2,14] are still little known. They are

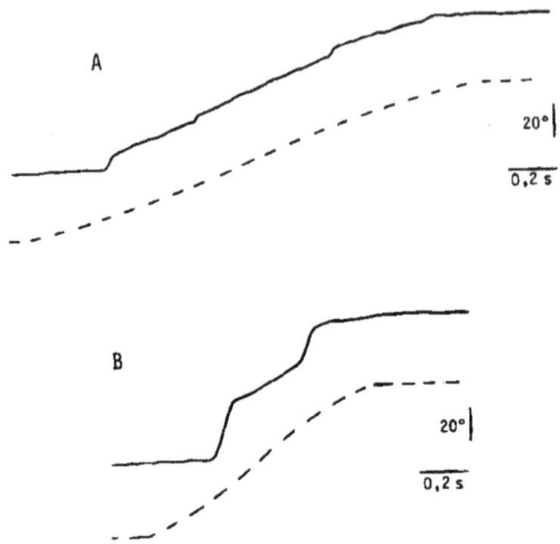

Figure 5 - Examples of eye position during tracking (continuous line) of moving targets (dashed lines). **A.** The target moves from 45° left to 45° right with maximal speed of 60°/s and tracking is mainly performed by smooth pursuit movement with small corrective saccades. **B.** The target moves as above but has maximal speed of 120°/s: most of the tracking is made of two large saccades.

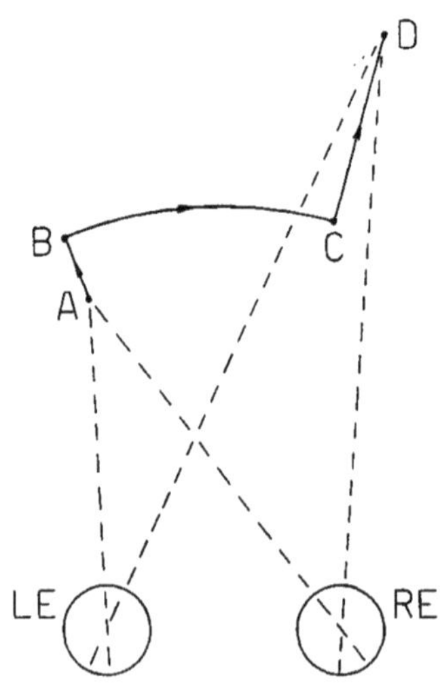

Figure 6 - Typical combination of version and vergence during rapid gaze orientation. due to shortest latency, the eye movement begins with vergence (A-B). Then a rapid versional saccade shifts the line of sight to the new direction (B-C). Final vergence brings the eyes on the new fixation point (C-D).

present in frontal-eyed animals, serve image fusion and stereopsis, have a latency of about 160 ms, and are slower and less accurate than other foveation movements, *divergence* being slower than *convergence*.

Specific stimuli for vergence are both retinal image disparity, eliciting *fusional vergence*, and image blur, eliciting *accomodative vergence* in association with lens accomodation. Fusional vergence results in reduction of the disparity that induced it, i.e. it's closed-loop controlled. Accomodative vergence has no effect on its own cause, then it is basically open-loop controlled. In physiologic conditions the two mechanisms work together and this should result in almost closed-loop control.

In every-day life, vergence and version movements are used together. An example of saccade-vergence coordination is given in Fig. 6.

COORDINATION, HIERARCHICAL CONTROL, AND PLASTICITY

Head is the moving platform supporting the oculomotor apparatus and, in turn, is supported by a further moving structure, the body. This situation has two major consequences. On the one hand, head and body movements may be detrimental to oculomotor control. On the other hand, coordinated body, head, and eye movements may extend the possibilities of gaze control.

During fixation of still objects, gaze stabilization against head movement is ensured by the cooperative action of vestibulo-ocular and optokinetic (and possibly cervico-ocular) reflexes, but the presence of stabilizing reflexes may be counterproductive in other conditions. For instance, when pursuing by head *and* eye movements a target moving to the right in front of a structured background, the eyes would be driven to the left by both the vestibulo-ocular reflex, stimulated by head rotation, and the optokinetic reflex, stimulated by the retinal slip of the background image. Actually it doesn't happen so and the eyes can easily pursue the target, provided its motion isn't too fast. This suggests that *in these and similar conditions* reflex movements are suppressed, or at least reduced, to allow easy pursuit. How this is obtained is still a matter for speculation[15].

The above example shows that conflicts easily may arise among the different oculomotor subsystems in the fulfilment of their tasks. Some of them are likely to be "automatically" solved by system organization and the presence of stabilizing reflexes. In other cases, conflict solution requires a hierarchical control to define goals, and consequent priorities, and to coordinate EM with movement of other body segments, in particular the head. Hierarchical control means a multi-level or multi-layer structure in which the higher levels/layers can make decision and intervene and modify the behaviour of the lower ones in order to meet particular needs or to cope with particular situations. The organization and the neural substrate of this structure are still unknown. Cerebellum is likely to play a major role and coordinate different motor systems[5].

Hierarchical control may also justify the fact that OMS can modify its behaviour in particular situations (plasticity)[16]. Plasticity includes: learning of repetitive patterns of target motion with consequent improved performance; progressive suppression of reflex response when repeatedly stimulated in conditions where it is purposeless (e.g. vestibular nystagmus in the dark); performance adaptation to particular working conditions (e.g. vestibular response enhancement or reduction induced by wearing magnifying or reducing goggles, respectively; reversal induced by reversing prisms); recovery after lesions.

EYE MOVEMENTS AND VISION

It is commonly admitted that EMs serve vision, but, in spite of such a general agreement, relationships between eye movement and vision are still poorly understood. Many studies dealt with the influence of vision conditions (luminance, contrast, target dimension, retinal location of target image, background, etc.) on OMS performance (various reviews are available, e.g. [4,2,13,17,18]). In the present context, it seems more interesting to examine how eye movements influence vision, a topic with a long and controversial history. Nowadays it's generally accepted that EMs are necessary to vision but say little about processing of visual information and, in particular, cognitive processes. A few facts are listed below. For a more comprehensive picture the reader is referred to various reviews available in the literature (e.g. [4,17] and in particular [19]).

Physiologic nystagmus prevents retinal image from fading. Its main role seems image refreshing to counterbalance the effect of the phasic behaviour of the retinal receptors. It's not clear whether it plays other roles in image processing, e.g. edge and contour detection.

During saccades, vision is at least partially suppressed (*saccadic suppression*) and this should be related with *retinal blur* resulting from the high retinal image velocity produced by saccade. It probably has a functional role and it isn't a passive consequence of blur, because it begins before saccade onset. It might be a means to prevent possible undesirable effects on perceptual stability of the environment.

It is admitted that *smooth pursuit* contributes to object motion perception and effects of retinal image velocity on perceived stimulus velocity have been described during both smooth pursuit and *optokinetic nystagmus*. Nevertheless retinal configuration change seems to provide much more effective information[20].

Binocular vision supporting stereopsis needs *vergence* movements, however vergence *in se* seems to give only minor, if any, contribution to depth perception[14].

Saccades are used to scan figures and scenes, however recognition of complex figures is possible even if eye movement is prevented[21]. Then saccades seem to be mainly involved in detailed analysis of the visual scene, but the exploration strategies are almost impossible to discover because many factors (expectation, emotional involvement, context, task) may influence scan patterns[9] and make their interpretation hard.

In summary, it seems appropriate to conclude that "empirical evidence from eye movement research, by itself, has been insufficient to specify models for the perceptual and cognitive processes involved in the visual exploration of the world"[21].

REFERENCES

1. R.A. Moses (ed.) *Adler's physiology of the eye*, Mosby, Saint Louis, 1970.
2. D.A. Robinson, in *Handbook of Physiology*, sect. 1, vol. II, part 2, V.B. Brook, ed., American Physiological Society, Bethesda (1981), p. 1275.
3. J.A. Büttner-Ennever (ed.). *Neuroanatomy of the Oculomotor System*, Elsevier, Amsterdam, 1988.
4. R.J. Leigh, D.S. Zee. *The Neurology of Eye Movements*, 2nd ed., Davis, Philadelphia, 1991.
5. W.A. MacKay and J.T. Murphy, *Progr. Neurobiol.* 13: 361 (1979).
6. R.W. Baloh, V. Honrubia. *Clinical Neurophysiology of the Vestibular System*, Davis, Philadelphia, 1979.
7. V.J. Wilson, G. Melvill Jones. *Mammalian Vestibular Physiology*, Plenum Press, New York, 1979.
8. R.H. Wurtz and M.E. Goldberg (ed.s). *The neurobiology of saccadic eye movements*, Elsevier, Amsterdam, 1989.
9. A.L. Yarbus. *Eye movements and vision*. Plenum Press, New York, 1967.
10. J.K. O'Regan, in *Eye Movements and their Role in Visual and Cognitive Processes*, E. Kowler, ed., Elsevier, Amsterdam (1990), p. 395.
11. M.R. Clark, L.W. Stark, *IEEE Trans.* AC-20: 345 (1975).
12. L.R. Young, S. Yasui, *Science* 190: 906 (1975)
13. E. Kowler, in *Eye Movements and their Role in Visual and Cognitive Processes*, E. Kowler, ed., Elsevier, Amsterdam (1990), p. 1.
14. H. Collewijn, C.J. Erkelens, in *Eye Movements and their Role in Visual and Cognitive Processes*, E. Kowler, ed., Elsevier, Amsterdam (1990), p. 213.
15. V.P. Laurutis, D.A. Robinson, *J. Physiol. (Lond.)* 373: 209 (1986).
16. A. Berthoz, G. Melvill Jones (ed.s) *Adaptive mechanisms in gaze control*. Elsevier, Amsterdam, 1985.
17. R. H.S. Carpenter. *Movements of the Eyes*, Pion, London, 1977.
18. W. Becker, in *The neurobiology of saccadic eye movements*, R.H. Wurtz and M.E. Goldberg, ed.s, Elsevier, Amsterdam (1989), p. 13.
19. E. Kowler (ed.). *Eye Movements and their Role in Visual and Cognitive Processes*. Elsevier, Amsterdam, 1990.
20. H. Wallach, in *Eye Movements and their Role in Visual and Cognitive Processes*, E. Kowler, ed., Elsevier, Amsterdam (1990), p. 289.
21. P. Viviani, in *Eye Movements and their Role in Visual and Cognitive Processes*, E. Kowler, ed., Elsevier, Amsterdam (1990), p. 353.

FILTERING OF THE INPUT IMAGE AND VISUAL PERCEPTION OF GEOMETRICAL FIGURES*

Vito Di Maio
Institute of Cybernetics of Italian National Research Council, Via Toiano 6, I-80072 Arco Felice (NA), Italy

1. INTRODUCTION

In the recent years, in co-operation with Petr Lánský, we have developed a model to explain some findings in visual perception of area of geometrical figures. This model is based on the idea that visual stimuli are spatially filtered before any cognitive evaluation takes place. This means essentially that the stimulus really evaluated is a transformation of the physical stimulus. This concept agrees with several psychophysical and with electrophysiological evidences [18,19]. The problem of what are the structures and/or the mechanisms that produce the spatial filtering of the input image, remains unresolved. It seems that retinal ganglion cells are the best candidates to this function [16,17] but, recently, some retinal models take into account the role of some neurotransmitters (Dopamine) released by D-amacrine cells in modulating spatial frequency by changing the electrical properties of other retinal cells [4]. Although some authors believe that the filtering along the optical pathway is negligible [17], my personal opinion is that this contribution cannot be completely neglected. I think that several filtering structures and mechanisms act during the early processes of vision and the contribution of all of them is important to the total filtering of the input image. The nature of the filtering mechanisms in the visual pathway, however, is unimportant in our model and so I will avoid this problem by considering simply a general band-pass filter resulting from the linear combination of all the filters acting in the visual system. Although not explicitly declared, some different opinions circulate on the role of the filtering activity of the visual system. Some authors, in fact, believe that they produce essentially a corruption of the input image rendering more difficult the work of the visual cortex in recovering from it the essential informations [33]. Others, believe that the filtering processes can be useful mechanisms for object recognition by the brain [24] that can be successfully employed also to built up algorithms for artificial vision [4,14-15,21,24,27,35-36]. Tough the latter hypothesis is currently more accepted, some of our results (although interpretable in different ways) seem to confirm partially the former idea [8]. However, it is not in the porpoises of this paper to lead to a philosophical dispute about this problem. What is for granted, however, is that the input image is spatially filtered and this is all I need to explain the model. Before to go on, we need to understand what is meaning for spatial filtering of the input image and so we need at least a short explanation of the basic model.

---

* *To my dougther **FRANCESCA**, that positively changed my life since her birth (17th March 1994), and to my wife **MARIAROSA** that gave me this wonderful golden gift.*

2. BASIC MODEL

By definition, the simplest object that can be perceived is a dot and so I will start just from the perception of a dot. The smallest number of parameters to define a dot in a stimulus plane are: its position (x, y co-ordinates) and its brightness (*i.e.*, a numeric value related to the intensity of the radiant energy at that position in the stimulus plane). Parameters also important, as well as the contrast [31] and the colour are constant in our experiments and unimportant for our model and then they will not be considered here.

The input intensity function describing the point P_d of co-ordinates x_d, y_d can be a delta function the values of which in any point P of co-ordinates x, y in the stimulus plain is

$$I(p) = M\delta(p - p_d) \tag{1}$$

Where $\delta(.)$ is the delta function and M is a numeric value related to the brightness. If we have several dots (a dot pattern composed of n dots for example) the input intensity function will be

$$I(\overline{p}) = \sum_{i=1}^{n} M_i \delta(p - p_i) \tag{2}$$

where P_i is the i dot of the pattern and M_i is the numeric value corresponding to its brightness. The convolution of equation (1) or (2) with an appropriate filtering function can be a good way to describe the filtering of the input image [22,24,26,32,34-36] although some authors believe that the Fourier transformation is a better system [17]. The circular structure of the retinal receptive fields suggests that a good filtering function for this porpoises could be, for example, a 2D Gaussian the form of which is

$$G(p) = e^{-\frac{|p|^2}{2\pi\sigma^2}} \tag{3}$$

Where σ is a spatial constant related to the band-pass properties of the filter [26,32,35-36]. Since the $\delta(.)$ is a particular Gaussian function, the convolution of (2) and (3) will be another Gaussian and so we can write.

$$F(p) = I(p) \otimes G(p) = \sum_{i=1}^{n} M_i e^{-\frac{|p-p_i|^2}{2\pi\sigma^2}} = M_i G(p - p_i) \tag{4}$$

where $\otimes$ stays for convolution [27]. Figure 1A, 1C and 1E shown respectively the graph of this function, its contour plot and a 1D profile for the case of a single dot. This function is normally called Image Function (IF) of the input image [6-13,22,25-26,32,34-36]. This is the simplest example and several other can be done by using different filtering functions. The simple convolution of the input intensity function with the Gaussian function, however, fails in explaining some aspect of vision as well as some electrophysiological evidences related to the on-centre/off-surround (and *vice versa*) structure of the visual receptive fields [24]. According to Marr [24], a better operator as filtering function is the second derivative of the above Gaussian (*i.e.*, the Laplacian operator) that is of the form

$$\nabla^2 G(p) = \frac{-1}{\pi\sigma^4}\left(1 - \frac{|p|^2}{2\sigma^2}\right)e^{-\frac{|p|^2}{2\sigma^2}} \tag{5}$$

The corresponding IF obtained by the convolution of this function with equation (2) will be

$$F(p) = I(p) \otimes \nabla^2 G(p) = \sum_{i=1}^{n} M_i \nabla^2 G(p - p_i) \tag{6}$$

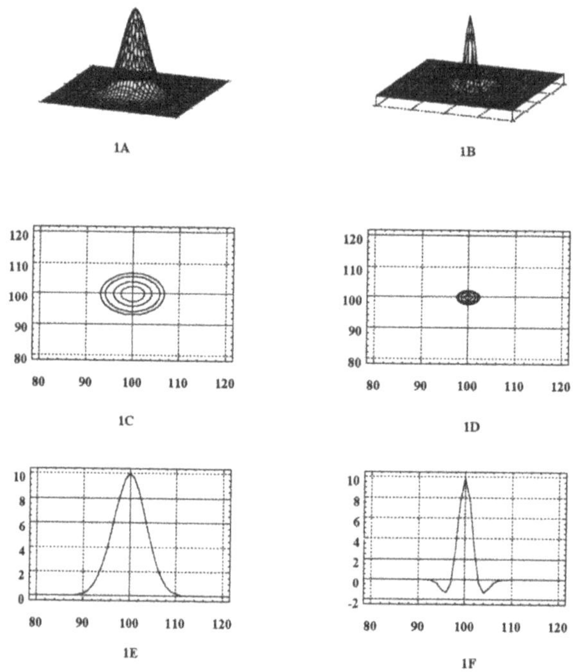

Figure 1- Image Function of a single dot: 1A Gaussian Form; 1B Laplacian Form; 1C and 1D respectively contour plots of 1A and 1B (for the Laplacian only the positive values are drawn in 1D); 1E and 1F are the 1D profiles of 1A and 1B.

Having both a positive and a negative part, this function well agrees with the structure of the receptive fields of the visual system and permits shape identification by the identification of its zero-crossing [24]. Figure 1B shows the IF of a dot by using the convolution of equation (1) and (6) and figure 1D and 1F show respectively its contour plot and 1D profile.

By using this function (or its modifications) several different algorithms for shape detection have been proposed as for example MIRAGE [35-36], MIDAAS [21] and HDC [15]. Other authors, however, have shown that this operator cannot satisfy completely for several reasons (as for example low ability in discriminating between edges and lines) and proposed different models as for example the model based onto the local energy distribution detected by identifying the converging phases of the Fourier components [28]. The Gabor function, however, seems more appropriate not only in describing the receptive fields' characteristics but also more coherent to what seems to be the role of the visual cortex in the process of vision [30]. For our porpoises, however, we do not need to go inside these details since just a common property of the filtered images is all that I need to explain our experimental results.

Being a line segment $\check{S}$ a dot pattern organized according to some geometrical rules, its IF can be directly obtained by the IF of a dot pattern by applying equations (4) or (6) expanding in one dimension the limiting properties by which the delta function is defined [3]. Since this represents the infinitesimal summation of the IFs of all the dots forming the line, for a general filtering function $H(p)$, we can write

$$F_1(p) = \int_{l_0}^{l_1} H(p - g(l))dl \tag{7}$$

where $g(l), l_0 \le l \le l_1$, is the parametric description of the line segment in the stimulus plane. A different way to obtain the IF of the line segment $\check{S}$ is to consider all the dots forming the line segment as having the same perceptual value and hence the same value of IF [35]. The IF at any point in the stimulus plain is then obtained by the IF of the closest dot belonging to the line segment [25,35]

$$F(p) = H(p - p_l) \tag{8}$$

The same consideration can be done for a polygon $\Re$ of m sides. The IF of $\Re$ can be either computed using equation (8) by the IF of the closest dot belonging to $\Re$, or by equation (7) by summing the IF of all the m sides forming $\Re$.

$$F_1(p) = \sum_{j=1}^{m} F_{1j}(p) \tag{9}$$

This last form of IF could give some problems of interpretation [26]. Due to the summation effect of the IFs of dots, the maximal values of IF would not necessarily correspond to the geometrical co-ordinates. In fact, in correspondence of the angles of the polygon there will be a higher density of dots producing a higher value of the IF in those points as well as a possible shift of positions of the maximal values with respect to the geometrical co-ordinates [25,26]. The size of this shift mainly depends on the amplitude of the angles of the polygon being higher for angles with lower amplitude [10-12]. This problem is avoided by using equation (8) [35]. In both cases, however, our model will explain the obtained results. In practice, when we have to work with concrete and complex figures, the parametric description of which can be of some difficulty, the IF is really computed using a discrete 2D input matrix $\boldsymbol{M}$ describing the stimulus plane (for example in pixels). The values of $\boldsymbol{M}$ will be a representation of the brightness of the image at any point (x, y) of surface of the plane considered for the figure representation. The application of the filtering function to such a descriptive matrix will give a resulting matrix $\boldsymbol{F}$ that corresponds to the IF of $\boldsymbol{M}$.

What is the physical meaning of the IF? It is simply that the radiant energy reflected (for example by a single dot) looses its characteristic of a pulse $[\delta(.)]$. The energy is dispersed around a central value corresponding to the geometrical co-ordinates of the dot. The distribution of this energy can be regulated in the model by varying M and σ [26].

The shape of the IFs resulting form of the above equations is important for our model. It is evident from the IF of a dot (Fig. 1) that the values all around the maximum are not so different from the maximum itself (see Fig. 1C, 1D and 1E, 1F). This characteristic is observable for all the above mentioned filtering functions (compare Fig. 1A-1B, 1C-1D, 1E-1F). Our assumption is that some values of the IF, different from the maximum and placed around it, can affect visual perception of the figure and as a consequence the evaluation of its geometrical properties. What is important in this respect, however, is not the absolute values of IFs able to affect the perception of figures but their distance from the maximal value.

3. MODEL FOR AREA PERCEPTION

3.1. General remarks

The problem of area perception has been addressed several times in the past with different and often contradictory results [37]. The larger part of the contributions concerning this aspect of visual perception has been given in the cognitive domain [1-2,29]. Piaget [29], regarding the judgement of quantities, divided the subjects into pre-operational (less then 6 year old) and post-operational (the rest). The pre-operationals do not integrated quantity so they usually judge only one dimension. The classical example is give by the experiment where subjects are asked to look a glass thin and tall, full of water. The water is then transferred in a glass large and short. Pre-operationals evaluates that in the first case there is more water than in the second one while post-operationals say that the quantity is the same in the two cases. By similar consideration holds the conclusion that pre-operational subjects fail more in judging area of geometrical figures while the post-operational practically gives correct answers. Anderson and Cuneo [1-2] rejected Piaget's theory and proposed that the two groups use different strategies in judgement of area of geometrical figures. In experiments of area evaluation of rectangles they used a fixed scale and rectangular cookies of different areas asking the subjects to align the cookies on the scale according to their area. The conclusions were that pre-operationals use an additive rule in estimating the area of rectangles (*i.e.* a model that follows the Height + Width rule) while post-operationals follow a multiplicative rule (*i.e.* a model that follows the Height x Width rule). The error committed in this case by the post operational are judged as due to the so-called cognitive algebra and to the relative model based on the Integration Information theory [1]. This theory cannot explain our results of area comparison of square and rectangle [25]. Being the geometrical areas of both the figures given by the multiplicative rule, according to the Anderson's model we would not expect systematic error in area comparison while we found the overestimation of the rectangle's area [25]. These considerations aimed our experiments and the need of a different model to explain area perception of geometrical figures. Before considering the model of Anderson we started our experiments with area comparison of square and circle. So let me start by these experiments.

3.2. Area comparison of square and circle

In this experiment we have presented simultaneously on the screen of an oscilloscope driven by a PDP11 computer a square and a circle. One of the two figures was fixed while the area of the other could be varied by the subject by rotating the knob of a potentiometer. Subjects were asked to match the area of the fixed figure by modifying the size of the adjustable one. Our results shown that the area of the square was overestimated with respect to that of the circle [8]. Now, let me to refer to the general model to explain this finding. Because we have used figures of the same brightness and for the sake of simplicity let me assume that the maximal value of the IF is the same and uniform in both the figures as for equation (8). If any value of IF placed at a distance *d* from the maximal value can affect the visual perception of area of these two figures, then, to compute the area of the perceived figures we need to add such *d* both to the side (a) of the square and to the radius (r) of the circle. The perceived areas of the circle (A_c) and square (A_s) will be respectively

$$A_c = \pi(r+d)^2 \qquad A_s = (a+2d)^2 \tag{13}$$

When the two geometrical areas are the same, the perceived areas are different and this is evident by a simple numeric example. If we have a square and a circle of the same area equal to 100 squared units (*i.e.*, A_c=A_s=100), then the side of the square will be 10 and the radius of the circle will be 5.64 approximately. For a d=0.5 unit in (13) the area of the square will be 11^2=121 squared units while that of the circle will be 118.38 squared units approximately. The perceived area of the square is larger than that of the circle and then the observed results are explained. By equating the right-end sides of A_c and A_s in (13) and resolving for d, this value can be computed [6-7]. However, since our experiments cannot completely exclude the influence of other factors on the value of *d*, its evaluation has to be considered mainly from the qualitative point of view.

3.3. Area comparison of square and rectangle

Three experiments have been done for area comparison of square and rectangle; two with complete figures [25] and one with interpolated figures [8]. The methods have been the same in all the experiments [13]. The procedural difference of these experiments with the previous one was only that we used the screen of a computer in spite of that of the oscilloscope. In this case the subjects fitted the area of the fixed figure by adjusting that of the adjustable one by pressing the Plus and Minus Keys on the keyboard. Once the subject judged the two areas equal he/she pressed the Enter Key. The first two experiments were done respectively with unfilled and filled figures [25]. In the first case the figures were presented as bright contours on completely black background. In the second experiment the figures were completely bright. The results were that the area of the rectangle is overestimated independently of whether the figures were filled or unfilled. In addition, we have found no significant differences between the results of the two experiments. Let me show also in this case the effect of the filtering by a simple numeric example, using the same value of d as in the previous case. For the two geometrical areas of 100 squared units, the perceived area of the square will equal to 121 squared units (see previous example). A rectangle of area A_r=100 can have the sides (b and c) of several different lengths. One of these configurations can be for example b=16 and c=6.25. Adding the same d =0.5 both to b and c the perceived area of the rectangle will be

$$A_r = (b+2d)(c+2d) = 17x7.5 = 127.5 \tag{14}$$

The perceived area of the rectangle then is larger then that of the square of the same geometrical area and so our observed results are fully explained. This can be shown for any configuration of the sides of the rectangle. Also in this case, equating the perceived areas of square and rectangle and resolving for d its value can be computed [25] but, the same considerations made for the previous case holds also for the present one. This hypothesis is in good agreement also with a model of area perception of rectangle proposed by King [20], if we assume that the super ordinate system he has proposed to explain his results is not produced by the geometrical co-ordinates of the figure but by their IFs.

The observed effect of overestimation disappeared in the experiment of area comparison of rectangle and square when we used interpolated figures [8]. This experiment was exactly as in the previous case and made by the same computer program [13] with the only difference that the figures were shown only by the corners. This result fully agrees with our model. No sides are present, no IFs of sides, no d influencing the perception of area [8].

3.4. A rule for area perception of geometrical figures.

The above results combined with very easy computations based on the above model lead to the conclusion that a rule for area comparison of geometrical figures can be given. The model, in fact, predicts that when the areas of two geometrical figures are compared, the figure that for a given fixed area has the longer perimeter is overestimated with respect to the other.

4 OTHER VISUAL ILLUSIONS

4.1. Müller-Lyer and Paggendorff illusions

Finally, let me show that the present criteria can be applied also to explain other visual illusions. Müller-Lyer illusion is a visual illusion where one line is overestimated when compared with another. The difference between the two lines is that in one case they end with arrows pointing inward while in the other the arrows point outward (I will define the former as figure A and the latter as figure B). The length of the figures A is overestimated with respect to the length of the figure B. The illusion is maintained also for a 3D configuration and in stereoscopic vision [28]. Ginsburg [18], however, showed that at least 80% of the magnitude of this illusion can be accounted by filtering of the input image. The results obtained in our experiments on interpolated square and rectangle as well as some results we have obtained in the study of orientation in random dot pattern by using interpolated line to match the direction [23] suggested us to try to demonstrate the filtering contribution on this illusion by using interpolated Müller-Lyer patterns [9-12]. The basic idea was that, if the filtering of the input image can be responsible of Müller-Lyer illusion then it has to be a boundary effect in correspondence of the tips of the arrows. This can be demonstrated by doing the difference of the IFs of figure A form that of figure B. This operation shows that the resulting function is always positive outside the tips [10]. This means that our *d* (i.e. the distance of a value of the IF able to be perceived as ending point of the lines in this case) is different in the two figures being always more distant from the tip in figure A than in figure B. Our experimental idea was that if we reduce the contribution of the lines forming the arrows to the total IF around the tips, the illusion has to decrease. For this reason, we have performed two experiments where interpolated Müller-Lyer patterns were presented [9-12]. In these experiments, the lines forming the arrows are to be interpolated between dots; in one experiment we used arrows with a fixed angle [10-11] and in the other with variable angles [9-10]. The size of illusion decreased with the number of dots forming the arrows [9-12] as the model predicts [10-13]. The size of the illusion decrease also increasing the distance between the dots forming the lines of the arrows [10] further confirming that reducing their contribution to the IF around the tips, decreases the magnitude of the illusion. This is due because when there is no influence of dots around the tips the IFs of the two central lines of the Müller-Lyer patterns have equal length. For those configurations where only two extreme dots identified the line forming the arrows we have found no illusion. This would stress the conclusion that the filtering of the input image can totally explain Müller-Lyer illusion. However, the findings of many authors suggested the presence of top-down interactions [5,28,37] and, in addition, this configuration is an extreme case. The conclusions would be, in any case, that the conclusions of Ginsburg [18] are confirmed by our results [10]. Our model also predicts that Paggendorf's illusion can be reduced or disappear when interpolated figures are used [12].

4. CONCLUSIONS

The results of the experiments I have reviewed above seem to confirm that one of the most important factor to have into account when geometrical characteristics of figures are evaluated during psychophysical experiments is the filtering of the input image. This conclusion holds independently of the filtering model proposes. For any filtering function it is easy to show that around the maximum of the IF produced there are some values that can be perceived in spite of the maximum. This conclusion induced us to predict the above models for area perception and to explain Müller-Lyer illusion. In addition our experiments with interpolated figures shown that when the filtering contribution is lacking the illusions decrease [8-12]. This has been predicted also for other illusion as for example the Paggendorf illusion [12]. However a wrong conclusion from the present paper would be that filtering can explain everything in visual perception. My idea is strongly different. I want only stress that in considering visual phenomena one cannot neglect the filtering contribution to the illusion. So, the model proposed by Anderson and Cuneo for area perception of rectangle [1-2] can agree very well with our results with interpolated figures were filtering is absent or negligible but not with real figures where filtering is an important factor. This example can be true also for all the model that consider the cognitive (top-down interactions) aspects of the illusion completely neglecting the possible role of the filtering of the input image. Models taking into account all the aspects of the mechanisms underlying the vision are necessary for the good understanding of visual perception in general and even more for visual perception of geometrical figures.

5. ACKNOWLEDGEMENTS

Many tanks to the director of the school on "Biocybernetics of Vision" Cloe Taddei-Ferretti for her kind invitation and to have been patient with me during the years.

6. REFERENCES

1. N.H., Anderson (1981). *Foundation of Information Integration Theory*. New York., Academic Press,
2. N.H., Anderson & D. Cuneo (1978). The Height + Width rule in children Judgement of quantity. *J. Exp. Psychol., Gen.* 107 379-387.
3. R., Bamler (1989). *Mehrdimensionale Linear Systeme*. Berlin, Springer-Verlag.
4. E., Boussard & J-F., Vibert (1994). Dopaminergic neuromodulation Brings a dynamical plasticity to the retina. In: *Neural Information Processing System 6* J.D. Cowan, G. Tesauro & J. Alspector (eds), San Francisco, Morgan Kaufmann Publishers, pp 559-565
5. P., DeLucia (1993). A quantitative analysis of illusion magnitude predicted by several averaging theories of the Müller-Lyer illusion, *Percept. and Psychophys.* 53 498-504.
6. V., Di Maio (1994). La visione nell'uomo. In: *Cibernetica Biologica*. F. Allocati, C. Musio & C. Taddei-Ferretti (eds), Torino, Rosemberg & Fellier, pp 183-201.
7. V., Di Maio & P., Lánský (1990). Area perception in simple geometrical figures. *Percept. & Motor Skills* 70 459-466.
8. V., Di Maio & P., Lánský (1994). Perception of area of interpolated figures *Cybernet.& Syst.* **25** 567-579.
9. V., Di Maio, & P., Lánský (1994). Influence of number of dots on the perception of interpolated Müller-Lyer patterns. *Perception* 23 (supplement) p. 14.
10. V., Di Maio, & P., Lánský, (1994). Partial explanation of Müller-Lyer illusion by the filtering of the input image (submitted for publication).
11. V., Di Maio, & P. Lánský, (1994) Müller-Lyer Illusion and the Image Function. In: *Cybernetics and Systems* '94. R. Trappl (ed) Singapore, World Scientific, pp 793-799.
12. V., Di Maio, M., Frucci & P., Lánský (1992). The Image Function and some visual Illusion *Cybernetics and Systems Research* '92. R. Trappl (ed),Singapore, World Scientific, pp 805-812.
13. V., Di Maio, M., Frucci, & P., Lánský (1994). A computer program to study area perception of squares and rectangles. *Comp. Method. and Programs in Biomedicine* 41 263-270
14. M.A., Garcia-Perez (1988). Space-variant visual processing: spatial limited visual channels. *Spatial Vision* 3 129-142.
15. M.A., Garcia-Perez (1988). HDC: a subroutine library for image processing using Burt's hierarchical discrete correlation. *Behav. Res. Method., Instrum. & Computers* 20 317-329.

16. A.P., Ginsburg (1984). Visual form perception based on biological filtering. In: *Sensory Experiences, Adaptation and Perception*, L. Spilman & B. R. Wooten (eds.) Hilldasale, NJ: Elbaum., pp. 53-72,
17. A.P., Ginsburg, (1986). Spatial filtering and visual form perception. In: *Handbook of Perception and Human Performance*, K. R. Boff, L. Kaufman & J.P. Thomas (eds.), New York Wiley,.pp 34.1-34.41
18. D.H., Hubel & T.N., Wiesel (1962).Receptive fields, binocular interactions and functional architecture in the cat's visual cortex. *J. Physiol., London* 160 106-154.
19. D.H., Hubel & T.N., Wiesel (1977). Architecture of macaque monkey visual cortex. *Proc. R. Soc. Lond.* 198 1-59.
20. D.L., King (1990). A large rectangle delays the perception of a separate small rectangle, *Percept. & psychophys.* 47 369-378.
21. F., Kingdom and B., Moulden (1992). A multi-channel approach to brightness coding. *Vision Res.* 32, 1565-1582.
22. J.J., Koenderink (1984). The structure of the images. *Biol. Cybernet.* **50**, 363-370
23. P., Lánský, J.,Mates, N., Yakimoff & V., Di Maio (1990). Matching the orientation of dot patterns with real, interpolated, and extrapolated lines. *Spatial Vision* 5 219-230.
24. D., Marr (1982). *Vision* , New York, W. H. Freeman and Company
25. J., Mates, V., Di Maio & P., Lánský (1992). A model for area perception. *Spatial Vision* 6 101-116.
26. J., Mates, P., Lánský & N., Yakimoff (1990). Models for the perception of random dot patterns. *Biol. Cybernet.* 63 71-80.
27. M.C., Morrone & D.C., Burr (1988). Feature detection in human vision: a phase dependent energy model. *Proc. of the Royal Soc. of London*: B, 235, 221-245.
28. R., Nijhawan (1991). Three-dimansional Müller-Lyer illusion. *Percept. & Psychophys.* **49** 333-341.
29. J. Piaget, B., Inhelder & A., Szeminska, (1948). *La Géométrie Spontanée de l'Enfant. Paris, Presses* Universitaires de France.
30. , Pribram (1991). *Brain and Perception: Holonomy and Structure in Figural Processing.* Hillsdale, NJ, Lawrence Erlbaum Associates.
31. R., Shapley & C., Enroth-Cugell (1984). Visual adaptation and retinal gain controls. In: *Progress in Retinal Research*, N.N. Osbourne & G. G. Chader (Eds), Oxford, Pergamon Press pp 263-346.
32. J.T.S., Smits & P.G., Vos (1986). A model for the perception of curves in dot figures: the role of local salience of "virtual lines". *Biol. Cybernet.* 54 407-416.
33. A.W., Snyder and Srinivasan, M.V. (1979). Human psychophysics: functional interpretation for contrast sensitivity versus spatial frequency curve. *Biol. Cybernet.* 32 9-17.
34. P.G., Vos, M.P., Van Oeffelen, H.J, Tibosch & J. Allik (1988). Interaction between area and numerosity. *Psychol. Res.* 50, 148-154.
35. R.J., Watt (1988). Visual Processing: Computational, *Psychophysical and Cognitive Research*. London: Lawrence Erlbaum.
36. R.J., Watt & M.J., Morgan (1985). A theory of primitive spatial code in human vision. *Vision Res.* 11, 1661-1674.
37. L., Zusne (1970) *Visual Perception of Form.* New York, Academic Press.

PIGEON'S BINOCULAR FIELD: A BEHAVIOURAL ESTIMATION

Giovanni Cesaretti, Claudia Kusmic and Daniela Musumeci
Dipartimento di Fisiologia & Biochimica, via S. Zeno 31, 56127 Pisa, Italy

Panoramic vision supports hunted animals allowing them to monitor a wide portion of the surrounding environment as well as binocular vision enhances the performance of hunting animals demanded for fine judgements[1, 2]. Among Aves, birds of prey possess a wide binocular field in which two eyed vision probably increases the accuracy in visual motor tasks, such as to gauge the position of limbs and talons during the grasping of prey in the final stage of the catching[3]. Granivourous birds like pigeons[4], that mainly use the beak to grasp seeds on the ground possess a rather narrow binocular field particularly useful, however, in those behaviours requiring a fine coordination between their effector -the beak- and the target, as pecking and feeding do. Pecking behavioural studies using classical discrimination paradigms as well as studies on feeding indicate that pigeons show a proper visuomotor coordination related to the distance and the size of the seed or the target. Pecking responses are very stereotyped and characterised by fast saccadic movements of the head interspersed with fixation stops[5-8]. All the Authors agree on the ovoid shape of the pigeon's binocular window; the estimate of its extension, however, varies from study to study depending on the animal breed and the technique applied (retinal or optical measurements). It ranges from 18° to 41° along the horizontal axis and from 80° to 135° along the vertical axis with the peak width at ± 10° around the beak[1,9-13]. Until now a very few attempts[14, 15] have been made to map the functional borders of pigeon's binocular field by investigating behavioural performances that can be improved by the use of binocular vision. In the present work we investigated a pecking behaviour recording reaction times (RTs) to visual stimuli displayed at increasing distances from a fixation point on a wide portion of the frontal field. In addition, we analysed the different spatial-temporal relationship between the head of the animal and the target position. The experimental conditions employed resembled the natural ones occurring during the feeding behaviour at seeds on the ground.

Methods. Three pigeons, maintained at 80% of their free feeding weight were required to produce fast responses to a small black spot displayed over a 45° outward sloped VGA 9" b/w monitor placed 10 cm above the floor of a modified Skinner box. The spot (3,8 mm diameter) could randomly occur on one of 17 positions. Spot positions lay at increasing distances (3.8, 7.6, 15.2 and 22.8 mm) along two Cartesian axes centred on the "0" point (x 320, y 210 pixel) as shown in figure 1. Birds got the stimulus (spot or white background) displayed by pecking a small starter stimulus (black square) always displayed in the same position of the "0" point (behavioural fixation). Stimuli remained on the screen one second. A single peck to the spot caused a single pellet reward, while a peck to the blank produced a 3 s black-out. Pigeons had to peck the positive stimulus and to refrain from pecking the negative one according to a Go-

Nogo procedure. Spots, blank screen, starter and grey-background during the intertrial interval were 640x480 16 colour .PCX files. Pecks were detected by a microphone and fed into a 80286 AT computer that drove the set-up and collected RT (i.e. the time elapsing between the pecking to the starter and the pecking to the positive stimulus) by means of a dedicated software.

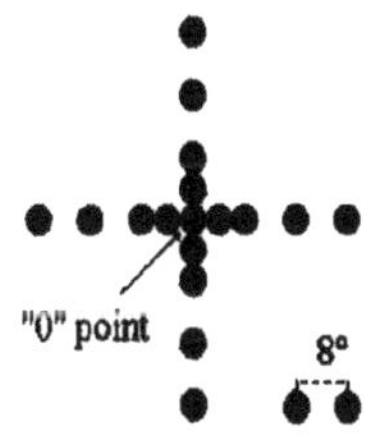

Figure 1. Calibration is for an eye looking at from 61.2 mm

A b/w PAL TV camera acquired and recorded on the same frame a side view, an overhead view of the pigeon (via a 45° mirror mounted on the Skinner box ceiling) and a 50 Hz digital counter. Images were successively analysed frame-by-frame at 20 ms rate by means of a digitizer card. Passive markers (white spot) were painted around the eye, at the tip of beak and at the centre of the skull to estimate the head position during the image analysis. Successive frames were digitised, and the marks identified as VGA coordinates. Pigeons performed a daily session of 600 trials divided in 5 blocks of 120 trials each, in which either 60 Go and 60 Nogo stimuli were displayed according to a Gellermann sequence. In the first block ("warm-up" block) the spot occurred always on the "0" point. In the successive four blocks spots occurred on five positions: the "0" point and 4 positions (up, down, left and right) at the same distance from the "0". Blocks differed each other in the spot distance and they randomly occurred within the session. When pigeons reached a steady performance level greater than 85% of correct responses in all blocks, then, at least 7 consecutive sessions run to provide RTs and images for data analysis.

Results. RTs of correct responses were separately collected for each spot position and median values calculated for each pigeon. Median RT obtained on the "0" point was the fastest value recorded in all sessions (the average across the subjects was 193.5 ms) and it increased not significantly of about 30 ms for positions up to 7.6 mm apart on the axes. RT measured on spot placed at or farther than 15.2 mm was significantly slower of about 170 ms ($p<0.001$) than that on the "0" point (figure 2). Image analysis indicated that the pecking response of pigeons to a visual target is composed of three well defined stages: 1) a *withdrawal* after the pecking to the starter; 2) a *stimulus fixation*; 3) *a motor output*. In addition, remarkable differences existed between the stages depending on the position of the target within the frontal field. During the stage 1 pigeon drew back after the pecking to the starter. It was a very stereotyped event lasting about 51 ms ± 10 and the final position of the head was unchanged from trial to trial. Stage 2, the phase of stimulus fixation, was a single step period for spot falling up to 7.6 mm apart from the "0" point and for nogo stimuli as well. The time of fixation was about 87 ms ± 15 for the "0" point and raised up significantly to 108 ms ± 14 for spot at 7.6 mm. In contrast, stage 2 was a three steps level for discrimination of spots lying from 15.2 to 22.8 mm apart on the axes. The first step, named first fixation or stage 2a, lasted about 115 ms. The position of the head was highly constant among the birds (mean ± SD in VGA coordinates: eye x =264 ± 6, y =313 ± 3; beak x =336 ± 1, y =356 ± 5) and the distance between the beak tip and the centre of the "0" point measured 15 mm, wherever the stimulus occurred. After the first fixation, a head rotation/translation towards the stimulus resulted (stage 2b). Finally, a second fixation step (stage 2c) occurred lasting about 107 ms ± 20 (figure 3). Stage 3, the motor output, was a standard behaviour going on for 60 ms ± 7.

Discussion. Data from RTs indicate that in the frontal field of pigeons it is possible to draw two rough and ready concentric areas with different speed of response. Inside to the inner area RTs lightly increase at increasing spot distance. RTs rise heavily on spots falling into the outer area. One may expect slower RTs on spots farther from the centre due to the longer distance to run along in producing the response and to the different retinal areas involved as well. This could be perhaps the case for stimuli falling up to 7.6 mm apart from the centre. It cannot, however, justify the long time required for the response to stimuli in the outer area of the frontal field. How could pigeon spend such a time? The remarkable slowing down of RTs on the outer area is due to the second fixation (stages 2b and 2c). According to Goodale (1983), the last fixation is the stage in which the pigeon, once taken the decision of pecking and keeping the target within its binocular field, obtains information about position, size and depth of the target. Assuming that pigeon profits by binocular vision to estimate its egocentric distance from the target and the size of it, then, it is reasonable to believe that stimulus positions requiring only one fixation stop fall within the actual binocular field and that stimulus positions requiring two fixation stops lie outside of it.

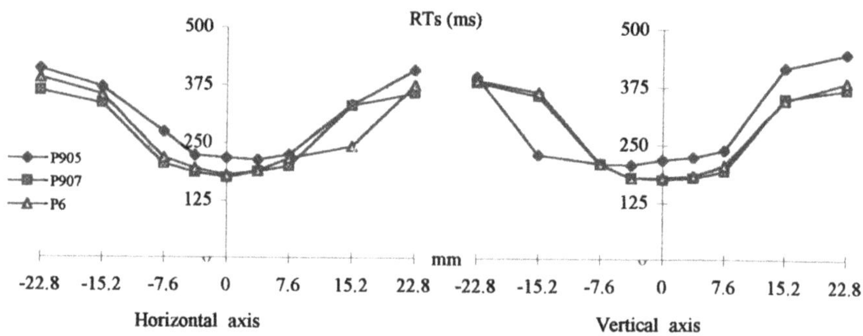

Figure 2. Individual median RTs on different spot positions. Negative values indicate left and down directions on the horizontal and the vertical axes, respectively.

Considering that spot position requiring only one fixation stop subtend a visual angle of about 23.5°, we believe that this is the measure of the pigeon's functional binocular field.

Our estimation comply with the retinoscopic measurements reporting a binocular field 24° wide on the horizontal axis. On the contrary, as the vertical axis is concerned, our data indicate that its extension is comparable to that of the other axis. It derives that the shape of the functional binocular field is approximately a circle centred on the eye-beak axis.

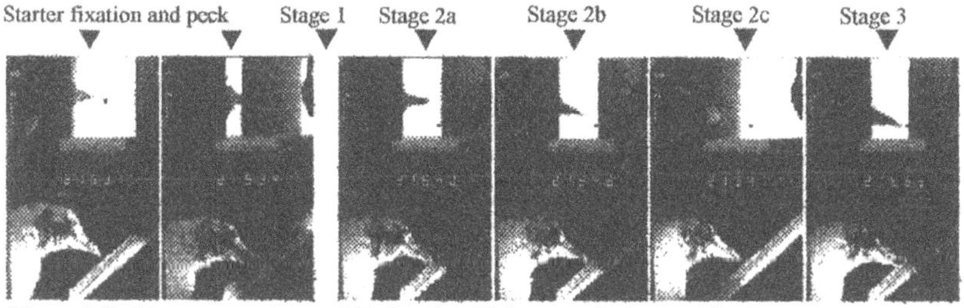

Figure 3. Pecking sequence to a spot displayed at -22.8 mm on the horizontal axis.

References

1) G. L. Walls (1942, Reprinted 1963). *The Vertebrate Eye and its Adaptative Radiation*. Hafner. New York, London, p. XIV, -785.

2) R. K. Jones and D. Lee (1981). Why two eyes are better than one: the two views of binocular vision. *J. Exp. Psychol.: Human Perc. and Perf.*, 7, 30-340.

3) R. S. Payne (1971). Acoustic location of prey by barn owl (*Tyto alba*). *J. Exp. Biol.*, 54, 535-573 .

4) W. J. Donovan (1978). Structure and function of the pigeon visual system *Physiol. Psychol.*, 6, 403-437.

5) C. Martinoya, J. Le Houezec and S. Bloch (1984). Pigeon's eyes convergence during feeding: evidence for frontal binocular fixation in a lateral-eyed bird. *Neuroscience Letters*, 45, 335-339.

6) Smith, R. F. Topography of the food-reinforced key and the source of 30-millisecond interresponse times. *J. Exp. Anal. Behav.*, 21 541-551 (1974).

7) R. Jager and H. P. Zeigler (1991). Visual field organization and peck localization in the pigeon (*Columba livia*). *Behav. Brain Res.*, 45, 65-69.

8) M. A. Goodale (1983). Visually guided pecking in the pigeon (*Columba livia*). *Brain, Behav. Evol.* 22 22-41.

9) B. P. Hayes, W. Hodos, A. L. Holden and J. C. Low (1987). The projection of the visual field upon the retina of the pigeon. *Vision Res.*, 27, 31-40.

10) G. R. Martin and S. R. Young (1983). The retinal binocular field of the pigeon (*Columba livia*: English racing homer), *Vision Res*, 23, 911-915.

11) C. Martinoya, J. Rey and S. Bloch (1981). Limits of the pigeon's binocular field and direction for best binocular viewing. *Vision Res.*, 21 1197-1200.

12) S. A. McFadden (1989). Eye design for depth and distance perception in the pigeon: an observer orientated perspective. *Int. J. Comp. Psych.,3 101-129.*

13) H-O. Nalbach, F. Wolf-Oberhollenzer and K. Kirschfeld (1990). The pigeon's eye viewed through an ophthalmoscopic microscope: orientation of retinal landmarks and significance of eye movements. *Vision Res.*, 30, 529-540.

14) C. Kusmic, P. Lombardi and D. Musumeci (1991). Monocular equivalence in pigeons: comparison of reaction time distributions to a stimulus randomly displayed within the binocular visual field. *Europ. J. Neurosci.*,77, 1324.

15) D. M. Wilkie and L. M. Saksida (1994). The spatial distribution of pigeons' target detection. *Behav. Brain Res.*, 60, 137-140.

This work was supported by M.U.R.S.T.. The dedicated software Te.Re.S.A. is available by R. Agnesi, Progemi s.p.a., via Caviglia, 11, Milano 20123, ITALIA.

Neural network model of *Hydra* photoresponsive behaviour

S. Chillemi[1], M. Riani[2], E. Simonotto[2],
C. Musio[3] and C. Taddei Ferretti[3]
[1] *Istituto di Biofisica, CNR, Pisa, Italy*
[2] *Dipartimento di Fisica, Universitá di Genova, Italy*
[3] *Istituto di Cibernetica, CNR, Arco Felice (NA), Italy*

The behaviour of *Hydra* is characterized by a continuos alternance of two sub-behaviour states. It is obtained by the alternate activity of the ectodermal and endodermal effector systems, the myofibrilis of which are arranged ortogonally, *i.e.* along the animal long axis and circularly such as the two sub-behaviour states, body shortenings (duration about 30 *sec.*) and body elongations (duration about 120 *sec.*), arise. The shortenings begin at the distal end of the body and spread as a wave toward the proximal one, the elongations present the opposite pattern. The total behavioural period (shortening and elongation) of each animal presents a low variability around a mean.

A nervous net lyes between the ectodermal and endodermal layers: it is responsible for the pacemaking of the shortening/elongation activity at the distal/proximal end of the animal, as well as for the spread of the two different activity waves toward the opposite ends (although the wave spread is carried out also by direct propagation along one or the other dermal layer). The net is composed by about 10.000 multipolar neurons [1] *quasi*-isotropically distributed everywhere: a slight increase of the neuron density is presented at the two ends of the animal with respect to the body column, and more at the proximal end (where body elongation, *i.e.* the longest sub-behavioural state, arises), such as the net could be considered composed of two sub-nets slightly differentiating each one into the other [2].

It has been hypothesized[3,4]:

i) that the long time (of the order of minutes) of each sub-behavioural states arises from the global activity of a sub-net, whose neurons have much shorter times of pacemaking activity and exert mutual inhibition of the neighbour neurons to which they are functionally linked [5];

ii) that the periodic alternations between the two sub-behavioural states result from the mutual inhibition of the two subnets, the activity of each one arising as a post-inhibitory rebound after the end of the time course of the inhibition received by the other subnet [6].

We propose, as a model of the Hydra alternating behavioural states, an autoassociative neural network that was previously introduced to model the multistability in human perception [7,8].

The network is made up of two populations of neurons. Each neuron is described by a continuos-valued variable (values between 0 and 1); the activity of each population is linked to the two behavioural states of the Hydra. Experimentally, the two behavioural states exclude each other; consequently the excitation of one population must exert an inhibiting influence on the opponent one.

The dynamical evolution of the neural network is time-discrete and is given by the set of equations:

$$f_i(t+\tau) = L\left(\left(\sum_{j=1}^{n_a+n_b} W_{ij} f_j(t) + g_i\right)\sigma_i(t)\right) \tag{1}$$

where f_i is the activation of the i-th network neuron, W_{ij} are the connection matrix elements, g_i is a costant input and $\sigma_i(t)$ is the habituation process that lowers the input sensibilty of a neuron after it has reached its maximum fire rate.

Preliminary results of numerical simulations are presented. We have simulated small networks, made up of some tens of neurons. We have considered networks whose populations have different numbers of neurons.

We have found conditions to ensure that the two populations periodically alternate activation peaks. A key role in this process is played by the habituation process which, by lowering the activity of the excited population, favours the recovery of the inhibited one. The limit cycle is robust with respect to noise applied both at the synaptic connections and at the inputs of each network neuron. Rarely, due to the noise effect, one population does not show any peak between the peaks of the opponent one.

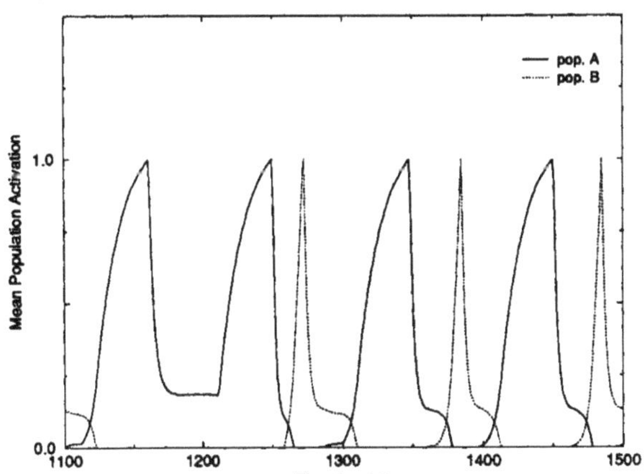

Figure 1: Dynamical evolution of the mean activities of the two populations.

We have measured the duration time of activation, that is, the duration time in which one population is more activated than the opponent one. The population composed by a greater number of neurons has mean activation time longer than the other population; the return time distribution, that is the histogram of the duration time in which one population is activated followed by the period of time in which the other one is activated, shows a marked peak with a small dispersion; secondary peaks are clearly visible though they are less pronounced.

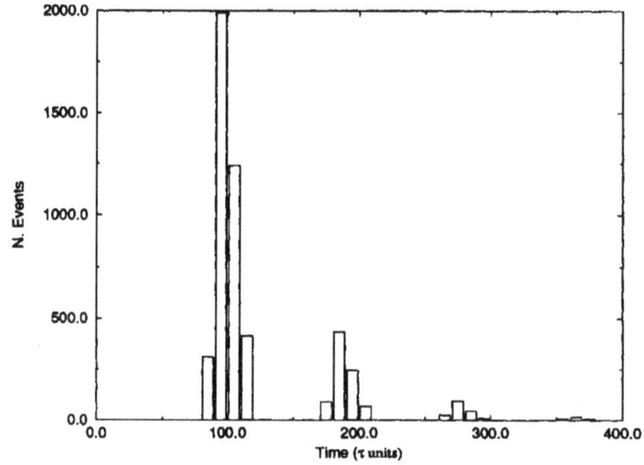

Figure 2: Typical duration time histogram measured in a simulation of the neural network model

This model is able to reproduce qualitatively some of the phenomenological aspects of the Hydra behaviour. We will verify, by an extensive computer simulation (e.g., using networks made up of a greater number of neurons), if this can be made quantitative.

References

[1] A.L. Burnett, A. L. and N.A. Diehl (1964), The nervous system of *Hydra*, I. Types, distribution and origin of nerve elements, *J. Exp. Zool.*, **157**, 237-250.

[2] P. Tardent and C. Weber (1976), A qualitative and quantitative inventory of neuron cells in *Hydra attenuata* Pall. In: *Coelenterate ecology and behaviour* (G. O. Mackie, ed.) 501-512, New York, London: Plenum,.

[3] C. Taddei-Ferretti and S. Chillemi (1987), Modulation of *Hydra attenuata* rithmic activity. V. A revised interpretation. *Biol. Cybern.*, **56**, 225-235.

[4] C. Taddei-Ferretti, S. Chillemi and C. Musio (1992), A descriptive model of the interactions in the nervous network of *Hydra*. In: *Structure: From Physics to General Systems*, (M. Marinaro and G. Scarpetta, eds), Singapore, New Jersey, London, Hong Kong: World Scientific, vol. II, 113-130.

[5] T. Pavlidis (1969), Population of interacting oscillators and circadian rhythms, *J. Theor. Biol.*, **22**, 418-436.

[6] L.D. Harmon (1964), Neuromimes: action of a reciprocally inhibitory pair, *Science*, **146**, 1323-1325.

[7] M. Riani, F. Masulli and E. Simonotto (1991), Perceptual alternation of ambiguos patterns: a model based on artificial neural network. Proc. of SPIE **1469**, 166-177.

[8] F. Masulli, M. Riani and E. Simonotto (1990), Modelling multistable perception by multilayer networks. In: *Third Italian workshop on parallel architectures and neural networks* (E. R. Caianiello, ed.), Singapore, New Jersey, London, Hong Kong: World Scientific.

LOCAL CIRCUITS UNDERLYING THE FUNCTION OF THE VERTEBRATE INNER RETINA

GREG MAGUIRE

Department of Neurobioloy & Anatomy, University of Texas-Houston Medical School, Houston, TX 77225, USA

e-mail: gmaguire@gsbs.gs.uth.tmc.edu

Transient amacrine cells were first reported by Werblin and Dowling [1]. These are the first transient neurons in the visual system, and are important for the feature extracting properties of retinal ganglion cells. The transient amacrine cell appears important to movement detection and directional selectivity of ganglion cells in most retinas tested, from frogs to monkeys [2,3,4].

Sustained amacrine cells were not initially reported by Werblin and Dowling tl] in the mudpuppy, but were later discovered and shown to be a high percentage of the amacrine cell population in salamanders by Maguire et al [5]. That sustained amacrine cells exist has important implications for how we model the retina, including the modeling of the receptive field center-surround and of the trigger features, such as directional selectivity, of ganglion cells (GCs). Thus, even today, studies that "take apart" the retina, identifying its elements and the connectivity of the elements, are important for conceptuali~ng even the most basic functional properties of the retina [6]. Putting the retina back together into a realistic biological model is dependent on knowing much more about the relatively poorly defined amacrine cell population. The studies described here are an attempt to better define the retinal amacrine cells in the tiger salamander retina, but we are now beginning to successfully record from amacrine cells in the primate retinal slice using patch clamp methods (Maguire, unpublished observations). These methods, along with those of patch recordings from GCs in the mammalian retinal whole mount [7] and patch recordings of GCs in the primate retinal slice [8], should provide precise descriptions of the synaptic interactions in the primate inner retina and lead to powerful structural and predictive models of primate visual processes at the level of the retina.

Sustained and Transient Amacrine Cells

The salamander retina is relatively simple compared to that of fish, cat. or primate for example.

However, even in the salamander there are many types of bipolar and amacrine cells. Many of the amacrine cells can be classified into two groups: neurons that respond with sustained or transient classical spiking to a light step [5].

The transient amacrine cells usually respond to the onset and offset of a light step with a burst of spikes (Figure 1A), whereas the sustained cells respond with continuous spiking during the step (Figure 1B shows a sustained ON amacrine cell). Other studies have shown the transient amacrine cells to stain for glycine or GABA antibodies [9], while the sustained cells are often GABA positive [5].

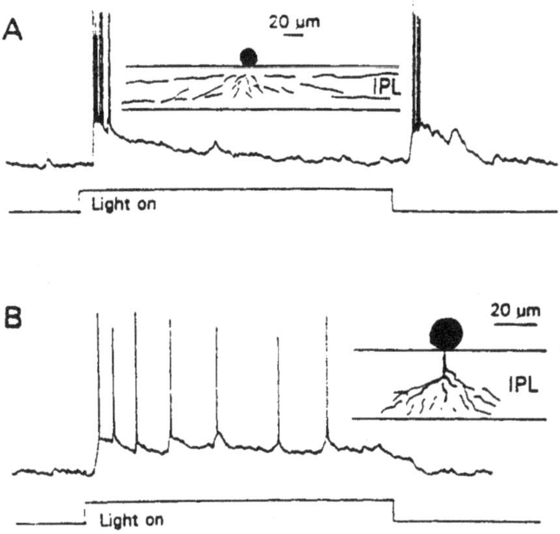

Figure 1. A) Voltage response of a wide-field transient amacrine cell. A burst of spikes is present at light-on and -off. Inset shows the "wide-field" spread of processes in the inner plexiform layer. **B)** Voltage response of sustained amacrine cell. Continuous, regular spiking is present throughout the light step. Inset shows the narrow lateral spread of processes in the inner plexiform layer.

Because the origins of the classical center-surround organization of the GC receptive field at the level of the bipolar/horizontal cell has been questioned [10,11], another possible candidate, the GABAergic sustained amacrine cell can now be considered as a candidate for supplying the antagonistic surround. This is because GABA acting, either presynaptic at the bipolar cell terminal or postsynaptic at the GC dendrites, through a sign-inverting synapse, would oppose the sign-conserving (glutamatergic; see Massey and Maguire, [12]) synaptic transmission from the bipolar cell terminal to the GC dendrites. Both synapses act in a sustained fashion and can be mutually antagonistic [13].

The light responses of the bipolar cells in the slice are sustained [14] and they presumably release glutamate onto the sustained amacrine cells in a tonic fashion. The glutamate receptors in the sustained cells have a large non-desensitizing component, and underlie the the sustained EPSC [15]. In contrast, the transient amacrine cells respond to the sustained release of glutamate from the bipolar cells with a transient EPSC. Underlying this transient EPSC are glutamate receptors that rapidly desensitize in the presence of tonic glutmate release [15]. Control of the glutamate release from the bipolaar cell terminals, and consequently control of the rate of desensitization of the glutamate receptors on the amacrine cells, can be achieved through activation of $GABA_B$ receptors in the bipolar axon terminal [16]. Further control of the amacrine cell's glutamate receptors may be achieved through postsynaptic $GABA_B$ receptors on the amacrine cells. These synaptic interactions are diagrammed in Fig. 2A.

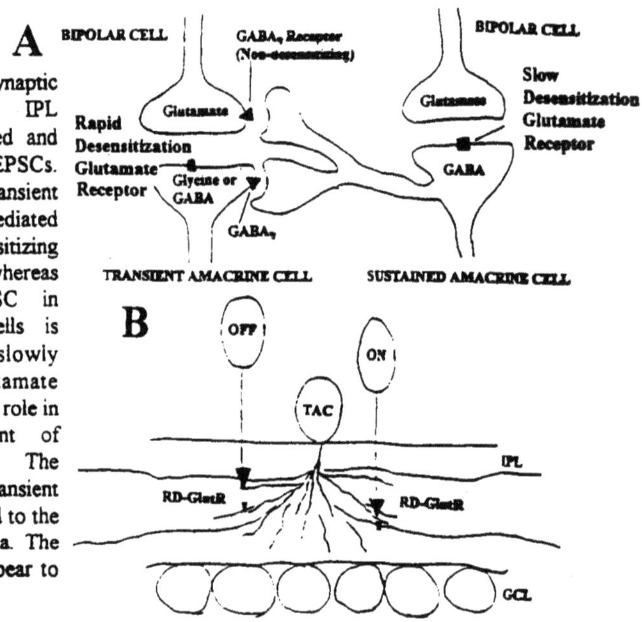

Figure 2. A) Synaptic interactions in the IPL responsible for sustained and transient amacrine cell EPSCs. The transient EPSC in transient amacrine cells is mediated through rapidly desensitizing glutamate receptors, whereas the sustained EPSC in sustained amacrine cells is mediated by slowly desensitizing glutamate receptors. GABA plays a role in regulating the amount of desensitization. B) The excitatory inputs to the transient amacrine cell are confined to the processes near the soma. The peripheral processes appear to be output sites.

The excitatory receptive field for the transient amacrine cells may be narrow. This is because only the processes close into the soma express glutamate receptors. Peripheral portions of the processes do not express functional glutamate receptors. Thus, ON and OFF excitatory inputs are confined to a narrow lateral extent, even though the processes of the transient amacrine cell extend laterally much farther. These characteristics are similar to the non-linear subunits recorded in retinal ganglion cells [17], that have ON-OFF non-linear excitatory responses with small receptive fields. Although these transient amacrine cells appear inhibitory [9], they are thought to provide excitation to the ganglion cells through disinhibition of other sustained, inhibitory amacrine cells [18]. Formalized computational models are now being developed to determine whether these amacrine cell interactions can account for the receptive field characteristics of the various ganglion cell types in the vertebrate retina.

References

1. Werblin, F. and Dowling, J.E. (1969) Organization of the retina of the mudpuppy, II. Intracellular recording. J. Neurophysiology 32:339-355.
2. Maguire, G., Connaughton, V., Wu, S. and Werblin, F. (1994) Physiological characterization of biochemically and morphologically identified sustained amacrine cells in the tiger salamander retina. J. Neurophysiology, In Press.
3. Maturana, H.R., Lettvin, J.Y., Pitts, W.H. and McCulloch, W.S. (1960) Anatomy and physiology of vision in the frog (Rana pipiens). J. Gen. Physiol. (Suppl) 43:129-175.
4. Werblin, F. (1970) Responses of retinal cells to moving spots: Intracellular recording in Necturus maculosis. J. Neurophysiology 33: 342-350.
5. DeMonasterio, F.M. and Gouras, P. (1975) Functional properties of ganglion cells of the rhesus monkey retina. J. Physiology 251: 17-195.
6. Beaudot, W. and Herault, J. (1994) A neurobiological and psychophysical consistent model of the vertebrate retina. Perception, In Press.
7. Ratto, G.M., Lambardi, L., Bisti, S., Cervetto, L. and Chalupa, L.M. (1994) Voltage-clamp and current-clamp recordings from retinal ganglion cells. Soc. Neuroscience Abstracts 20: 1574.
8. Zhou, Z.J., Marshak, D.W. and Fain, G.L. (1994) Amino acid receptors of midget and parasol ganglion cells in primate retina. Proc. Natl. Acad. Sci. USA 91: 4907- 4911.
9. Yang, C.Y., Lukasiewicz, P.D., Maguire, G.W., Werblin, F. and Yazulla, S. (1991) GABAergic, glycinergic, dopaminergic, and glucagonergic amacrine cells in the tiger salamander retina: morphology, physiology and neurotransmitter identification. J. Comp. Neurology 312: 19-32.
10. Boycott, B.B., Hopkins, J.M. and Sperling, H.G. (1987) Cone connections of the horizontal cells of the rhesus monkey's retina. Proc. R. Soc. Lond. B 229: 345-379.
11. Kolb, H. and Nelson, R. (1984) Neural architecture of the cat retina. Prog. Retinal Res. 3: 21-60.
12. Massey, S. and Maguire, G. (1994) The role of glutamate in retinal circuits: In: Excitatory Amino Acids and Synaptic Transmission, Wheal, H. and Thomson, A., eds. Academic Press, London, UK, In Press.
13. Maguire, G., Lukasiewicz, P. and Werblin, F. (1989) Amacrine cell interactions underlying the response to change in the tiger salamander retina. J. Neuroscience 9: 72-735.

VISUAL PERCEPTION AND COGNITIVE PROCESSES

VISUAL FIELD ASYMMETRIES AND LOCAL ANISOTROPIES IN PATTERN DISCRIMINATION: A SIGN OF CORTICAL ASYMMETRIES.

Adriana Fiorentini and Nicoletta Berardi
Istituto di Neurofisiologia C.N.R., Via S.Zeno 51, Pisa, Italy

Visual performance varies considerably from the center to the periphery of the visual field. Changes of performance in various detection or discrimination tasks may reveal how structural and functional properties of visual neural mechanisms vary across the visual field, and whether the limitations of visual performance with increasing eccentricity are imposed at a retinal level or in the brain, either by local or global properties of cortical neural networks.

Anisotropy of spatial resolution in extrafoveal vision

Visual acuity is highest in the center (fovea) and decreases progressively in the periphery of the visual field. In normal eyes the foveal acuity is about 40 - 50 cycles/deg in photopic vision. In the very center of the fovea the limit may be imposed by the optical qualities of the eye [1] but optical factors are not responsible for the decline of visual acuity in the region surrounding the fovea. Among the neural factors that impose a limit to visual acuity outside the fovea, the spacing of the photoreceptors (cones) plays the major role within several degrees from the center of the fovea, whereas more peripherally, because of the convergence of cones onto the retinal neurons of second and third order, the limit is imposed by the density of the ganglion cells [2,3]. This is the case both in man and monkey (see the comparison of the monkey behavioural acuity with cone and ganglion cell density distributions in Merigan and Katz [4].

Contrast sensitivity for sinusoidal gratings also diminishes at increasing distances from the fovea. If however the size of the stimulus is increased progressively, and scaled in proportion to retinal sampling, then the periphery is not inferior to the fovea (see Drasdo [5] for a review).

Another interesting fact emerges at retinal eccentricities exceeding 20 degrees: visual acuity measured with high contrast gratings at various positions along the vertical, horizontal or oblique meridians of the visual field varies locally according to the orientation of the grating, and is highest for gratings oriented radially (i.e. pointing to the fovea) and lowest for gratings having the opposite orientation (tangential to the local meridian) (Fig. 1) [6]. A possible explanation for this orientational anisotropy may be found in the morphological and functional properties of retinal ganglion cells. In the cat and monkey, a large proportion of the ganglion cell receptive fields are only approximately circular, being somewhat elongated in the radial direction [7-9]. This is reflected in an orientational bias observed in the distribution of cell receptive fields in the visual cortex: cells preferring radial orientations are more numerous than those preferring the opposite orientation [10, 11].

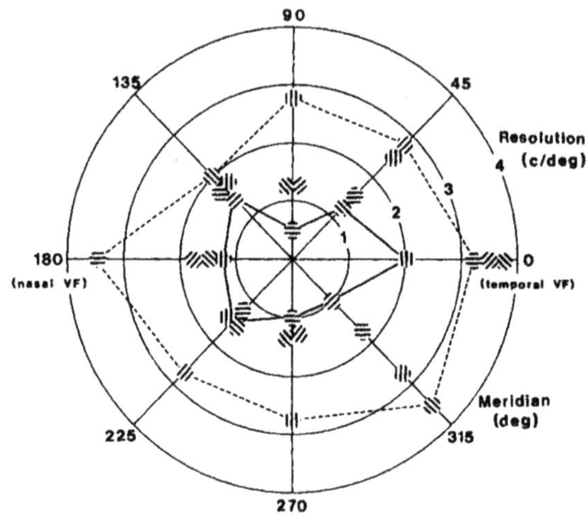

Fig. 1 - Visual acuity (in cycles/deg) measured with gratings of different orientations located at 25 deg eccentricity on different meridians of the visual field of the right eye of a human subject. The distance from the center indicates spatial resolution in cycles/deg, the radii of the graph indicate the various meridians. The acuities relative to gratings oriented radially (parallel to the local meridian) are those connected by a dashed line, the acuities for perpendicular orientations are those connected with a solid line (Reproduced from Rovamo et al. [6], with permission).

In summary, the limits in spatial resolution across the visual field are largely accounted for by retinal factors. The role of central limiting factors, however, may become predominant for other visual tasks, like the so-called hyperacuity tasks. Some hyperacuities (e.g. vernier acuity and stereoacuity) exibit foveal thresholds of only a few seconds of arc, largely exceeding the limit imposed by the spacing of foveal cones. This superior performance has been ascribed to cortical factors [12], although single retinal ganglion cells may show hyperresolving response properties [13].

Hyperacuities decline more steeply than visual acuity in peripheral vision. The change in peripheral hyperacuity with increasing eccentricity parallels the change of the reciprocal of the cortical magnification factor M [14] as estimated for man [15] and monkey [16]. M is the extent of the surface of the visual cortex (in mm) onto which 1 deg of visual angle is represented. Scaling the stimulus size with 1/M allows the stimulus to be projected onto a constant amount of visual cortex, irrespective of its position in the visual field, and therefore may compensate for the progressive compression of the cortical representation of the peripheral visual field (but see Drasdo[5]).

Several types of hyperacuities present a local anisotropy in peripheral vision, similar to the anisotropy described for grating acuity. For instance, at eccentricities exceeding 20 deg, the discrimination of curvature in nearly.straight lines is most accurate for lines oriented towards the fovea [17]. Between 5 and 10 deg from the fovea, the precision of bisecting the space between two reference dots with a third dot is highest if the reference dots lie on a circle centered on the fovea (isoeccentric dots) and lowest for dots aligned radially [18]. Anisotropy in peripheral vision has been reported for a number of other visual tasks [19-21], including motion detection [22-23].

Visual field asymmetries

In the paracentral region of the visual field the decline in visual performance for most resolution or hyperresolution tasks is more or less the same along the various meridians. This is the case for instance for grating acuity [24] and grating contrast sensitivity [19]. At larger eccentricities differences are found between the nasal and the temporal monocular fields: visual acuity declines less steeply with eccentricity in the temporal than in the nasal visual field both in man and monkey [4, 24, 25] probably because of differences in ganglion cell densities between the nasal and temporal retina [2]. Asymmetries are observed for other visual tasks between the peripheral nasal and temporal fields [20] or between the upper and lower visual field [26] or both [23]. Asymmetries between the upper and lower visual field projections are observed in the primary visual cortex of the monkey, where more cortex is devoted to the representation of lower than upper fields [27]. Asymmetries between the left and right hemifields, related to right or left hemisphere superiority, will be dealt with in a following chapter.

Discrimination of spatial phase in peripheral vision

The discrimination of a particular type of patterns, i.e. compound gratings resulting from the superposition of two sinusoids of different spatial frequency (for instance spatial frequencies f and 3f, or f and 2f) with different spatial phase relationships (Fig. 2), provides interesting examples of visual field asymmetry and orientational anisotropy. The performance in a two-alternative forced-choice task, where the observer is required to discriminate between two gratings differing only for the relative spatial phase of their harmonics, is very accurate in the fovea [28] but declines very rapidly with increasing eccentricity [29-33], possibly reflecting positional uncertainty in extrafoveal vision [32]. Aperiodical patterns (bars with ramp-like luminance profile corresponding to one period of f + 2f, or f + 3f compound gratings with mirror symmetric profile) are also more difficult to discriminate peripherally than foveally.

The discrimination of spatial phase in these either periodical or non-periodical patterns presents both a visual field asymmetry and an orientational anysotropy [33-35]. Discrimination is better performed in the left than in the right visual hemifield and in the lower than in the upper hemifield. This is true both in binocular vision and in monocular vision with either eye. Therefore the left-right asymmetry must reflect a hemispheric difference in the performance of this task, that will be further described in a following chapter.

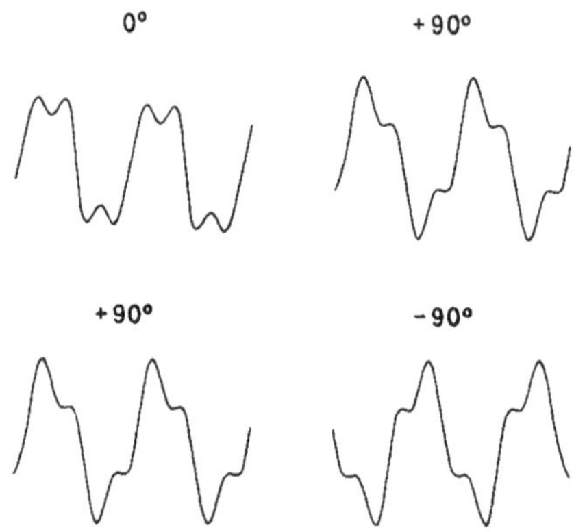

Fig. 2 - Luminance profiles of patterns obtained by the sum a two sinusoids of spatial frequency f and 3f and relative amplitudes 1 and 1/3, with different phase relationships. The discrimination task may imply either two patterns with mirror symmetric profiles (bottom) or two patterns non mirror symmetric (top).

The orientational anisotropy shows up at a few degrees from the fovea[33,35]: patterns oriented radially (pointing towards the fovea) are better discriminated than patterns in the opposite orientation (tangential to the local meridian). For instance, horizontal patterns located at 2 deg from the fovea along the horizontal meridian are better discriminated than vertical patterns, whereas the opposite is true for patterns located at 2 deg from the fovea along the vertical meridian (Fig. 3). In foveal vision horizontal and vertical patterns are discriminated equally well.

This orientational anisotropy recalls that observed in various hyperacuity tasks (see above). For the discrimination of spatial phase, however, another interesting fact has been observed: at eccentricities larger that 4 - 5 deg the orientational anisotropy vanishes for patterns displaced along the horizontal meridian, while it is preserved along the vertical meridian (Fig. 4). In contrast to the anisotropies observed in other visual tasks, the anisotropic properties of spatial phase discrimination are radially symmetric in the visual field only within a few degrees from the fovea. This indicates that in addition to the local orientational anisotropies, which are radially symmetric, there are functional asymmetries in the representation of the vertical and horizontal meridians in the human visual cortex.

Among the possible factors that may account for the orientational anisotropy of visual hyperacuities perhaps the most obvious is the anisotropy of the representation in striate cortex of the vertical meridian of the visual field. In the monkey, the magnification factor measured along the projection of a line oriented radially on the vertical meridian in the peripheral visual field is larger than the magnification factor along a contour of constant eccentricity. For the horizontal meridian, on the contrary, the representation is substantially isotropic [27].

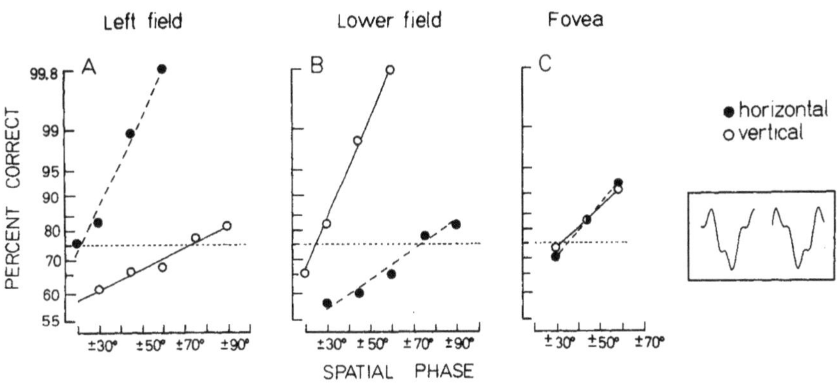

Fig. 3 - Percentage of correct responses in a 2 alternative forced choice task for the discrimination of aperiodical patterns (see insert) having a mirror symmetric luminance profile with various relative phase differences (abscissa), oriented either horizontally or vertically and located at 2 deg eccentricity in the left (A) or lower (B) hemifield or viewed foveally (C). Stimulus size 1 x 1 deg.

The anisotropy of the cortical map is probably related to the pattern of ocular dominance columns. These form an approximately circularly symmetrical pattern in the cortical representation of the fovea, whereas outside the fovea they tend to be oriented perpendicularly to the border between cortical areas V1 and V2, where the vertical meridian is represented. Along the cortical projection of the horizontal meridian, the ocular dominance columns are much more randomly oriented (Fig.5). [36-38].

There is evidence that the fundamental processing modules (hypercolumns) along the V1 - V2 border in the monkey striate cortex are elongated by a factor of 2 in the direction perpendicular to the ocular dominance columns.

Recent findings obtained in the monkey by real-time optical imaging moreover reveal a marked anisotropy of long-range lateral interactions in the striate cortex: the cortical activity elicited by a localized retinal stimulus spreads laterally through a distance twice as large in the direction perpendicular to the ocular dominance columns, than parallel to the columns [39].

It is possible that there is a difference in the processing of spatial phase according to whether the representation of the stimuli on the visual cortex crosses the border of the ocular dominance columns or is oriented parallel to the border [40]. If so, the difference between the

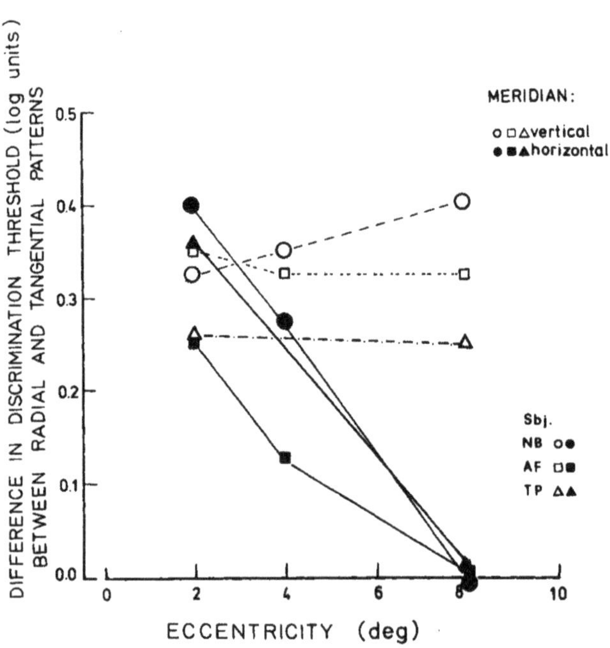

Fig. 4 - Difference between the log contrast thresholds for the discrimination of aperiodical mirror-symmetric patterns oriented parallel to the meridian or perpendicular to the meridian, and located at different eccentricities along the lower vertical (open symbols) or the left horizontal (closed symbols) meridian, for two subjects. (Reproduced from Berardi and Fiorentini [33], with permission).

anisotropies observed along the vertical and the horizontal meridian (Fig. 4) would suggest that the arrangement of ocular dominance columns in the human visual cortex is not dissimilar to that of non-human primates. It has been reported that in the human cortex the columns are oriented perpendicular to the border between visual areas V1 and V2 as in the monkey [41].

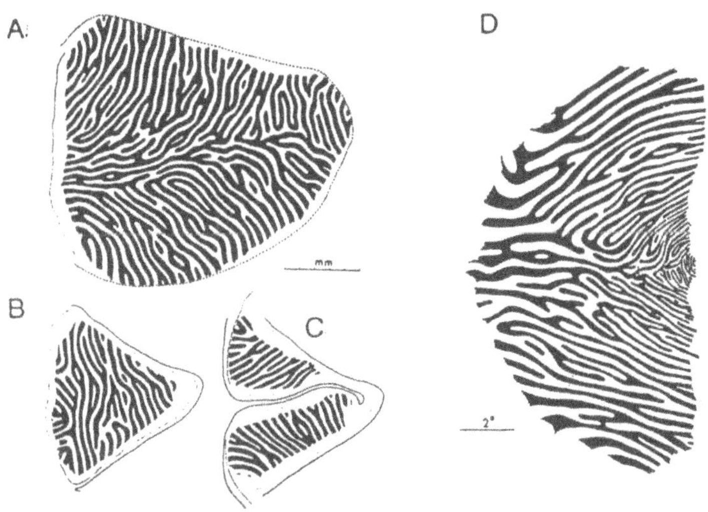

Fig. 5 - A: Pattern of ocular dominance columns as they appear on the flattened surface of the striate visual cortex of a macaque monkey right occipital lobe. The region of visual field represented is from the fovea (at the extreme right) to about 9 deg eccentricity. The vertical meridian is represented along the dashed line, the horizontal meridian, on the line from F that bisects the figure. Note that the ocular dominance columns are perpendicular to the contour representing the vertical meridian, while are more irregularly oriented near the horizontal meridian, especially in the parafoveal region. B: Visual field reconstruction of the pattern of ocular dominance columns shown in A. Since the portion of striate cortex represented in A is from the right hemisphere, the projection is into the left visual field (Reproduced from Hubel and Freeman [37], with permission).

References

1 - J.Hirsh and C.A.Curcio (1989). The spatial resolution capacity of human foveal retina. *Vision Res.* **9,** 1095-1101.

2 - C.A.Curcio, K.R.Sloan, O.Packer, A.E.Hendrickson and R.E.Kalina (1987). Distribution of cones in human and monkey retina: individual variability and radial asymmetry *Science.* **236** 579-582.

3 - V.H.Perry and A.Cowey (1985). The ganglion cell and cone distributions in the monkey's retina: Implications for central magnification factors. *Vision Res.* **25** 1795-1810.

4 - W.H.Merigan and L.M.Katz (1990). Spatial resolution across the macaque retina. *Vision Res.* **30** 985-991.

5 - N.Drasdo (1991). Neural substrates and threshold gradients in peripheral vision. In: *Limits of Vision*, J.J.Kulikowski, V. Walsh and J.J.Murray (eds.). London, MacMillan Press. pp 251-265.

6 - J.Rovamo, V.Virsu, P.Laurinen and L.Hyvarinen (1982). Resolution of gratings oriented along and across meridians in peripheral vision. *Invest.Ophthalmol.Vis.Sci.* **23** 666-670.

7 - W.R.Levic and L.N.Thibos (1982). Analysis of orientation bias in cat retina. *J. Physiol.* **329** 243-261.

8 - A.G.Leventhal and J.D.Schall (1983). Structural basis of orientation sensitivity of cat retinal ganglion cells. *J. compar. Neurol* **220** 465-475.

9 - J.D.Schall, V.H.Perry and A.G.Leventhal (1986). Retinal ganglion cell dendritic fields in Old-World monkeys are oriented radially. *Brain Res.* **368** 18-23.

10 - J.D.Schall, D.J.Vitek and A.G.Leventhal (1986). Retinal constraint on orientation specificity in cat visual cortex. *J. Neuroscience* **6** 832-836.

11 - R.Bauer and B.M.Dow (1989). Complementary global maps for orientation coding in upper and lower layers of the monkey's striate cortex. *Exp. Brain Res.* **76** 503-509.

12 - G.Westheimer (1982). The spatial grain of the perifoveal visual field. *Vision Res.* **22** 157-162.

13 - R.Shapley and J.Victor (1986). Hyperacuity in cat retinal ganglion cells. *Science* **231** 999-1002.

14 - D.M.Levi, S.A.Klein and A.P.Aitsebaomo (1985). Vernier acuity, crowding and cortical magnification. *Vision Res.* **25** 963-977.

15 - A.Cowey and E.T.Rolls (1974). Human cortical magnification factor and its relation to visual acuity. *Exp. Brain Res.* **21** 447-454.

16 - B.M.Dow, R.G.Snyder, R.G.Vautin, R.G. and R.Bauer (1981). Magnification factor and receptive field size in foveal striate cortex of the monkey. *Expl. Brain Res.* **44** 213-228.

17 - M.Fahle (1986). Curvature detection in the visual field and a possible physiological correlate. *Exp. Brain Res.* **63** 113-124.

18 - Y.I.Yap, D.M.Levi and S.A.Klein (1987). Peripheral hyperacuity: isoeccentric bisection is better than radial bisection. *J. opt. Soc. Am.* **4A** 1562-1567.

19 - J.S.Pointer and R.F.Hess (1989). The contrast sensitivity gradient across the human visual field: with emphasis on the low spatial frequency range. *Vision Res.* **29** 1133-1151.

20 - R.P.Scobey and P.L.E.van Kan (1991). A horizontal stripe of displacement sensitivity in the human visual system. *Vision Res.* **31** 99-109.

21 - A.Toet and D.M.Levi (1992). The two-dimensional shape of spatial interaction zones in the parafovea. *Vision Res.* **32** 1349-1357.

22 - F.H.McColgin (1960). Movement thresholds in peripheral vision. *J. Optical Soc. Am.* **50** 774-779.

23 - W.A.van de Grind, J.J.Koenderink, A.J.van Doorn, M.V.Milders and H.Voerman (1993). Inomogeneity and anysotropies for motion detection in the monocular visual field of human observers. *Vision Res.*. **33**, 1089-1107.

24 - J.Rovamo and V.Virsu (1979). An estimation and application of the cortical magnification factor. *Exp. Brain Res.* **37** 495-510.

25 - C.Marzi and M.R.Di Stefano (1981). Hemiretinal differences in visual perception. *Doc.Ophthalm.Proc.Series* **30** 273-278.

26 - C.M.Murasugi and I.P.Howard (1989). Up-down asymmetry in human vertical optokinetic nystagmus and afternystagmus: contribution of the central and peripheral retinae. *Exp. Brain Res.* **77** 183-192.

27 - D.C.Van Essen, W.T.Newsome and J.H.R.Maunsell (1984). The visual field representation in striate cortex of the macaque monkey: asymmetries, anisotropies and individual variability. *Vision Res.* **24** 429-448.

28 - D.C.Burr (1980). Sensitivity to spatial phase. *Vision Res.* **20** 391-396.

29 - O.Braddick (1981). Is spatial phase degraded in peripheral vision and visual pathology? In: *Pathophysiology of the Visual System*, Maffei L. (Ed.). Documenta Ophthalmologica Proceedings Series, no. 30. The Hague, W.Junk. pp 255-262.

30 - I.Rentschler and B.Treutwein (1985). Loss of spatial phase relationships in extrafoveal vision. *Nature* **313** 308-310.

31 - P.J.Bennett and M.S.Banks (1987). Sensitivity loss in odd-symmetric mechanisms underlies phase anomalies in peripheral vision. *Nature* **326** 873-876.

32 - M.C.Morrone, D.C.Burr and D.Spinelli (1989). Discrimination of spatial phase in central and peripheral vision. *Vision Res.* **29** 433-445.

33 - N.Berardi and A.Fiorentini (1991). Visual field asymmetries in pattern discrimination: a sign of asymmetry in cortical visual field representation? *Vision Res.* **31** 1831-1836.

34 - A.Fiorentini and N.Berardi (1984). Right hemisphere superiority in the discrimination of spatial phase. *Perception.* **13**, 695-708.

35 - P.J.Bennett and M.S.Banks (1991). The effect of contrast, spatial scale and orientation on foveal and peripheral phase discrimination. *Vision Res.* **31**, 1759-1786.

36 - S.LeVay, D.H.Hubel and T.N.Wiesel (1975). The pattern of ocular dominance columns in macaque visual cortex revealed by a reduced silver stain. *J. compar. Neurol.* **159** 559-576.

37 - D.H.Hubel and D.G.Freeman (1977). Projection into the visual field of ocular dominance columns in macaque monkey. *Brain Res.* **122** 336-343.

38 - R.B.H.Tootell, E.Switkes, M.S.Silverman and S.L.Hamilton (1988). Functional anatomy of macaque striate cortex. II.Retinotopic organization. *J. Neurosci.* **8** 1531-1568.

39 - A.Grinvald, E.E.Lieke, R.D.Frostig and R.Hildesheim (1994). Cortical point-spread function and long-range lateral interactions revealed by real-time optical imaging of

macaque monkey primary visual cortex. *J.Neuroscience* **14** 2545-2568.

40 - J.C.Horton and E.T.Hedley-Whyte (1984). Mapping of cytochrome oxidase patches and ocular dominance columns in human visual cortex. *Phil. Trans. R. Soc. Lond. B* **304** 255-272.

NEURONAL RESPONSE PLASTICITY

SHAUL HOCHSTEIN
and
VOLODYA YAKOVLEV

Neurobiology Department, Institute of Life Sciences
& Neural Computation Center
Hebrew University,
Jerusalem, 91904, Israel

ABSTRACT

The study of neural plasticity is central to our understanding of the mind since information acquisition, storage and retrieval, and, of course, adaptation, learning, and change, are essential ingredients for confronting an ever changing environment. Physiologists attempt to find the synaptic transformations, and neural circuits and network mechanisms underlying memory and learning. Single cell recording in alert, behaving primates has advanced greatly this endeavor, and the sites of attentional effects of various kinds have been found. In addition, long-lasting neural activity, of the type which may subserve short-term memory, has been identified and is being studied. We have studied the changing characteristics of this activity as the monkey learns a new, match-to-sample, task. We find an increase in this delay activity, as well as in the activity in response to the performance- and reward-contingent stimulus.

1. Introduction

> "It becomes all men who deliberate on dubious matters, to be influenced neither by hatred, affection, anger, nor pity. The mind, when such feelings obstruct its view, cannot easily see what is right, nor has any human-being consulted, at the same moment, his passions and his interest. When the mind is freely exerted, its reasoning is sound; but passion, if it gain possession of it, becomes its tyrant, and reason is powerless."
>
> Gaius Julius Caesar, On the treatment of the Conspirators

With these words Julius Caesar described the need for attentional mechanisms within the brain. One of the most powerful aspects of brain function is the ability to selectively attend to specific stimuli, and to adapt this attention to new situations, that is, to learn. Experience is incorporated through learning processes and affects future actions via memory mechanisms. The study of the physiology of this cerebral plasticity has occupied neuroscientists for many generations. Only today, however, with an increased understanding of the neural networks underlying information representation within the brain, and a more solid theoretical basis concerning the synaptic transformations which may serve as long term memory store and retrieval, have we begun to attack learning phenomena directly.

The concept of the receptive field was coined by Keffer Hartline, and has since been expanded greatly. This tool for the study of neuronal response characteristics may be seen as a measure of the neuronal wiring, forming the input to the cell under study, and the basic summation processes within the neuron. Today, measuring the simultaneous activity of neighboring neurons, we may use the joint PST histogram of the units as a multi-unit receptive field [10]. This is a good predictor for processing within the cortical area. Most importantly, changes in the receptive field characteristics are a main measure of changes in neuronal summation as a function of behavioral state and learning.

Hubel and Wiesel, studying ocular dominance columns and orientation columns in V1 of developing kittens and primates, found receptive fields were extremely plastic during, and only during, an early critical period. Thereafter, receptive fields seemed to be quite inflexible. With the advent of single-unit recording techniques from the alert, behaving monkey, and the simultaneous exploration of multiple cortical areas subserving the processing of the different visual dimensions (including form, color and motion, and the representation of where and what), work began on the study of plasticity in higher cortical areas.

2. Attention in Extra-striate Cortical Areas

There are important differences between the receptive field properties in low-level and high-level cortical areas: Low-level neuron receptive fields are small, position sensitive, simple dimension dependent (orientation, size/spatial frequency, color, contrast, 1-dimensional

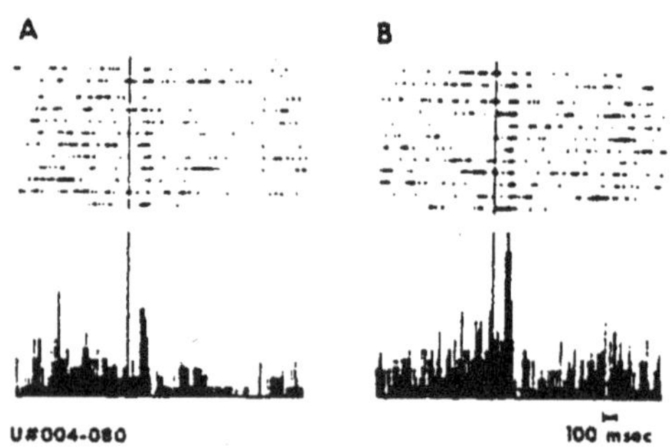

Figure 1.

Component motion direction), and are under bottom-up control. In contrast, high-level neuron receptive fields are large, position insensitive, depend on "complex" stimulation patterns (relative spatial frequency, integrated pattern motion, faces, fractal patterns, etc.), display size-constancy, color-constancy, contrast integration, and are under top-down task-relevant attentional control.

The formative studies of Wurtz, Goldberg and Robinson [11] implicated parietal cortical areas in processes of attention. The responses of these neurons to visual stimuli are enhanced when the stimuli are the subject of attention, and have behavioral significance. For example, Figure 1 demonstrates responses from an area 7 unit when the monkey performs the fixation task (A: respond to central fixation dimming; disregard peripheral stimulus) or the attention task (B: respond to either a central or peripheral spot

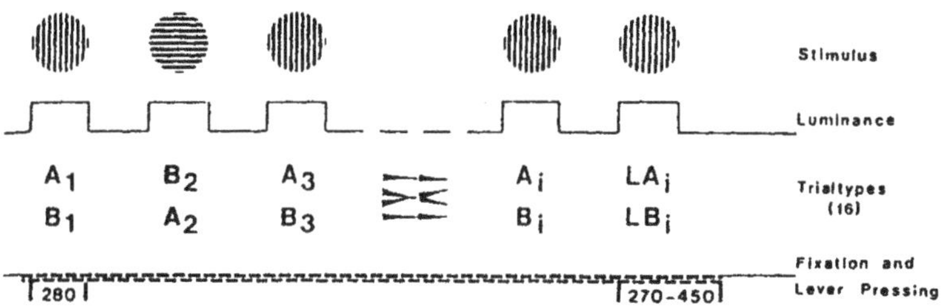

Figure 2.

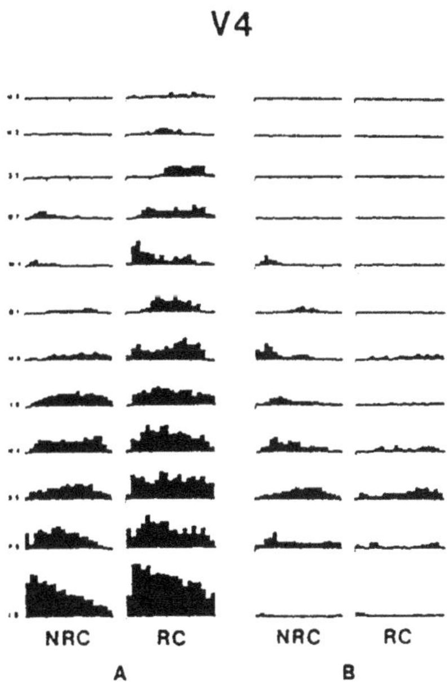

Figure 3.

dimming; attend both spots; see [2]). Note the enhancement due to attention, even though no eye movement was made to the peripheral spot, and the monkey did not know which spot would dim. This enhancement is quite different from that found in Superior Colliculus or Frontal Eye Fields, where enhancement is only seen when it precedes eye movement to the attended spot, or in V1-V3 where enhancement is due to the monkey's being generally alert.

Haenny, and Schiller [5] and Haenny, Maunsell and Schiller [6] studied similar phenomena in area V4, and found behavioral context to be an important determinant of response magnitude. For example, in one task, described in Figure 2, the monkey had to attend to a alternating series of oriented gratings, and respond (by lever press) only when two successive gratings were of the same orientation. As shown in Figure 3, the responses of V4 neurons were markedly enhanced when the stimulus was reward contingent (RC), i.e. was the second of the non-alternating pair, compared to the response to the non-reward-contingent stimulus, the first of this same pair.

Maunsell and Hochstein [7] also studied "dimensional attention" effects in V4, i.e. the dependence of the response tuning curves on the visual dimension being attended to.

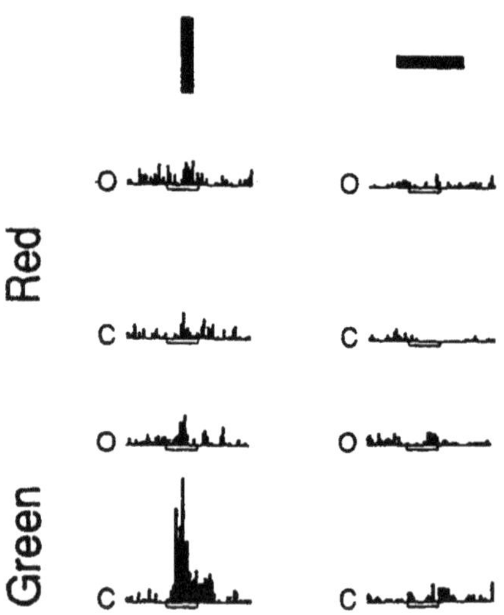

Figure 4.

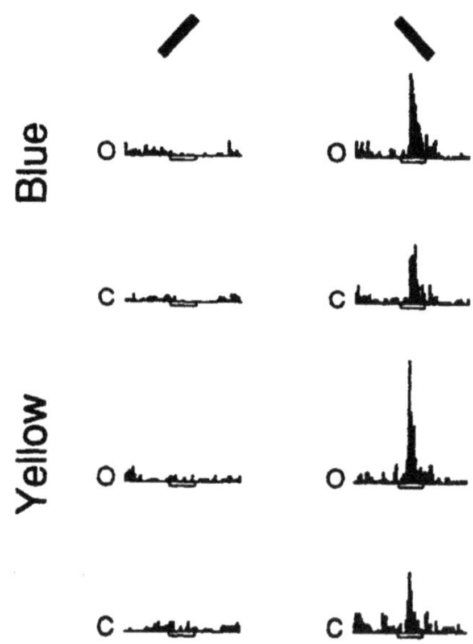

Figure 5.

The monkey had to chose the colored, oriented grating which matched either the color of a solid colored disc cue, or the orientation of a black and white oriented grating cue. We tested the orientation and color tuning curves in either condition, i.e. when the monkey was attending to the color or the orientation of the presented gratings. A significant proportion of V4 neurons had response modulations dependent on this dimensional attention [7]. Interestingly, this modulation could be enhancement or diminution of the response in the attended condition. For example, the neuron whose responses are demonstrated in Figure 4 responded to a green grating, but only when attending to the color of the grating. When the monkey was matching orientations, the response to the same stimulus was greatly reduced, and was not different than the response to the red stimulus. Figure 5 demonstrates responses of another neuron which responded selectively to left oblique compared to right oblique stimuli, but with a greatly enhanced response when matching orientations rather than colors.

Recently, Moran and Desimone [9] showed that receptive field sizes in V4 and IT are flexible. When attention is paid to a sub-region of the receptive field, it shrinks down to encompass only that area being attended to. On the other hand, when attention is outside the receptive field, responses are unaffected. Thus, attention to one stimulus has an inhibitory effect on responses to non-attended stimuli within the receptive field of the same neuron. This effect is found both in V4 and in IT. The higher area effect is *de nova* since when the stimulus is outside the V4 receptive field (and therefore is absent there) it still appears in IT.

In general, however, findings of response dependence on attention, and response plasticity with learning, have been quite disappointing. The effects have been weak, and statistical, rather than robust and dramatic, and investigators had to work hard to design the paradigms to study the effects. Why are these effects so hard to find and study? Learning-induced changes in cortical neuron characteristics may be more prevalent in still "higher" cortical areas; the changes may indeed be subtle, since small variations may be what the system is able to utilize; and the changes may reside in network dynamics, rather than in single unit activity.

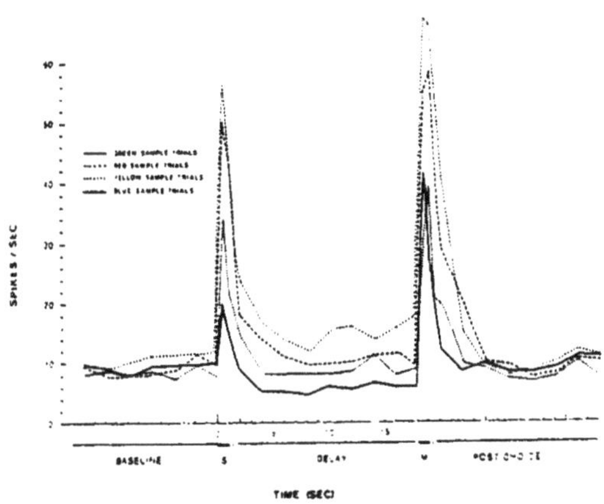

Figure 6.

3. Infero-temporal Cortex and Memory

One particularly fruitful domain of study has been the neurons of inferotemporal cortex (IT). Neurons in this region respond when the monkey views a complex pattern or object. It was found that monkeys with lesioned IT have difficulty in learning new object recognition tasks, and fail at match-to-sample tasks. Fuster and Jervey [4] found that when monkeys perform a visual memory (match to sample) task, their IT neurons fire during the period over which they must remember the preceding stimulus. For example, Figure 6 demonstrates average response firing rates during the sample stimulus (S), during the delay period (DELAY) and during the match stimulus (M). Not only is the match stimulus response enhanced relative to the response to the sample stimulus, but the firing level during the delay period is also enhanced relative to the firing at baseline, i.e. before the sample stimulus. Furthermore, the firing levels are correlated with each other in the sense that the order of response magnitudes for different colors are the same for these three periods.

Since these IT neurons are clearly involved in processes of memory of complex forms, it was natural to look for their participation in perceptual learning, as well. Miyashita and Chang [8] found dramatic neuronal correlates of pictorial memory in primate IT. Since they used new pictures (fractal patterns) which the monkey had obviously never seen before, their training clearly produced learning of these patterns. As seen in Fig. 7, activity patterns correspond to neural network type memory representations

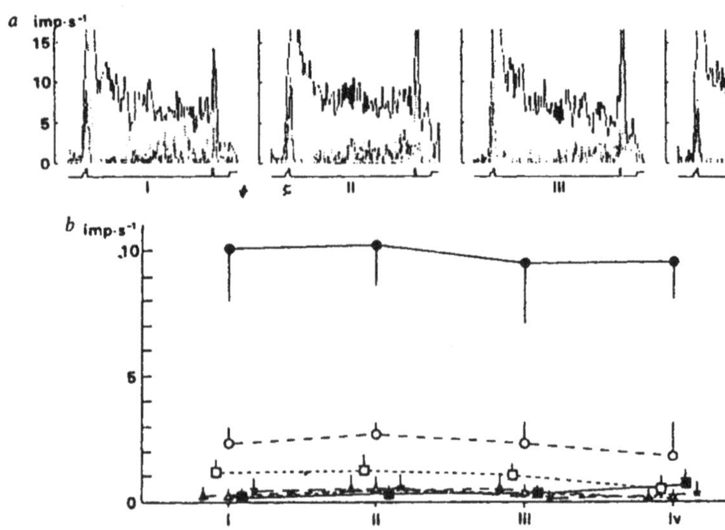

Figure 7.

and specific correlations were found between the neurons responding to each pattern, related to the order of presentation of the stimuli during the training phase of the experiment. Learning is thus a very dramatic part of the activities of neurons in higher cortical areas, and learning induced patterns may be seen if they are looked for in the right way.

4. Learning effects in IT Cortex

These studies only tested the "before" and "after" learning conditions. We have now been looking at learning-induced changes in patterns of IT neuron activities by studying the neuronal activities during the learning process, itself [12]. Again, effects are seen only when the correct tools are used for the study [1].

Multiple single-unit recordings were taken from IT cortex of a macaque monkey performing a standard match-to-sample task. The stimuli were a bright solid circle and rectangle, of equal area and energy. We studied the responses of IT cells in all stages of learning: from the level of chance performance to 90% correct responses, as seen in the learning curve of Figure 8. For analysis the learning period was divided into 3 stages: I. chance performance (sessions 1-12); II. ~ 60% correct responses (13-33); and III. > 60% correct responses (34-60).

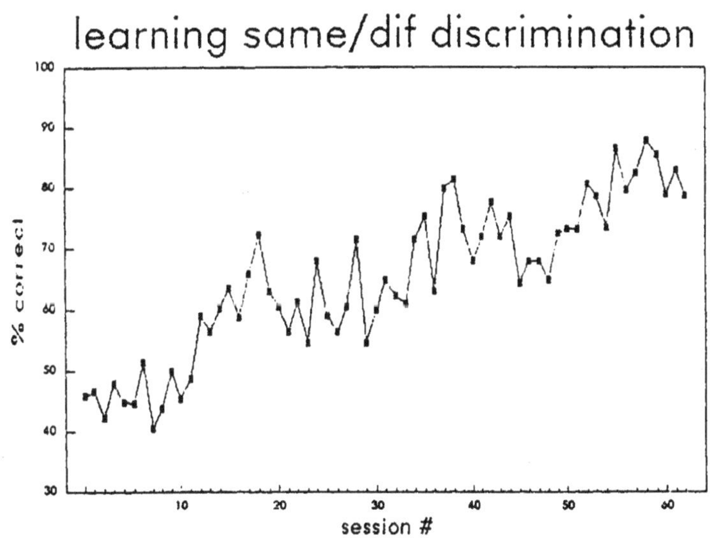

Figure 8.

IT cell activity, recorded from those neurons which were responsive to our stimuli, can be classified into the follow types: (1) Some of the responsive cells are unselective, responding similarly to the rectangle and the circle. The response during the first, sample, stimulus was often stronger then during the second, test stimulus. (2) Some cells were

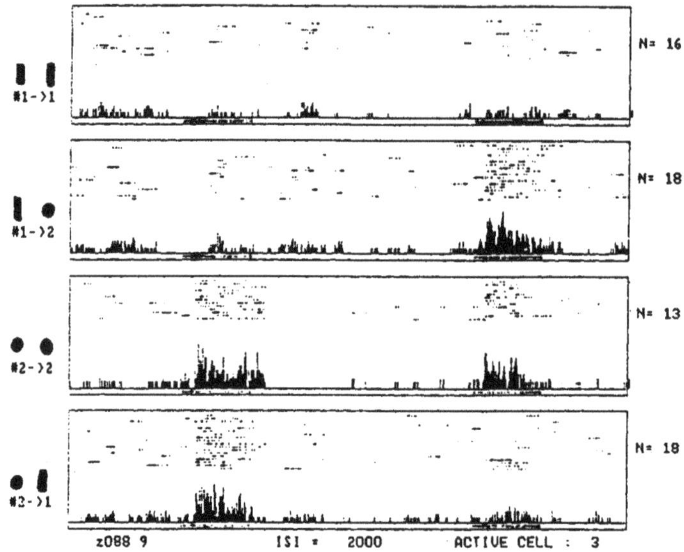

Figure 9.

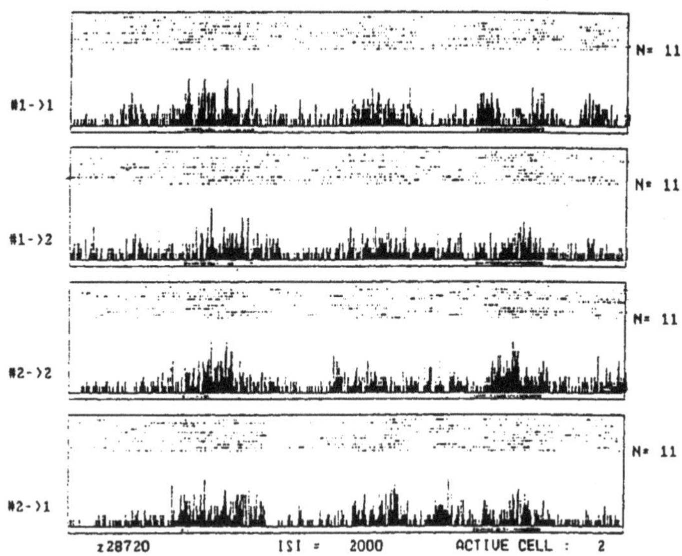

Figure 10.

selective to one of the stimuli, responding to the rectangle and ignoring the circle or vice versa, as in Figure 9. (3) Some cells were active during the interstimulus interval, as in Figure 10, often only following one of the two stimuli, and therefore with an activity which may reflect memory of this preceding stimulus, as needed for correct performance of the task.

We analyzed how the activity of IT cells changed depending on the temporal position of the stimulus, that is whether it was the first (sample) or second (match) stimulus. In addition, how neuron activity changed during the ISI period, compared to background activity. We show here only the data for the SAME situation (that is, for match stimulus the same as the sample stimulus) to demonstrate how the response depends on temporal position of the stimulus.

Figure 11 (squares) displays the ratio of activity in terms of spike frequency during the first stimulus relative to during the second stimulus if the stimuli were both circles or both rectangles, for each stage of learning. Zero means that the responses are equal. During the 1st stage the values are above 0 (the response to the sample stimulus is stronger) and the values are lower, or negative, during the next stages (strengthening of response to the test stimulus).

The diamonds of Figure 11 display the activity of IT cells in the same trials during the ISI. During the first Stage the average ratio of ISI activity to spontaneous activity is below 0 (activity during ISI is suppressed). However, in the next stages the distribution is above 0 (during stages II and III, IT cells are activated during ISI). Finally, in the 1st stage, responses of IT cells to the test stimulus were suppressed for the match condition compared to the non-match condition (triangles) while in later stages the match response grew in size and presumably in importance

The most dramatic changes in activity of IT cells take place as the monkey progressed from stage I to stage II. We suggest that during the first stage the sensory

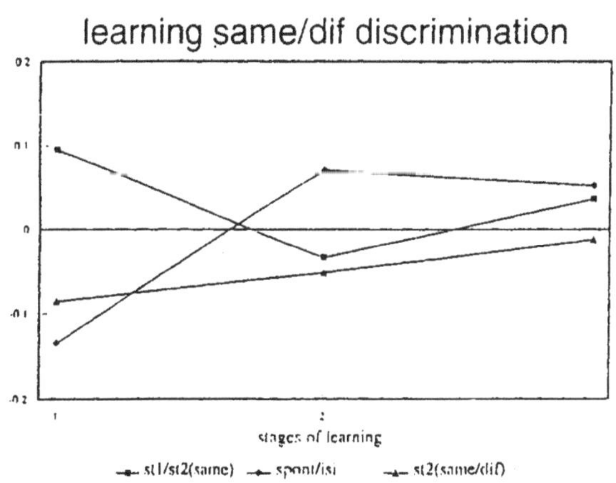

Figure 11.

reaction dominated IT activity, and most of the activity was connected to the shape of the stimuli, but during later stages, IT neuron activity during the ISI takes on an important role. If activity during the ISI reflects the process of remembering the sample stimulus, then these data confirm the idea that for successful learning the crucial point is creation of neural assembles with memory properties.

5. Acknowledgments

We thank Daniel Amit for helpful comments throughout this study. This work was supported by the U.S.-Israel Binational Science Foundation (BSF) and the Israel Science Foundation of the Israel Academy of Sciences and Humanities.

6. References

[1] Amit, D.J. (1993) In defense of single electrode recording. Network, 3, 385.

[2] Bushnell, M.C., Goldberg, M.E. and Robinson, D.L. (1981) Behavioral enhancement of visual responses in monkey cerebral cortex. I. Modulation in posterior parietal cortex related to selective attention. Jour. Neurophysiology, 46: 755-772.

[3] Desimone, R., Schein, S.J., Moran J. & Ungerleider, L.G. (1985) Contour, color and shape analysis beyond the striate cortex. Vision Research, 25, 441-452.

[4] Fuster, J.M. & Jervey, J.P. (1982) Neuronal firing in the inferotemporal cortex of the monkey in a visual memory task. Jour. Neuroscience 2, 361-75.

[5] Haenny, P.E. and Schiller, P.H. (1988) State dependent activity in monkey visual cortex. I. Single cell activity in V1 and V4 dependence on visual tasks. Experimental Brain Research, 69: 225-244.

[6] Haenny, P.E., Maunsell, J.H., Schiller, P.H. (1988) State dependent activity in monkey visual cortex. II. Extraretinal factors in V4. Exp. Brain. Res. 69, 245-259.

[7] Maunsell, J.H.R. and Hochstein, S. (1991). Effects of behavioral state on the stimulus selectivity of neurons in area V4 of the macaque monkey. pp 447-470. In: Channels in the Visual System. B. Blum, ed. Freund, London.

[8] Miyashita, Y & Chang, H.S. (1988) Neuronal correlate of pictorial short term memory in the primate temporal cortex. Nature 331, 68-70.

[9] Moran, J. & Desimone, R. (1985). Selective attention gates visual processing in the extrastriate cortex. Science, 229, 782-785.

[10] Stern, E., Aertsen, A., Vaadia, E. and Hochstein, S. (1993). Stimulus encoding by multidimensional receptive fields in single cells and cell populations in V1 of awake monkey. Neural Information Processing Systems, 5: 377-384.

[11] Wurtz, R., Goldberg M., & Robinson D.L. (1980). Behavioral modulation of visual responses in the monkey: Stimulus selection for attention and movement. Prog. Psychobiol. & Physiol. Psychol. 9, 43-83.

[12] Yakovlev, V. & Hochstein, S. (1994) Learning a visual task as reflected in activity dynamics of macaque IT neurons, Soc Neurosci Abstr, 20.

PERCEPTUAL LEARNING SPECIFICITY

SHAUL HOCHSTEIN
and
MERAV AHISSAR

Neurobiology Department, Institute of Life Sciences
& Neural Computation Center
Hebrew University,
Jerusalem, 91904, Israel

ABSTRACT

The phenomenon of transfer of learning in early perceptual tasks is used to study the mechanism(s) and site(s) of learning. Psychophysical studies found that perceptual learning is specific to stimulus attributes in various visual tasks, including in discriminations of phase in stereopsis, relative phase in complex gratings, orientation, motion direction, pattern or texture, and in vernier acuity, feature search, array orientation identification, and Gabor-patch detection. Improvement in performance does not transfer to different spatial locations, orientations, sizes, spatial frequencies, or even, in some cases, across eyes. This specificity has implications for the site of perceptual learning since neurons at lower levels of the visual pathway are highly specialized for position, size, orientation, spatial frequency, and eye of origin while neurons in higher anatomical visual areas generalize over these variables and are sensitive to more complex aspects of the stimuli. Thus, stimulus-specific learning- induced changes seem to occur at an early cortical processing stage, perhaps in primary visual cortex. However, these studies of perceptual learning do not rule out the possibility that there might be both stimulus-specificity and also top-down control. We used a paradigm whereby two independent tasks could be performed on the same set of visual stimuli. Following practice on one task, we tested consequent improvement in the other task, keeping the retinal stimuli constant. We found that perceptual learning is specific to the trained task. No substantial improvement was found for attributes which were not attentively practiced. Interestingly, improvement of detection did transfer across eyes, to homologous positions near fixation in the other hemifield, to mirror symmetric orientations and to expanded images. These generalizations may reflect initial stages of spatial invariances.

1. Introduction

There are three necessary elements in the investigation of the cognitive processes of perception: **sensation**, the study of the representation of perceptual information in the brain; **memory and learning**, inquiry into the way these sensory representations and the processes of their encoding are changed by experience; and, **transfer**, measurement of the degree to which learning-induced changes affect new and different sensations [21,27]. The understanding of these three levels of perception are wholly intertwined and any comprehensive theory must account for them all.

Yet, sensory physiologists have spent many decades studying, through the use of single neuron recording, the representation of sensory information in columns of detector neurons, without much concern for where and how this information is stored, let alone generalized. Similarly, cognitive psychologists have discussed different types of memory traces (short and long term; implicit and explicit) without always considering the limitations which the neuronal basis of memory puts on these processes, and only occasionally testing their theories by measuring the degree of transfer of newly acquired capabilities.

2. Early Perceptual Learning

Learning is a strange and wonderful thing. We start out lacking something which we acquire and later have. What is the process of this acquisition? Where in the brain does it occur? What is the nature of the changes which serve as the underlying mechanism of learning?

Because these questions are so hard to answer, there has been quite a neglect of the whole field of learning. Electrophysiologists study the "steady state" - How could they do otherwise, working in an anesthetized animal, which is probably incapable of learning (though we are not even sure of that!), or working with an alert behaving animal which is purposely over-trained in the performance of the behavioral task so that the experimenter may record from a population of neurons, sequentially over a period of weeks or months, while the effective state of the monkey is unchanged - i.e. no learning takes place.

Psychologists, too, have turned their backs on learning as a field of study. They rather talk of memory - explicit and implicit, long term or working - the encoding of information, the characteristics of internal memory traces, and the mechanisms for memory retrieval and recall.

Learning is a process of change. Having learned something about a stimulus "A", I will respond to a **re-viewing** of "A" differently - more quickly, perhaps; with a different response, perhaps. What really interests us is: What is the mechanism of this change? What change(s) in the brain underlie the changed response? Does learning take place at the site of very low-level, stimulus-locked, sensory cortical processing? Or, is it part of a higher level, strategy-related, task-dependent, perceptual or behavioral process?

3. Specificity and Transfer

3.1. Introduction

As a tool for the study of learning, we are using the phenomenon of learning transfer. In a way, this research strategy goes back to the original Kefer Hartline method of the receptive field. The receptive field of a sensory neuron is defined as the collection of stimuli - the volume of stimulus space - which affects the firing rate of the neuron. The learning transfer study strategy is based on the hypothesis that, learning will not transfer beyond the (union of) receptive field(s) of the neuron(s) in which the learning takes place.

The classical experiment in this field of early perceptual learning was performed by Adriana Fiorentini and Nicoletta Berardi nearly 15 years ago [10]. See also their papers in this volume [12,13]. They found that perceptual learning was specific for orientation and spatial frequency. In our terminology, then, and in terms of their classical experiment: If learning takes place in spatial frequency tuned neurons, then training on a discrimination task using complex grating stimuli in one frequency band will not transfer to subsequent testing of performance using similar gratings in another frequency channel. One of the beauties of this technique is that following the successes of the spatial frequency channel bandwidth definitions using threshold and adaptation studies, here was a new psychophysical technique for measuring channel bandwidth and one didn't know at the outset if the same width would be measured.

Why not? Why isn't it obvious that a task which depends on spatial frequency tuned units would be learned without significant transfer to performance on stimuli of another spatial frequency? Because, it might be that what is learned is a performance strategy, not a sensory adroitness. It might be that learning takes place in the higher, behavioral, areas of the cortex, in neurons whose "receptive fields" are not spatial frequency limited at all, but which collect information converging from numerous spatial frequency channels.

In fact, the hierarchical model of pathways of visual processing, with information climbing from V1 to V2,3,4, MT, IT, etc., sees the receptive fields of neurons within these areas as gradually increasing in size, both in terms of spatial extent and in terms of simple dimensional tuning bandwidths, in exchange for other, higher-order selectivities. Face cell responses, for example, should be "invariant" for variations in spatial frequency, intensity, contrast, color, and maybe orientation, being specific to the identity of the complex pattern to which it responds, namely a specific face. So, if these are the cells involved in learning a face discrimination task, then learning may very well transfer to other size or other contrast faces, though maybe not to new faces.

Thus, Fiorentini & Berardi's results are not obvious at all. They lead us to the conclusion that learning is a very low level phenomenon, taking place where neurons have receptive fields which have narrow spatial frequency bandwidths.

In fact, these results themselves were a renewing of an old interest in perceptual learning abilities. In 1858, A.W. Volkman studying two-point touch discrimination found that the just-noticeable-difference could be reduced to half [29]. Figure 1 demonstrates these results. The improvement in performance with training (continuous curve) was limited to the trained finger (striped bars show lack of improvement in another finger), though transfer was found to the homologous finger on the other hand (black bars).

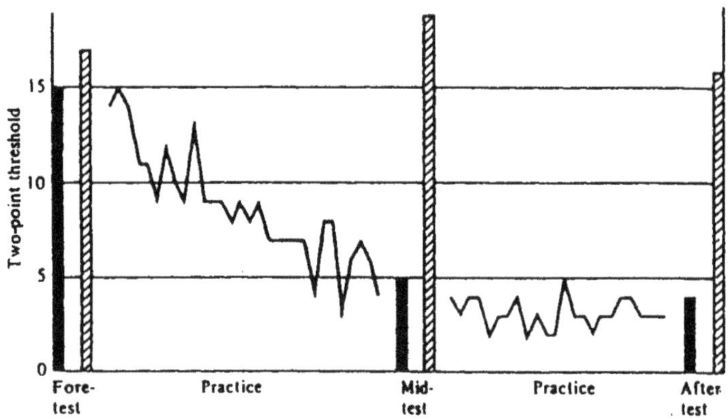

Figure 1.

3.2. Transfer and Specificity for Numerous Perceptual Tasks

Recently there has been a surge of interest in the field of transfer of learning in early perceptual tasks. Psychophysical studies have found that perceptual learning is specific to stimulus attributes in various visual tasks. Table 1, on the following page, reviews these tasks and specificities, and some are demonstrated in Figure 2, on the page following the next. For example, in all cases where oriented elements were used, learning was specific to the trained orientation, including discriminations of phase in stereopsis [25], relative phase in complex gratings [6,10,11,12,13], orientation [26, 28], texture [16,17], and vernier [9,23], (Fig 2 bottom left), and in feature search [1,2,3]. When sinusoidal gratings were used, improvement was specific to spatial frequency [10,11], (Figure 2, middle right).

Similarly, when testing discrimination of motion direction, learning was specific to speed and direction [4,5], (Figure 2, bottom right). When element size was greatly reduced, performance deteriorated towards original levels [1,2,3], (Figure 2, middle left). Some tests show specificity to eye [16], (Figure 2, top, left), while others don't [1,2,3,4,5,10,11].

In all tasks that tested specificity to retinal location, substantial specificity was found [4,5,9,22,23,24,26]. There was nearly no transfer of texture discrimination when moving from 3° above and to the right of fixation to the same distance below and to the left of fixation [16], and there was complete specificity when moving a 1° complex grating

Table 1. Studies of simple early perceptual learning task specificities

task	transfer characteristics	reference
touch position just-noticeable-difference	specific to location; transfers to homologous finger on other hand.	29
stereopsis	orientation specific.	25
complex grating relative phase discrimination	specific to orientation, spatial frequency, location (1°) except across hemifields; transfer across eyes.	6, 10, 11, 12, 13
orientation discrimination	no transfer from obliques to principal axes; some transfer to other obliques.	28
motion direction discrimination	specific to motion direction and speed; specific to retinal location.	4, 5
bar pattern discrimination	specific to retinal location.	22
texture discrimination	specific to background orientation, retinal location, trained eye.	16, 17
vernier acuity	specific to location and orientation.	9, 23
orientation discrimination	specific to location and orientation.	26
feature search in pop-out task	specific to orientation of target or background (but not to mirror image transformation), to retinal location (within 0.7°, but not across midline), to size for contraction (but not expansion). transfer across eyes.	1, 2, 3
array orientation discrimination	specific to array size (but not to element size).	1, 2, 3
Gabor patch detection	specific to location and eye.	24

patch from 1-2° from fixation to 2-3° from fixation (though there was complete transfer across the midline [6,10,11], (see Figure 2, top right, and papers in this volume [12,13]). We found specificity to 0.7° when moving away from fixation, but not when moving across the midline [1,3].

Thus, in each of these cases, the general conclusion of Fiorentini and Berardi was confirmed: there was no transfer of the learning effect beyond the receptive fields of the low-level neurons which we would have guessed would have acted as the "detectors" and the "discriminators" for these stimuli and for these tasks.

This specificity has implications for the site(s) of perceptual learning: There is both anatomical and physiological evidence for hierarchical information processing in the visual system. Neurons at lower levels of the visual pathway are highly specialized for position,

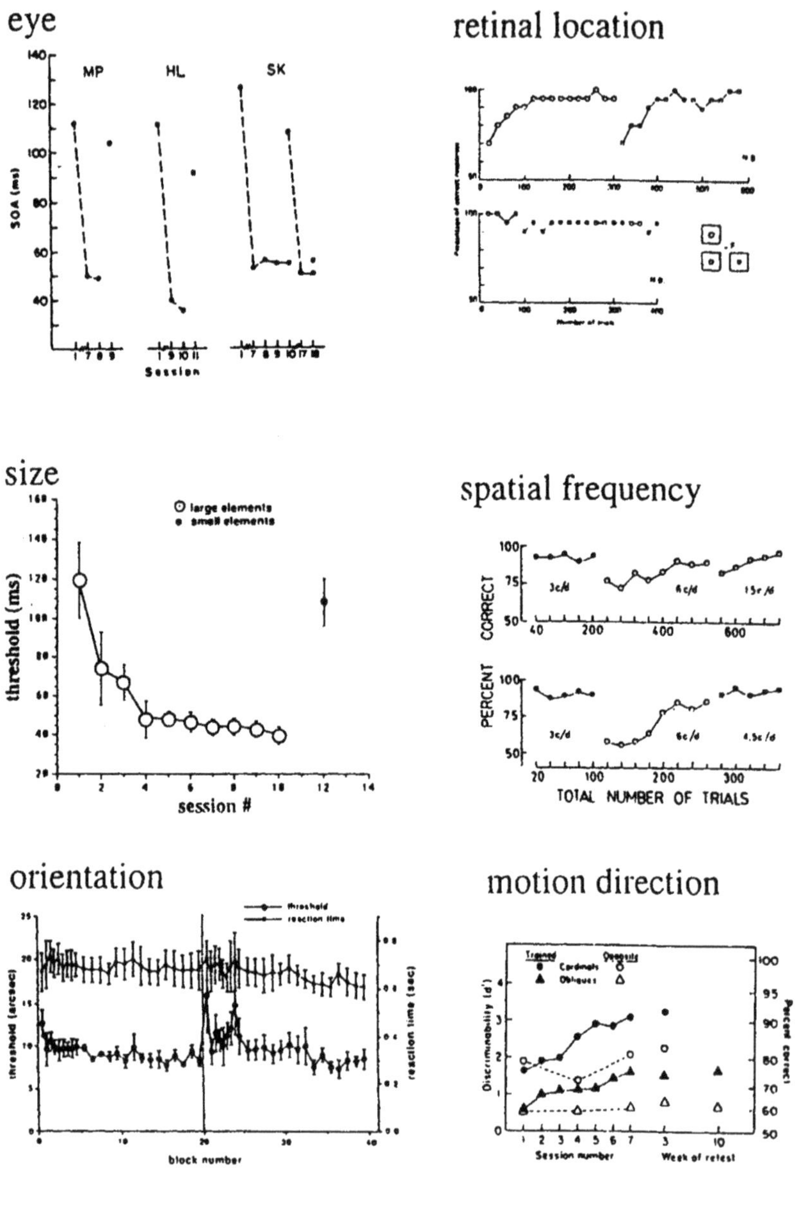
eye
MP
HL
SK
SOA (ms)
Session
retinal location
spatial frequency
PERCENT CORRECT
TOTAL NUMBER OF TRIALS
size
large elements
small elements
threshold (ms)
session #
orientation
threshold
reaction time
threshold (arcsec)
reaction time (sec)
block number
motion direction
Trained
Opposite
Cardinals
Obliques
Discriminability (d')
Percent correct
Session number
Week of retest

Figure 2.

size, orientation, spatial frequency, and eye of origin. Neurons in higher anatomical visual areas, however, generalize over these stimulus variables (see reviews [7,19]) and are sensitive to increasingly more complex aspects of the stimuli. These considerations imply that stimulus-specific learning-induced changes occur at an early cortical processing stage, perhaps in primary visual cortex, V1.

3.3. Attentional Control of Perceptual Learning

Does this mean that all learning is low level, and that learning is a matter of becoming acquainted with the stimulus at the stimulus-sensing level? In fact, these studies of perceptual learning did not rule out the possibility that there might be both stimulus-specificity and also "top-down" control. Computational considerations suggest that solving many complex image processing tasks in vision may involve an initial "bottom-up" processing stage followed by analysis at later stages which are top-down modulated, i.e. where information about task and prior knowledge may modify neural responses and computations [18]. This notion agrees with findings from behaving monkeys where the firing rates of neurons in higher visual areas were more strongly modulated by top-down task-dependent influences than neurons in early cortical areas [14,15,20,30], (for a review, see [8]). According to this simple hierarchical view, the reported stimulus specificity of perceptual learning that indicates a location early in the visual pathway implies that there should be little top-down control of these learning processes.

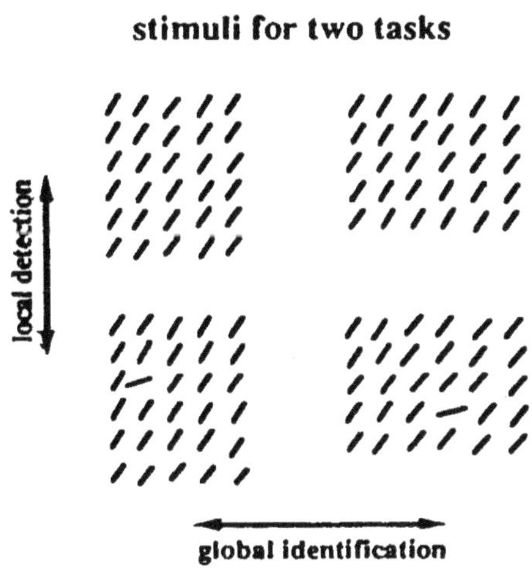

Figure 3.

To test the prediction that perceptual learning was not affected by influences from higher levels of cerebral cortex, we used a paradigm whereby two independent tasks could be performed on the same set of stimuli. The two tasks are demonstrated in Figure 3, on preceding page. Subjects were shown one of the arrays, chosen in random from the four arrays displayed in the figure. Subjects were asked to judge either the presence or absence of an odd element within the array ("local detection task" - differentiate between upper and lower arrays), or, the vertical or horizontal alignment of the array ("global identification task" - differentiate between left and right arrays). A mask stimulus (a square array of asterisk-like elements) was presented following a variable Stimulus-Onset-Asynchrony (SOA). From the resulting psychometric curve (performance vs. SOA), threshold SOA for 82% correct performance was determined and plotted as a function of training.

Following practice on one task, we tested the consequent improvement in the other task while keeping the retinal stimuli constant. Alternatively, we keep the task fixed while changing the physical stimuli. We found that while perceptual learning shows stimulus specificity, it is even more specific to the trained task. No substantial improvement was found for attributes which were not attentively practiced [1,2,3]. This result is shown in Figure 4, for two groups of 8 subjects. Learning curves show relative threshold change due to training with the "local" pop-out task, and subsequent training on the "global" array orientation task (C), or learning first the "global" task, and subsequently the "local" task (D). Learning in both cases is task specific, though there is some small degree of global to local task transfer. Thus, though learning is stimulus specific, as seen

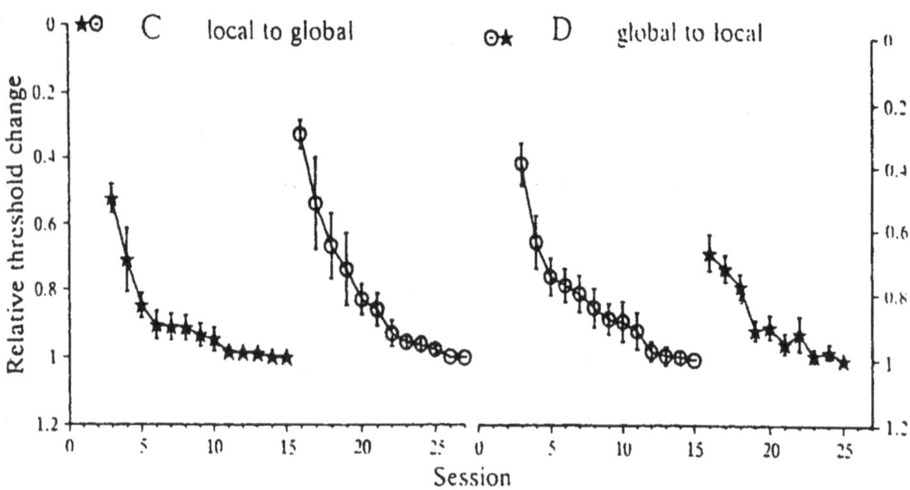

Figure 4.

in Figure 2, it is even more task specific, suggesting involvement of top-down attentional control mechanisms.

In conclusion, learning occurs only in those neurons that are both activated by the stimulus and at the same time are relevant for the performance of the task. Which neurons are indeed relevant is dictated by the mechanism or strategy used for task performance, and is signaled by a top-down attentional control to the low-level cortical site of learning.

4. Acknowledgments

We thank Anne Treisman and Howard Hock for helpful comments throughout this study. Many thanks to Cloe Taddei-Ferretti of the Instituto di Cibernetica, CNR, for organizing a wonderful Study Program at Casamicciola on the Isle of Ischia. This work was supported by the U.S.-Israel Binational Science Foundation (BSF) and the Israel Science Foundation of the Israel Academy of Sciences and Humanities.

5. References

[1] Ahissar, M. & Hochstein, S. (1993). The role of attention in early perceptual learning. Proceedings of the National Academy of Science U.S.A. 90, 5718-5722.

[2] Ahissar, M. & Hochstein, S. (1995a). Perceptual Learning. In: *Perceptual Constancies: Why Things Look as They Do.* V. Walsh & J. Kulikowski, Eds. Cambridge University Press, Cambridge.

[3] Ahissar, M. & Hochstein, S. (1994). How early is early vision? Evidence from perceptual learning. In: *Early Vision and Beyond.* T.V. Papathomas, C. Chubb, A. Gorea and E. Kowler, Eds. M.I.T. Press, Cambridge MA.

[4] Ball, K. & Sekuler, R. (1982). A specific and enduring improvement in visual motion discrimination. Science, 218,697-698.

[5] Ball, K. & Sekuler, R. (1987). Direction specific improvement in motion perception. Vision Research, 27, 953-965.

[6] Berardi, N. & Fiorentini, A. (1987). Interhemispheric transfer of visual information in humans: spatial characteristics. Journal of Physiology, 384, 633-47.

[7] Desimone, R., Schein, S.J., Moran J. & Ungerleider, L.G. (1985) Contour, color and shape analysis beyond the striate cortex. Vision Research 25, 441-452.

[8] Desimone, R., Wessinger, M., Thomas L, and Schneider, W. (1990). Cold Spring Harbor Symposia on Quantitative Biology, Volume LV. pp. 963-971.

[9] Fahle, M. & Edelman, S. (1993). Long term learning in vernier acuity: Effects of stimulus orientation and range, and of feedback. Vision Research, 33, 397-412.

[10] Fiorentini, A. & Berardi, N. (1980). Perceptual learning specific for orientation and spatial frequency. Nature (London), 287, 43-44.

[11] Fiorentini, A. & Berardi, N. (1981). Learning in grating waveform discrimination: specificity for orientation and spatial frequency. Vision Research, 21, 1149-1158.

[12] Fiorentini, A. & Berardi, N. (1995). Visual field asymmetries and local anisotropies in pattern discrimination: a sign of cortical asymmetries. In this volume.

[13] Fiorentini, A. & Berardi, N. (1995). Interhemispheric transfer of visual information: a cross-talk between the two cerebral hemispheres. In this volume.

[14] Haenny, P.E., Maunsell, J.H., Schiller, P.H. (1988) State dependent activity in monkey visual cortex. II. Visual and nonvisual factors in V4. Experimental Brain Research, 69, 245-259.

[15] Haenny, P. E. & Schiller, P. H. (1988). State dependent activity in monkey visual cortex. 1. Single cell activity in V1 and V4 dependence on visual tasks. Experimental Brain Research, 69, 225-244.

[16] Karni, A. & Sagi, D. (1991). Where practice makes perfect: Evidence for primary visual cortex plasticity. Proceedings National Academy of Science U.S.A., 88, 4966-4970.

[17] Karni, A. & Sagi, D. (1993). The time course of learning a visual skill. Nature 365, 250-252.

[18] Marr, D. (1982). *Vision*. Freeman & Company, New York.

[19] Maunsell, J. H. R. & Newsome, W. T. (1987) Visual processing in monkey extrastriate cortex. Annual Review of Neuroscience 10, 363-401.

[20] Moran, J. & Desimone, R. (1985). Selective attention gates visual processing in the extrastriate cortex. Science, 229, 782-785.

[21] Newell A. & Simon, H. (1972) Human Problem Solving. Prentice Hall, NJ.

[22] Nazir, T. & O'Regan, J.K. (1990). Some results on translation invariance in the human visual system. Spatial Vision 5, 81-100.

[23] Poggio, T., Fahle, M. & Edelman, S. (1992). Fast perceptual learning in visual hyperacuity. Science 256, 1018-1021.

[24] Polat, U. & Sagi, D. (1994). Spatial interactions in human vision: from near to far via experience dependent cascades of connections. Proceedings of the National Academy of Science USA, 91, 1206-1209.

[25] Ramachandran, V.S. & Braddick, O. (1973) Orientation specific learning in stereopsis. Perception 2, 371-376.

[26] Shiu, L. & Pashler, H. (1992). Improvement in line orientation discrimination is retinally local but dependent on cognitive set. Perception & Psychophysics, 52, 582-588.

[27] Treisman, A. Vieira, A. & Hayes, A. (1992). Automaticity and preattentive processing. American Journal of Psychology, 105, 341-362.

[28] Vogels, R. & Orban, G. A. (1985). The effect of practice on the oblique effect in line orientation judgments. Vision Research, 25, 1679-1687.

[29] Volkman, A.W. (1858). Ueber den Einflus der Uebung. Leipzig Berichte. Math.-phys. Classe, 10, 38-69.

[30] Wurtz, R., Goldberg M., & Robinson D.L. (1982). Brain mechanisms of visual attention. Scientific American 246(6), 100-107.

INTERHEMISPHERIC TRANSFER OF VISUAL INFORMATION: A CROSS-TALK BETWEEN THE TWO CEREBRAL HEMISPHERES

Adriana Fiorentini and Nicoletta Berardi
Istituto di Neurofisiologia C.N.R., Via S.Zeno 51, Pisa, Italy

Hemispheric lateralization is a well assessed property of the human brain: in the majority of subjects there is a left-hemisphere dominance for language functions and right hemisphere dominance for visuospatial functions. The most convincing evidence comes from patients with lateralized lesions. Aphasia and in general disturbances of language, are related in most cases to left hemisphere damage, while disturbances of spatial relations of objects or recognition of faces or other complex patterns are related to right hemisphere damage [1].

In normal subjects, the role of either hemisphere in visual tasks can be tested by presenting stimuli of short duration laterally, either in the left or in the right visual hemi-field, since either hemifield projects to the contralateral hemisphere. The findings obtained in normal subjects are in good agreement with those obtained in patients with lateralized lesions. Left visual field (right hemisphere) superiority has been found for the recognition of complex patterns, and in particular human faces, and for other visual tasks not implying a verbal labeling, as well as for visual spatial localization [2].

On the contrary, there is no hemispheric specialization for basic sensory processes underlying visual functions of contrast sensitivity and spatial resolution: visual tasks, like the detection of sinusoidal gratings of low contrast or the discrimination of spatial frequency differences between sinusoidal gratings of suprathreshold contrast are performed equally well in the left and in the right visual hemifield [3-5]. Fig. 1 shows monocular contrast sensitivity functions for horizontal sinusoidal gratings as a function of spatial frequency measured at 5 deg distance from the fixation point in the left and right visual field . No visual field preference is observed. There are only slight differences between the ipsilateral and contralateral visual field of either eye, suggesting a preference for the temporal hemifield (nasal retina), but these are significant only at high spatial frequencies, close to visual acuity [3]. If the gratings are oriented obliquely, rather than horizontally or vertically, resolution is somewhat poorer (oblique effect), but again there is no difference between the right and left hemifield [4].

Left hemifield (right hemisphere) advantage for spatial phase discrimination.

As reported in a previous chapter, the discrimination of spatial phase in compound gratings consisting of the sum of two harmonics (sinusoids of spatial frequency f and 3f or f and 2f) and viewed peripherally is more accurate for stimuli located in the left than in the right visual field (Fig. 2). If the discrimination of these compound gratings is interpreted in terms of sensitivity to spatial phase differences, then the left field advantage in this task

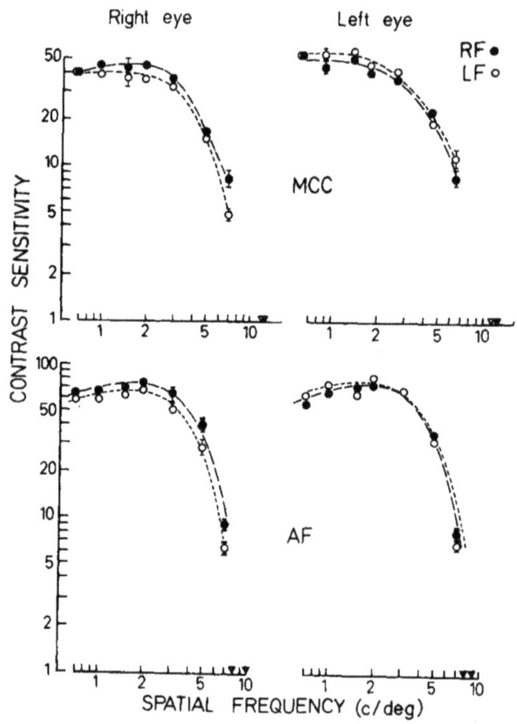

Fig. 1 - Contrast sensitivity functions for the right and left eye of two subjects in the left (open circles) and right (closed circles) visual field. The triangles indicate visual acuity. The inner edge of the stimulus grating (8 x 8 deg) was at 8 deg on the right or left of the fixation point. (Reprinted from ref.3 with permission).

indicates a right hemisphere superiority in spatial phase processing. Information on the relative phase of spatial frequency components of a complex visual stimulus is important for the recognition of complex forms [6]. Since no hemispheric specialization is observed either in contrast sensitivity and visual acuity, or in spatial frequency discrimination, the processing of spatial phase may be one of the earliest steps in the processing of the spatial organization of a complex stimulus at which a right hemisphere specialization is evident. If, alternatively, the limits in the discrimination of gratings having only two harmonic components in the peripheral visual field are interpreted in terms of positional uncertainty, rather than phase insensitivity [7], the left visual field advantage in the discrimination of complex gratings may represent an example of the superiority of the right hemisphere in the processing of visual spatial relationships.

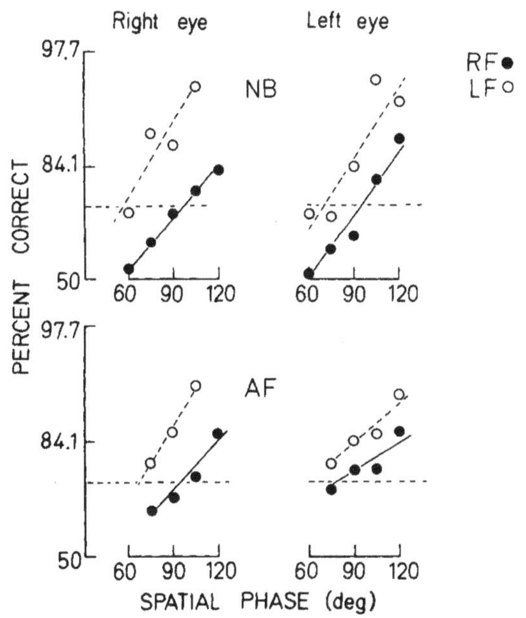

Fig. 2 - Psychometric functions for spatial phase discrimination of compound gratings consisting of f and 3f components, at 8 deg from the fixation point, in the left (open circles) and right (closed circles) visual field. One grating had the two components superimposed in the peak subtract phase (0 deg) and the other at a variable phase (abscissa). Spatial frequency f: 0.5 cycle/deg (AF), 1 cycle/deg (NB). (Reprinted with permission from ref. 3).

Interhemispheric transfer of visual information

Under normal conditions, the specificity of the two hemispheres for different visual tasks can be integrated by means of commissures connecting the two hemispheres, primarily by the *corpus callosum* and the anterior commissure. The anterior commissure, that is by far the smaller of the two commissures, connects non-visual areas, like the amigdalae and the olfactory bulbs, but in the monkey also contains fibers that connect regions of the visual temporal cortex.

In contrast, reciprocal callosal connections are present in the striate and all extrastriate visual cortical areas [8]. In cats and monkeys, the callosal connections in areas 17 and 18 of the visual cortex are limited to a narrow region on either side of the 17 - 18 border, where the

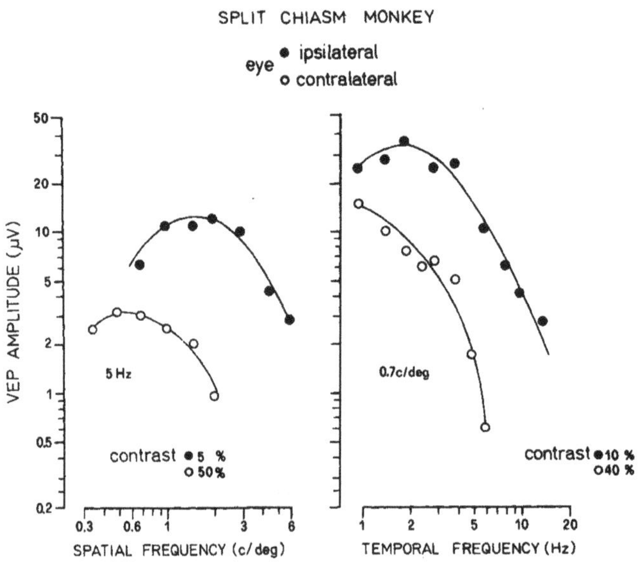

Fig. 3 - Visual Evoked Potentials recorded from one hemisphere at the V1 - V2 border, in a split-chiasm macaque monkey (where the contralaterally projecting retinal fibers are sectioned and the ipsilateral projecting fibers are preserved), in response to sinusoidal gratings of various spatial frequencies alternating in contrast at 5 Hz (left) or of fixed spatial frequency (0.7 cycle/deg) alternating at various temporal frequencies (right). The closed circles are responses to monocular stimulation of the eye ipsilateral to the recording hemisphere (that in split-chiasm animals represent activity arising from the direct geniculo-cortical input), the open circles are responses to stimulation of the contralateral eye, and represent activity reaching the recording hemisphere via the callosal fibers. Note that in the latter case the contrast of the stimulus grating was 4 or 5 times higher than in the former[20].

vertical meridian of the visual field is represented. Callosal neurons, i.e. neurons that send their axons contralaterally via the occipital portion (splenium) of the corpus callosum, have receptive fields that straddle the vertical meridian and thus contribute to extend into the ipsilateral hemifield the representation of the contralateral hemifield in either hemisphere [8-10]. Most receptive fields are binocular and the direct and callosal input have the same preferred orientation. The region covered by callosal receptive fields of areas 17 and 18, however, is limited to a few degrees from the vertical meridian of the visual field.

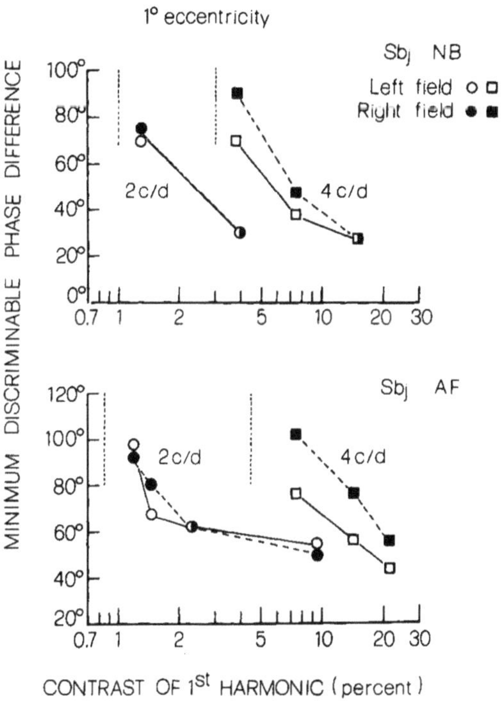

Fig. 4 - Minimum discriminable phase difference between compound gratings (consisting of two harmonics of spatial frequencies f and 3f, with contrast ratio 3 to 1) as a function of contrast of the f component. Stimuli separated 1 deg from the vertical meridian, either in the left (open symbols) or right (closed symbols) visual field. Spatial frequency f: 2 cycle/deg (circles), 4 cycle/deg (squares). (Replotted from ref. 25).

The transfer of visual information via the reciprocal callosal connections may subserve various roles. For instance, it may assure perceptual continuity across the vertical meridian of the visual field, by binding together the two halves of the visual field represented separately onto the two hemispheres [11,12]. It has also been suggested [13], that it may play a role in binocular vision: stimuli in front or behind the fixation point giving rise to a large crossed or uncrossed binocular disparity are imaged on both temporal or both nasal retinae. Sensitivity of single neurons to such large disparities may result from the convergence of a direct retinogeniculate input from the temporal retina of the ipsilateral eye and the callosal input from the nasal retina of the contralateral eye (neurons tuned to large crossed disparities), or viceversa (neurons tuned to large uncrossed disparities). Evidence in favour of this role of the corpus callosum is somewhat controversial, however, [14] and there is evidence that sensitivity to small disparities, subserving stereoacuity, does not require the contribution of interhemispheric connections [15].

Another possible role for the forebrain commissures is the interhemisferic transfer of learning of a visual task. In cats, there is evidence for the transfer of learning of shape discrimination mediated by the corpus callosum [see ref. 14 for review]. In monkeys, this role is shared by the anterior commissure that assures interconnections between the inferotemporal cortices of either hemisphere [16-18].

In spite of its importance for integrating the visual functions of the two hemispheres, the capacity of the callosal channel to trasmit visual information in the cat and monkey is limited both in the spatial- and in the temporal-frequency domain: compared with the direct geniculo-cortical pathway, the responsiveness of the callosal pathway to stimulus contrast is much reduced and especially the responses to high spatial and temporal frequencies are heavily attenuated [19-21] (Fig. 3).

Evidence for a limited spatial-frequency bandwidth of callosal connections in man is provided by psychophysical findings about the discrimination of spatial phase differences in compound gratings with two harmonic components. For gratings of relatively low fundamental frequency (2 cycle/deg or lower) a left field (right hemisphere) superiority in this task is observed when the stimuli are presented at a distance of several degrees from the vertical meridian (see above), but not for stimuli removed only 1 or 2 deg from the meridian, whatever the stimulus contrast. On the contrary, for stimuli of higher spatial frequency the visual field asymmetry in the performance of this task is preserved even at very short distances from the vertical meridian, unless the stimulus is of very high contrast (Fig. 4). The lack of lateralization for spatial phase discrimination in the region of the visual field close to the vertical midline can be ascribed to the transfer of information through the interhemispheric commissures, and in particular the corpus callosum. An explanation in terms of a dual representation of the vertical midline on both hemispheres seem to be very unlikely[22]. The psychophysical findings therefore agree with those obtained electrophysiologically in the cat and monkey, indicating an attenuation of high spatial frequencies in callosal transfer.

Interhemispheric transfer of perceptual learning

The discrimination of spatial phase in compound gratings with two harmonic components improves with practice and the effects of training are maintained for long [23-24]. These effects however are specific for the characteristics of the stimuli used for training (orientation and spatial frequency) and for their location in the visual field, as is typical of perceptual learning.

In particular, the learning effects in spatial phase discrimination acquired with stimuli presented at 5 deg eccentricity on one side of the vertical meridian are not preserved if the

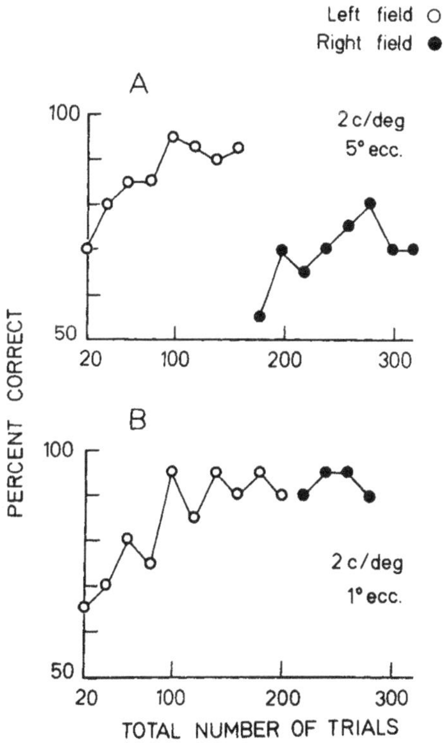

Fig. 5 - Percentage of correct responses in a forced choice task for the discrimination of compound gratings (f + 3f, with spatial phases in the gratings to be discriminated: 0 deg and 90 deg, respectively) displaced laterally 5 deg (A) or 1 deg (B) from the vertical meridian, either in the left (open circles) or right (closed circles) visual field. Each point represents a block of 20 trials. The stimulus was initially located in the left hemifield and after 160 (A) or 200 (B) trials it was displaced to the right field, at the same eccentricity. The improvement in performance (perceptual learning) transfers to the contralateral field in B, but not in A[25]. Note that the final level of performance is the same for the two visual fields in B, but is lower in the right than in the left field in A.

stimuli are presented at the same eccentricity on the other side of the meridian. Close to the meridian, on the contrary, the effects of learning can be trasferred from one visual hemifield to the other, provided that the fundamental spatial frequency of the gratings is relatively low (2 cycle/deg or less) (Fig. 5). The transfer occurs interocularly (training with one eye and testing with the other eye), but is specific for the orientation of the stimulus and is limited to areas of the left and right hemifields mirror-symmetric with respect to the vertical meridian, within 1 - 2 deg from the meridian. For higher spatial frequencies, no transfer is obtained interhemispherically, even at small eccentricities [25].

These findings are suggestive of a role of the forebrain commissures in mediating the interhemispheric transfer of perceptual learning. A contribution of the anterior commissure cannot be excluded. This possibility would be supported by the evidence of interhemispheric transfer of perceptual learning for spatial phase discrimination in a subject with callosal agenesis [26]. In normal subjects, however, a role of the anterior commissure seems unlike, since in non-human primates this commissure connects visual temporal areas with very large receptive fields extending far away from the vertical meridian and relatively insensitive to stimulus orientation. If the interhemispheric transfer of perceptual learning is mediated by the corpus callosum, then the conditions required for this transfer to take place suggest that in man, as in cats and monkeys, the callosal connections relevant for relatively early steps of visual processing are limited to a narrow strip on either side of the vertical meridian and that the filtering properties of the callosal channel in terms of spatial-frequency are also similar in human and non-human primates.

REFERENCES

[1] O.J.Gruesser and T.Landis (1991) Visual Agnosias. In: *Vision and Visual Dysfunction* J.R.Cronly-Dillon, ed. London, MacMillan Press.

[2] S.P.Springer and G.Deutsch (1981). *Left Brain, Right Brain.* Freeman, S.Francisco.

[3] A.Fiorentini and N.Berardi (1984) Right-hemisphere superiority in the discrimination of spatial phase. *Perception* **13** 695-708.

[4] F.L.Kitterle and R.S.Kaye (1985) Hemispheric symmetry in contrast and orientation sensitivity. *Percept. & Psychophys.* **37** 391-396.

[5] D.H.Peterzell, L.O.Harvey and C.D.Hardyck (1989) Spatial frequencies and the cerebral hemispheres: contrast sensitivity, visible persistence and letter classification. *Percept. & Psychophys.* **46** 443-455.

[6] L.N.Piotrowski and F.W.Campbell (1982) A demonstration of the visual importance and flexibility of spatial-frequency amplitude and phase. *Perception* **11** 337-346.

[7] D.C.Burr (1980) Sensitivity to spatial phase. *Vision Res.* **20** 391-396.

[8] D.C.Van Essen, W.T.Newsome and J.L.Bixby (1982) The pattern of interhemispheric connections and its relationship to extrastriate visual areas in the macaque monkey. *J. Neurosci.* **2** 265-283.

[9] G.Berlucchi (1981) Recent advances in the analysis of the neural substrates of

interhemispheric communication. In: *Brain Mechanisms and Perceptual awareness.* O.Pompeiano and C.Ajmone-Marsan eds. New York, Raven Press. pp 133-152.

[10] C.Blakemore, Y.Diao, M.Pu, Y.Wang and Y.Xiao (1983) Possible functions of the interhemispheric connexions between visual cortical areas in the cat. *J. Physiol.* **337** 331-349.

[11] D.H.Hubel and T.N.Wiesel (1967) Cortical and callosal connections concerned with the vertical meridian of visual fields in the cat. *J. Neurophysiol.* **30** 1561-1573.

[12] G.Berlucchi and G.Rizzolatti (1968) Binocularly driven neurons in the visual cortex of split chiasm cats. *Science, N.Y.* **159** 308-310.

[13] C.Blakemore (1970) Binocular depth perception and optic chiasm. *Vision Res.* **10** 43-47.

[14] G.Berlucchi (1990) Commissurotomy studies in animals. In: *Handbook of Neuro-psychology,* Vol. 4. F.Boller and J.Grafman eds. Amsterdam, Elsevier. pp 9-47.

[15] A.Cowey and F.Wilkinson (1991) The role of the corpus callosum and extrastriate cortex visual areas in stereoacuity in macaque monkeys. *Neuropsychologia* **29** 465-479.

[16] C.G.Gross, D.B.Bender and M.Mishkin (1977) Contributions of the corpus callosum and the anterior commissure to visual activation of inferior temporal neurons. *Brain Res.* **131** 227-239.

[17] L.Seacord, C.G.Gross and M.Mishkin (1979) Role of inferior temporal cortex in interhemispheric transfer. *Brain Res.* **167** 259-272.

[18] M.J.Eacott and D.Gaffan (1989) Interhemispheric transfer of visual learning in monkeys with intact optic chiasm. *Exper. Brain Res.* **74** 348-352.

[19] N.Berardi, S.Bisti and L.Maffei (1987) The transfer of visual information across the corpus callosum: spatial and temporal properties in the cat. *J. Physiol.* **384** 619-632.

[20] N.Berardi, S.Bisti, A.Fiorentini and L.Maffei (1988) The transfer of visual information across the corpus callosum in cats, monkeys and humans: spatial and temporal properties In: *Progress in Brain Research,* Vol. 75. T.P.Hicks and G.Benedek eds. Amsterdam, Elsevier. pp181-185

[21] A.Fiorentini and N.Berardi (1991) Limits in pattern discrimination: central and peripheral factors. In: *Limits of vision,* J.J.Kulikowski, V.Walsh and I.J.Murray, eds. London, MacMillan. pp 266-276.

[22] N.Berardi, A.Fiorentini and A.Gravina (1988) Functional dissociation of the hemispheres in the discrimination of complex gratings. *Vision Res.* **28** 491-496.

[23] A.Fiorentini and N.Berardi (1980) Perceptual learning specific for orientation and spatial frequency. *Nature* **287** 43-44.

[24] A.Fiorentini and N.Berardi (1981) Learning in grating waveform discrimination: specificity for orientation and spatial frequency. *Vision Res.* **21** 1149-1158.

[25] N.Berardi and A.Fiorentini (1987) Interhemispheric transfer of visual information in humans: spatial characteristics. *J. Physiol.* **384** 633-647.

[26] A.Fiorentini, N.Berardi, B.Falsini and V.Porciatti (1992) Interhemispheric transfer of visual perceptual learning in callosal agenesis. *Clin. Vision Sci.* **7** 133-141.

PARALLEL PATHWAYS: ANATOMO-PHYSIOLOGICAL AND PERCEPTUAL CHARACTERISTICS

C.A. Marzi and M. Girelli

Dipartimento di Scienze Neurologiche e della Visione, Sez. di Fisiologia, Lab. di Psicologia, Università di Verona, Strada Le Grazie 8, 37134 Verona, Italy.

That the retinal ganglion cells can be subdivided into different classes on the basis of their morphological, connectionistic and physiological characteristics is now common knowledge.

In spite of species differences in the nomenclature reflecting some morphological and functional differences, the basic picture emerged in the last decade is that the parallel organization of the visual system is an important trait common to all mammalian species [23].

In this article we review studies aimed at assessing whether selective lesions at subcortical sites of the parallel pathways in cats, monkeys and humans lead to selective visual impairments. A positive answer to this question would contribute a great deal to an understanding of the perceptual contribution of the parallel pathways to vision.

Effects of selective lesions of the alpha axons in the cat

The cat's retina contains at least three morphological types of ganglion cells with correspondingly different physiological properties [23]. One of the most conspicuous differences is cell

body and axon size: **alpha** ganglion cells, which correspond to the physiologically defined **Y**-type with transient response properties to brief light flashes have the largest size. They are followed by **beta** cells corresponding to the **X**- or sustained type and by a heterogeneous variety of small cells loosely termed as **gamma** corresponding to **W**-cells with sluggish response properties. A fundamental problem is to relate the functional properties of the different ganglion cells as assessed with single-cell recordings to selective perceptual roles. One direct way of sorting this problem out is to lesion each of the parallel pathways selectively and study the behavioural functions subserved by the remaining ones. In the cat this has been done by different authors with different methods. Burke et al. [2] have used a selective pressure block of optic nerve fibers in the cat to damage the large alpha axons. Degeneration of the Y-fibers was monitored histologically by examination of the optic nerve fibers caudal to the block and electrophysiologically by showing a loss of the short-latency response following optic nerve stimulation. Both controls indicated a substantial and in some case complete loss of large (>5 microns) axons. The only clear-cut impairment they found was a difficulty in discriminating fast visual motion; a function which they ascribe to the alpha fibers and the Y system. A different strategy has been used by Spear et al. [21] who used antibodies against large retinal ganglion cells and found an impairment in eye and head orientation responses to stimuli presented to the nasal visual hemifield.
In our laboratory we used an approach that takes advantage of the size segregation found by Guillery et al. [7] in the optic tract of the cat (OT) as well as of the monkey and other mammals (see [8] for a recent review including the developmental aspects of the segregation). Large alpha axons are segregated in the ventral sub-pial portion of the tract while middle-sized beta axons run more dorsally and deeply. The small gamma axons are instead represented throughout the whole dorsal-ventral extent of the OT. Interestingly enough, such segregation is not present in the optic nerve and appears to begin in the optic chiasma (OC). We developed a surgical technique whereby we expose the OC and with the aid of an operating

microscope and a sucking needle we aspirate the most ventral extent of the OT. Such a removal, when successful, allows to take away most of the alpha axons while leaving the vast majority of beta axons intact. A contingent of gamma fibres which belong to the phasic type [22] is inevitably severed along with alpha axons while leaving the tonic type of gammas intact (see Reese et al. [17] for methodological details). Operated cats and controls were tested on contrast sensitivity thresholds at various spatial frequencies and on flicker discrimination thresholds using various luminance levels. According to what is known from single-cell recordings from different class of retinal ganglion cells, cats with removal of alpha axons corresponding to the functional category of Y-cells should be relatively unimpaired in spatial resolution while they should be impaired in temporal resolution. In fact, this is not entirely what happens because we found in a study still in progress (Marzi et al, in preparation) that two animals with a practically complete removal of the alpha axons are impaired in both spatial and temporal resolution thresholds and this confirms in the cat evidence obtained in the monkey in favour of the idea that the functional dichotomy between the two main parallel pathways turns out not to be so sharp [14].

All in all, the picture emerging from selective lesions of the Y-system in the cat is only loosely in keeping with the single-cell recording data and a reasonable conclusion might be that there is a good deal of overlap between the two channels.

Effects of selective lesions of the P- or M- layers of the monkey Lateral Geniculate Nucleus (LGN)

In the primate's LGN the four dorsal layers are made up of small cells and therefore are named the Parvocellular layers while the two more ventral layers are composed of larger neurones and are, accordingly, named the Magnocellular layers. The parvocellular layers receive afferents from corresponding smaller (P) ganglion cells while the magnocellular layers receive input from the larger (M) ganglion cells; the specificity of ganglion cells classes perfectly matches that of their LGN targets both morphologically [11,15] and functionally [9,20].

In contrast to the cat in which the different classes of LGN cells are not segregated in different layers, in primates it is possible to selectively lesion one or the other pathway. This has been accomplished by Schiller et al. [19] and by Merigan and collegues [14] by placing small ibotenic acid lesions in either the parvo- or the magno-cellular layers of the LGN and by testing the animals psychophysically in the visual field area affected by the lesion. As is well known, ibotenic acid is an excitotoxin that damages cell bodies but spares fibers of passage. Another excitotoxin, that damage selectively the P-cells is acrylamide; this substance has been used extensively by Merigan's group and collegues [see 13] by means of oral daily dosing.

The results of these lesion studies show clear differences in the contributions of the P and M pathways to vision. For the most part, the results obtained correspond well with the response properties of the neurones of the two pathways. Thus, color vision appears to be subserved uniquely by the P- pathway; this result is consistent with the color opponency of P cells. Furthermore, the P-pathway appears to subserve acuity, but this is not determined by the physiological spatial resolution of P-cells that is similar to that of M-cells, but by the greater sampling density of retinal P ganglion cells (ratio P-M =8). In general, psychophysical assessments have shown that P-pathway lesions cause complementary effects to those of M lesions. They reduce luminance contrast sensitivity for stimuli of higher spatial and lower temporal frequency content. M- pathway lesions, by contrast, determine a large decrease in luminance contrast sensitivity at higher temporal frequency and lower spatial frequency. As a consequence, there is a reduced visibility of rapidly moving or rapidly flickering stimuli. Here there is a discrepancy between the results of Merigan's and Schiller's groups in that the latter found a large reduction of flicker resolution thresholds while the former found little impairment in flicker resolution for high contrast stimuli.

In summary, as for Y and X pathways of the cat, P- and M- pathways are anatomically and functionally separable although there is certainly some functional overlap. Their specializations concern basic visual properties such as sensitivity to spatial and temporal frequency; an important question is if this holds true for the human primate as well.

Effects of selective damage of the parallel pathways in humans

It has been shown by Rodieck et al. [18] that Golgi stained human retinae contain two different ganglion cell types. One group, parasol cells had larger dendritic arbors and somas as well as thicker axons than the other type (midget cells) with smaller dendritic fields. These types correspond to the alfa-beta distinction in the cat and to the M-P distinction in primates. In keeping with this distinction, we found recently that in the human OT not only there is a rather distinct population of large and smaller optic fibres but such types are segregated anatomically in much the same way as that has been found in cats and non-human primates [24]. It is important to mention that we have now preliminary evidence that size segregation in humans is present already at the OC especially on its lateral aspects. This is in accord with recent evidence obtained in ferrets [1,16]. A functional validation of a size segregation within the human OC can be found in unpublished evidence from Vingrys, Benes, and King-Smith (in King-Smith [10]). They found a selective loss of flicker modulation sensitivity at high temporal frequencies and a selective loss in contrast sensitivity at low spatial frequencies in a patient with a compression of the optic chiasma exerted by a pituitary adenoma. These results fit in well with the basic properties of the two systems and can be interpreted as a selective loss of the large achromatic fibres compressed by the tumour in the ventral portion of the optic chiasma.

To cast light on the possible perceptual correlates of the parallel pathways in humans, recently, we carried out a study on the putative neural bases of visual persistence [12]. Cells in the primate P and M pathways show large differences in the time course of their response to brief visual stimuli, with P-cells showing a longer-lasting tonic type of response and the M-cells showing a more transient response. An interesting question is whether iconic persistence is subserved by one or the other pathway or if both of them contribute to a different stage of persistence. As is well known, iconic persistence is a central form of persistence to be distinguished from after-images that are of retinal nature [5]. Iconic persistence plays an important role in most information-processing models of visual perception in that it represents a necessary stage prior to encoding in short-term memory. We sought to ascertain whether patients with a compression of the large fibers at the OC as a consequence of a pituitary adenoma might show an impairment in visual persistence. They were tested on a task of temporal integration of visual form similar to that employed in recent studies by Christman [4,5]. Such a task requires that a stimulus persists in the visual system until another stimulus appears so that together can give a meaningful percept. We used a standard LED array positioned with its center at 5 degrees either to the right or to the left of a fixation point. Lateralized presentations were used in order to be able to compare on each patient the temporal and the nasal hemifields in monocular vision. The dependent variables were the number of errors and vocal RT for identification of the stimulus. The procedure was the following: a pseudorandom subset of 2-3 elements of the seven-digit LED display was lit for 4 msec followed by a blank ISI of 15-60 msec range. Afterward, a second complementary subset of 2-3 elements was lit for another 4 msec. Displayed simultaneously or perceptually fused the two sets of elements formed one of four digits. The subject's task was to identify the digits at the various ISIs.

Patients were significantly impaired in accuracy of recognition in the temporal hemifields only at the shorter but not at longer ISIs. This suggests that iconic persistence can be dissociated into two components, one that is susceptible to compression of the ventral OC and the other that shows no impairment. The former may be subserved by neurones with transient response of the M- or Y-type, the latter may be subserved by the sustained response of the P- or

X pathways. The impairment found at short ISIs is likely to be related to a lack of the contribution of the M-input which, together with the P-input, subserves stimulus recognition in the initial phase of visual persistence. No impairment is seen at later ISIs because at later stages persistence is subserved by the P-input only. Interestingly enough, in some of the patients we were able to carry out pre-post surgical comparisons and found a remarkable degree of recovery following removal of the pituitary adenoma. This suggests that long-lasting OC compression, at variance with experimental glaucoma [6] or experimental pressure block [2] exerts reversible ill-effects on M-axons.

In conclusion, the selective lesion of the parallel pathways at subcortical sites has provided unequivocal evidence that the two systems, although clearly separated both anatomically and physiologically, are coactivated in mastering many perceptual tasks. It is only by careful psychophysical and cognitve dissection of the various components of task that one can find hopes of disentangling the specific contributions of the different pathways.

References

1 G.E., Baker (1990) Prechiasmatic reordering of fibre diameter classes in the retinofugal pathway of ferrets. *Eur. J. Neurosci.*, 2, 24-33.

2 W. Burke, L.J., Cottee, K., Hamilton, L., Kerr, C., Kyriacou, and M., Milosavljevic (1987). Function of the Y optic fibres in the cat: Do they contribute to acuity and ability to discriminate fast motion? *J.Physiol. (Lond.)*, 392, 35-50.

3 S., Christman (1987). Effects of perceptual quality on hemispheric asymmetries in visible persistence. *Perception and Psychophysics*, 41, 367-374.

4 S., Christman (1989) Temporal integration of form as a function of subject handedness and retinal locus of presentation. *Neuropsychologia*, 27, 1373-1382.

5 M. Coltheart (1980) Iconic memory and visible persistence. *Perception and Psychophysics*, 27, 183-228.

6 Y., Glowinski, H.A., Quigley, M.E., Pease (1993) Foveal ganglion cell loss is size dependent in experimental glaucoma. *Inv. Ophthalm. & Vis. Sci*, 34, 395-400.

7 R.W., Guillery, E.H., Polley, F., Torrealba (1982) The arrangement of axons according to fibre diameter in the optic tract of the cat. *J. Neurosci.*, 2, 714-721.

8 R.W., Guillery (1991) Rules that Govern the Development of the Pathways from the Eye to the Optic Tract in Mammals. In D. Man-Kit Lam and C.J. Shatz (Eds.) *Development of the Visual System. Proc.*

of the Retina Res. Foundation, Vol.3, pp. 153-171, MIT Press, Cambridge, Mass.
9 E., Kaplan and R., Shapley (1986) The primate retina contains two types of ganglion cells , with high and low contrast sensitivity. *Proc. Nat. Acad. Sci. Usa*, 83, 2755-2757.
10 P.E., King-Smith, (1991) Psychophysical Methods for the Investigation of Acquired Colour Vision Deficiencies. In: D.H. Foster (Ed.) "*Inherited and Acquired Colour Vision Deficiencies: Fundamental Aspects and Clinical Studies*", *Vision and Visual Dysfunction*, Vol.7, MacMillan Press, London, pp. 38-55.
11 A.G., Leventhal, R.W., Rodieck, and B., Dreher (1981) Retinal gangion cell classes in the old-world monkey: morphology and central projections. *Science*, 213, 1139-1142.
12 C.A., Marzi, M., Girelli, G., Tassinari, L., Cristofori, A., Talacchi, G., Marchini (1994) The role of parallel pathways in visible persistence. In C. Umilta' and M. Moscovitch (Eds.) "*Attention and Performance XV - Conscious and Nonconscious Information Processing*". MIT Press, Cambridge, Mass., pp. 110-121.
13 W.H.Merigan & T.A. Eskin (1986) Spatiotemporal vision of macaques with severe loss of Pb retinal ganglion cells. *Vis. Res.*, 26, 1751-1761.
14 W.H., Merigan, J.H.R., Maunsell (1993) How parallel are the primate visual pathways?. *Ann. Rev. Neurosci.*, 16, 369-402.
15 V.H., Perry, R., Oehler, and A. Cowey (1984) Retinal ganglion cells that project to the dorsal lateral geniculate nucleus in the macaque monkey. *Neuroscience*, 12, 1101-1123.
16 B.E.Reese, and G,E., Baker (1990) The course of fiber diameter classes through the chiasmatic region in the ferret. *Eur. J. Neurosci.*, 2, 34-49.
17 B.E., Reese, R,W., Guillery, C.A., Marzi, and G. Tassinari (1991) Position of axons in the cat's optic tract in relation to their retinal origin and chiasmatic pathway. *J. Comp. Neurol.*, 306, 539-553.
18 R.W., Rodieck, K.F., Binmoeller, J., Dineen (1985) Parasol and midget ganglion cells of the human retina. *J. Comp. Neurol.*, 233, 115-132.
19 P.H., Schiller, N.K., Logothetis, E.R., Charles (1990) Role of the color-opponent and broad-band channels in vision. *Visual Neurosci.*, 5, 321-346.
20 R., Shapley, and E. Kaplan (1989) Responses of magnocellular LGN neurones and M retinal ganglion cells to drifting heterochromatic gratings. *Inv. Ophthalmol. and Vis. Sci. Supplement*, 30, 323.
21 P.D., Spear, S., Miller, K., Vielhuber, S,E., Kornguth (1986) Visual field defects in cats with neonatal or adult immunological loss of retinal ganglion cells. *Brain Res.*, 368, 154-157.
22 L.R., Stanford (1987) W-cells in the cat retina: correlated morphological and physiological evidence for two distinct classes. *J. Neurophysiol.*, 57, 218-244.
23 J., Stone (1983) *Parallel Processing in the Visual System: the Classification of Retinal Ganglion Cells and its Impact on the Neurobiology of Vision*, Plenum Press, New York.
24 G., Tassinari, D., Campara, G., Balercia, M., Chilosi, G., Martignoni, C.A., Marzi (1994a) Magno- and parvocellular are segregated in the human optic tract. *NeuroReport*, 5, 1425-1428.

A MULTI-SCALE NETWORK MODEL OF BRIGHTNESS PERCEPTION

Ennio Mingolla

Center for Adaptive Systems and
Department of Cognitive and Neural Systems
Boston University
111 Cummington Street
Boston, Massachusetts 02215 USA

Abstract

A neural network for brightness perception is constructed in the tradition of filling-in theories. The model is developed to account for a wide variety of difficult data, including the classical phenomenon of Mach bands, low- and high-contrast missing fundamental and nonlinear contrast effects associated with sinusoidal luminance waves. The model builds upon previous work by Grossberg and colleagues on filling-in models that predict brightness perception through the interaction of boundary and feature signals. A new interpretation of feature signals through the explicit representation of contrast-driven and luminance-driven information is provided and directly addresses the issue of "absolute" brightness values. Simulations of the model implement a number of refinements already described in the development of Grossberg's (1987, 1994) Form-And-Color-And-DEpth (FACADE) theory, which though conceived as part of the theory, were not implemented in the simulations of Grossberg and Todorović (1988). These include: a) ON and OFF channels with separate filling-in domains; b) multiple spatial scales; c) computations for simple and complex cells; and d) boundary computations that engage a recurrent competitive circuit. Simulations of the present system of equations account for human perception of a wide variety of stimuli, including ones whose brightness contains shallow spatial gradients. This article briefly summarizes and comments upon joint research reported fully in Pessoa, Mingolla, and Neumann (in press).

A fundamental idea of the FACADE theory is that boundary signals -- in the Boundary Contour System (BCS) -- are used to generate filling-in compartments where featural quality (e.g., brightness in our case) is diffused, or spread. The final diffused activities in the Feature Contour System (FCS) correspond to the model's predicted brightness, which is the outcome of interactions between boundaries and featural quality, whereby boundaries control the process of filling-in by forming gates of variable resistance to diffusion. Boundary signals work to contain diffusion; large boundary values do not allow a featural value at a given spatial position to affect a neighboring one. In regions with zero boundary activity, featural quality is free to diffuse, while in regions containing spatially dense boundary signals, little diffusion of featural quality throughout a large area may occur. (See Figure 1.)

While it may seem natural to assume that boundary signals only exist in locations corresponding to discontinuities of luminance ("edges"), Grossberg and Mingolla (1987) showed that "boundary webs" can form in regions of luminance gradients. That is, signals sensitive to the local orientation of contrast gradients in the BCS can be generated not only (as is commonly assumed) in response to thin lines or edge contours, but also as a spatially dense plexus in response to shading, that is, gradients of luminance such as typically occur when surfaces that curve smoothly in depth are illuminated by a light source.

The significance of boundary webs for brightness perception is as follows. If there were no boundary webs, if instead boundaries existed only as sharp, focused contours, then the only realizable brightness percepts might be those for which homogeneous regions of a given brightness were separated by abrupt transitions (at edges) from other such regions of different brightnesses. The perception of a smooth, continuous change in brightness could then never occur within the model, thereby severely limiting its scope.

Fortunately, however, the formation of boundary webs is consistent with the specification of the BCS, as shown in computer simulations by Grossberg and Mingolla (1987). Moreover, the interaction of a smoothly modulated, spatially extended region of boundary signals with the nonlinear diffusion of the FCS can be shown to be capable of supporting a variety of outcomes in which the model's "perceived" brightness output is itself smoothly modulated, in agreement with the perceptual phenomena being modeled. Such results occur because the process of diffusion may be totally or partially blocked within extended regions, yielding a percept of spatially gradual changes in brightness.

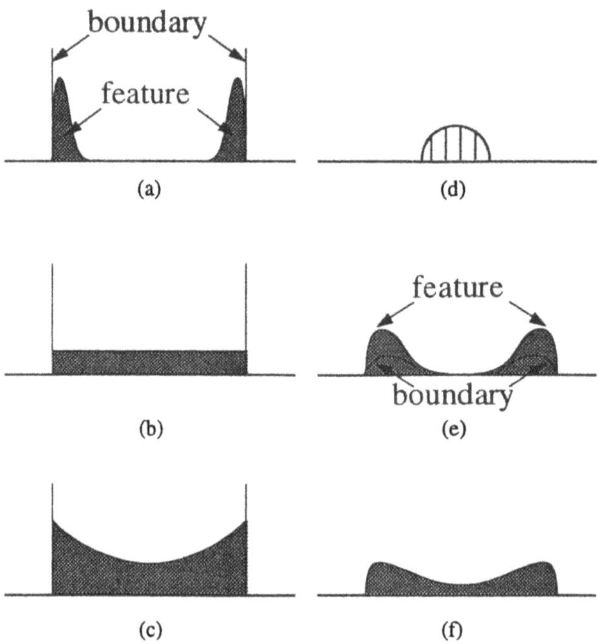

Figure 1: Two ways that filling-in can result in perceived smooth gradients of brightness. (a) This panel shows a one-dimensional cross-section of boundary and feature signals before diffusion. (b) Conservative diffusion by the heat equation would result in a flat featural profile at equilibrium. (c) Nonconservative diffusion by BCS/FCS mechanisms can strike a balance between sources (initial featural distributions) and sinks (passive decay of featural activity), resulting in a modulated distribution of brightness at equilibrium. (d) Continuous curves that denote boundary activity can be thought of as interpolated distributions between discrete network nodes (whose activity strengths are denoted by the lengths of vertical lines). (e) Spatially extended (fuzzy) distributions of boundary signals may overlap with distributions of featural signals. (f) Boundary signals can "trap" featural quality over extended areas, resulting in smooth modulation of activity.

Figure 1 diagrams two distinct ways in which a smooth or gradual modulation of perceived brightness may be realized as the outcome of the interaction of BCS and FCS mechanisms. The first is the nonconservative nature of the hypothesized diffusion itself, whereby a nonhomogeneous equilibrium may exist that balances the injection of new energy into the diffusion process (from earlier processing stages) with the tendency for activity of cells at the diffusion layer to spontaneously decay away. The second mechanism can be thought of as the spatial "trapping" of an entire signal profile from earlier processing stages by an extended region of strong BCS signals. Note that the need for the outcome of BCS/FCS interactions to be a smooth brightness profile for some inputs must be balanced against the need to generate flat plateaus separated by sharp transitions for other inputs; moreover, the model must itself make the determination of smooth vs. sharp somehow.

The results of any initial boundary filter will be some distribution of activity in a network of nodes with overlapping receptive fields. Some of these initial boundary signals will need to be sharpened, while others will need to be left fuzzy. Analysis indicates that stimuli with abrupt luminance transitions (e.g., luminance steps) generally require sharp boundary signals. Such boundaries create spatially abrupt barriers to diffusion between regions, resulting in discretely differing output levels (corresponding to the model's predicted brightness distribution). Stimuli containing smooth luminance modulations, on the other hand, typically require smoother boundaries to be able to trap (at least some of) the modulation that is present in the featural quality to create smoothly varying brightness distributions. How then can a model of brightness perception decide without a homunculus -- that is, the invocation of further rules invoked by yet other models or processes -- whether or not to sharpen its initial distribution of boundary signals?

What differentiates the situations requiring sharp and extended boundary signals? Consider a system where the input waveform is filtered by both ON and OFF center-surround operators. The solution originates from the observation that sharp transitions of luminance produce strong responses in both the ON and OFF channels. In other words, in the region surrounding an "edge," there will be strong on-activity (at the "light" region) and strong off-activity (at the "dark" region). On the other hand, waveforms with more gradual variations of luminance lead to a different distribution of ON and OFF responses. Specifically, the natural pairing of closely juxtaposed ON and OFF responses characteristic of luminance edges may be disrupted, so that a peak in, for example, ON activity may be relatively isolated. The details of such behavior depend critically on system parameters, and many important specifics are ignored in the present discussion. The essential steps in achievement of sharp vs. extended boundary signals is, however, illustrated in Figure 2.

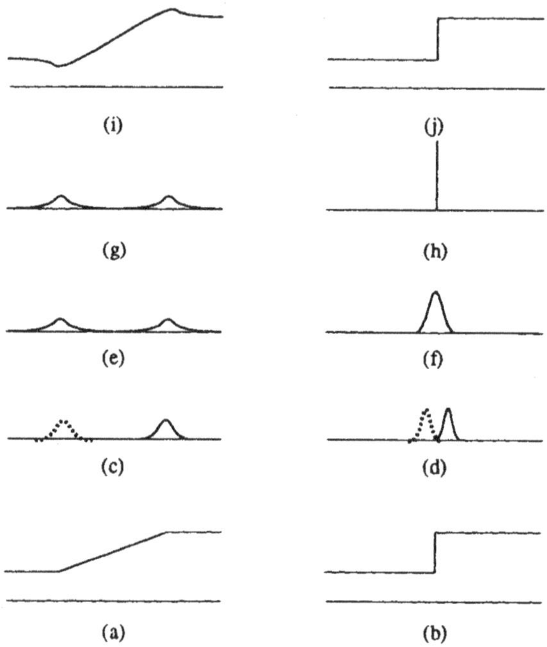

Figure 2: Mach bands for ramps vs. no Mach bands for step edges. Two luminance profiles are shown: (a) a ramp, and (b) a step. Panels (c) and (d) diagram the responses of OFF (dashed) and ON (solid) channels. Responses to the ramp are smaller in amplitude and further apart than the corresponding responses to the step. Panels (e) and (f) display a transformation, including blurring and pooling of ON and OFF signals from (c) and (d), respectively. Recurrent competition can then leave the distribution of (e) effectively unchanged, as shown in (g), while transforming the distribution of (f) into that of (h) by a "choice" (winner-take-all) transformation. Filling-in of featural signals is differentially affected by boundary signals of (g) and (h). The spatially extended boundaries that "survived" recurrent competition can trap visible featural signals, resulting in Mach bands (undershoots and overshoots), as shown in (i). Panel (j) shows how the spatially collapsed boundary distribution corresponding to the luminance step segregates the resulting brightness profile into two relatively flat plateaus, without Mach bands.

The preceding analysis indicates that ON and OFF responses can be used to guide the computation of boundaries: close spatial juxtaposition of "pre-boundary" signals generated by both ON and OFF signals should eventually lead to sharp boundaries while relatively isolated signals from either channel should produce spatially shallow boundaries. To realize this competence in the present model requires two stages. First, a nonlinear design for a contrast-sensitive "simple cell" (named by analogy to visual cortical cells sensitive to oriented contrast) is used. Simple cells are often diagrammed as elongated ellipses, with two subfields separated by the major axis. One half of the cell's receptive field is excited and the other half inhibited by increases in luminance. Our proposed circuit employs this typical arrangement, but with an important modification. Suppose that simple cells receive input from both ON and OFF center-surround cells of an earlier processing level. Then one half of each simple cell would be excited by ON input and the other half would be excited by OFF input. However, there is also a cross-inhibitory influence between ON and OFF subunits of the simple cell, whereby the summation of a small amount of ON signal to the appropriate side and an equal OFF signal to the appropriate side results in a much larger response than, for example, doubling the initial input to the ON side and setting the OFF side to zero.

As shown in Figure 3, the proposed mechanism favors the appropriate spatial juxtaposition of ON and OFF signals to the appropriate sides of the simple cell by "penalizing" input distributions that excite only one pathway. The penalty for the ON pathway is in the form of a tonic "self-inhibition" along the pathway that can be shut off only by a simultaneous signal for inhibition of the inhibitor (disinhibition) from the OFF pathway. The simple cell thereby acts as a "soft" edge detector, that is, a unit that is exquisitely sensitive to even small amplitude, sharp spatial transitions at edges -- by virtue of its amplified response to close ON/OFF juxtapositions -- but which also responds to smooth gradients of luminance, such as are found in the transition from "flat" to "ramp" portions of the luminance profiles of Mach band stimuli. (See Figure 2.)

The second mechanism for insuring that sharp luminance transitions lead to sharp boundaries, while shallow transitions lead to fuzzy boundaries, is the use of a recurrent competitive circuit for processing the output of the nonlinear simple cell. The competitive circuit, in current parlance, is capable of automatically switching -- that is, without a change of parameters -- between "winner take all" and "leader take most" behavior, or even, for relatively homogeneous input patterns, to have little effect at all on its input. Note that the case of "winner take all" obtains when sharp boundary transitions are desired (at luminance edges), while the "softer" outputs occur in just those cases where fuzzy boundaries are needed. (See Figure 4.) The design of the recurrent competitive circuit mimics that of Grossberg and Marshall (1989).

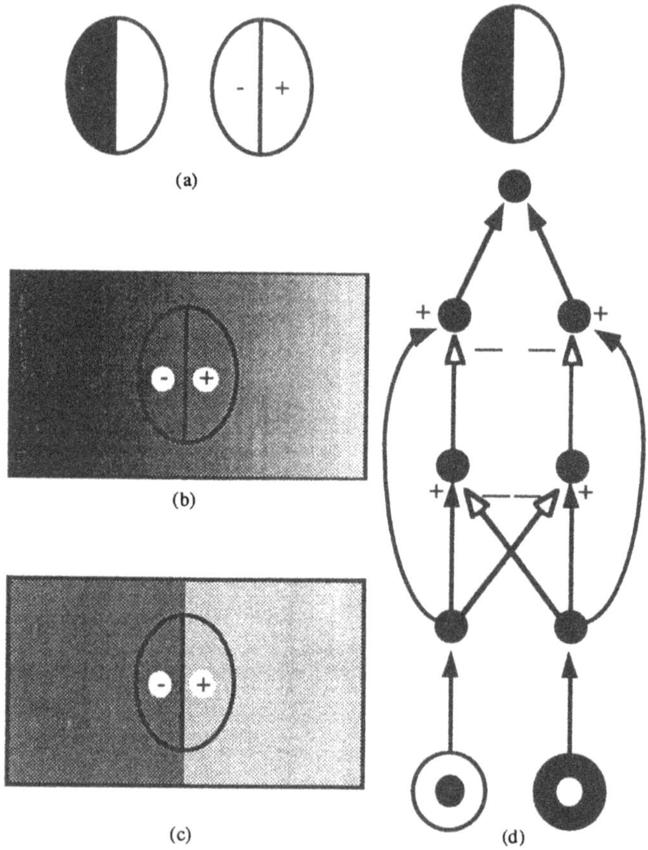

Figure 3: In (a) a simple cell is denoted by shading on the left and plus and minus symbols on the right. Such cells emit *some* response to smooth gradients of luminance, shown in (b). While simple cells are not purely "edge detectors," their response to step edges of luminance, shown in (c), is enhanced relative to that for smooth gradients. That is, even the same "energy" or "weighted luminance" within the two halves of the receptive field will have a different effect, depending on whether the two halves are separated by a luminance edge. Thus, given the preprocessing by concentric ON and OFF cells and the circuitry shown in (d), the simple cell of (c) responds more than that of (b), even for equal luminance energies.

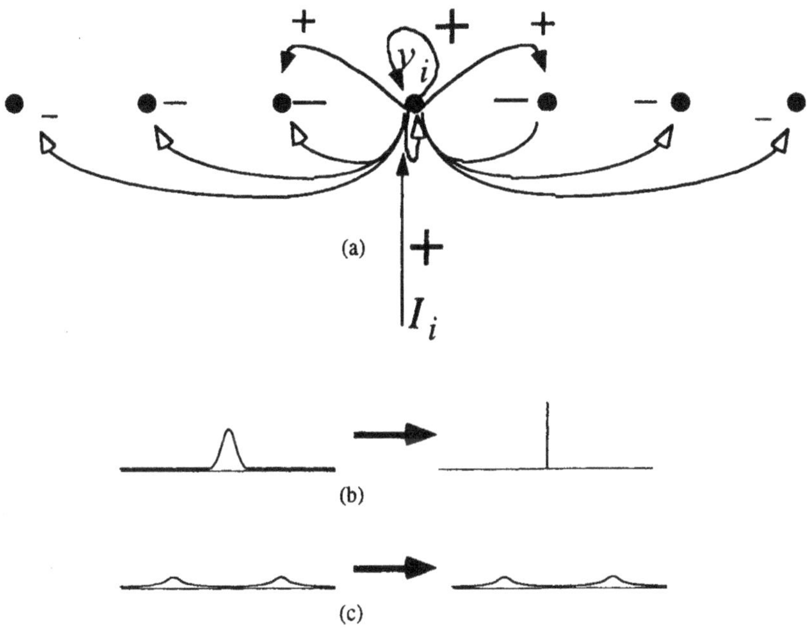

Figure 4: In a recurrent competitive field (a), strengths of lateral excitation (plus signs, solid arrow heads) and inhibition (minus signs, open arrow heads) fall off as a function of distance from a central node. Such a circuit can produce "switching behavior," whereby distributions of responses with relatively sharp peaks are further sharpened (b), while those distributions where all values are (absolutely) small or where many values very similar in size to neighboring ones are essentially left intact (c).

This chapter gives an intuitive description of only a few key mechanisms from a recently implemented model of brightness perception. For a full description of this model, including simulation results and an analysis of its relation to other models of brightness perception (e.g., Kingdom and Moulden, 1992; Morrone and Burr, 1988), readers should consult Pessoa *et al.* (in press). That work builds on an analysis articulated by Neumann (1993, 1994) that clarifies how the double-opponent subtraction of ON/OFF signals for brightness computation can be thought of

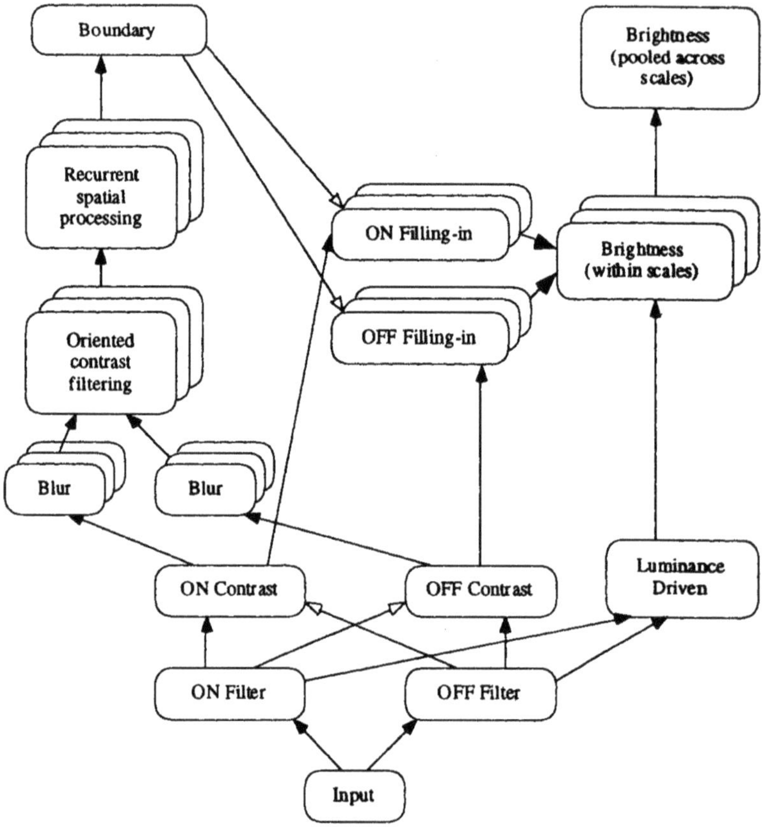

Figure 5. Model macrocircuit: Early processing by concentric ON and OFF filters is factored into ON contrast, OFF contrast, and luminance driven representations. Contrast signals drive the formation of boundary signals (left portion of diagram), which in turn constrain filling-in of ON and OFF featural information in several scales, which are then combined to yield a final percept. See text for details.

as "factored" into the computation in parallel of "contrast" and "luminance-derived" signals, as shown in the model macrocircuit of Figure 5. The current model combines and modifies the work of Grossberg and Todorović (1988) and of Neumann (1993, 1994) to provide an analysis of a range of brightness phenomena, especially those involving the automatic selection of processing to yield categorical differences between smooth and abrupt spatial transitions in brightness percepts. Finally, it must be noted that the reference to "multi-scale" processing in the title refers to the multiple copies of processing across spatial scales (from "small" to "large") that occurs in several stages of the model. Multiple scales are needed to rationalize how some percepts, including those of Mach bands, require a fusion of results from some scales that produce abrupt spatial discontinuities in brightness output and others that produce smooth transitions. The net result is a hybrid percept. It may be that the full range of brightness data has to date resisted a satisfactory parametric characterization by any model precisely because human brightness perception of even seemingly simple images requires complex interactions at multiple scales among nonlinear mechanisms whose principles have only recently begun to be uncovered.

References:

Grossberg, S. (1987). Cortical dynamics of three-dimensional form, color, and brightness perception: II. Binocular theory. *Perception & Psychophysics*, **41**, 117-158.

Grossberg, S. (1994). 3-D vision and figure-ground separation by visual cortex. *Perception & Psychophysics*, **55**, 48-120.

Grossberg, S. & Marshall, J. (1989). Stereo boundary fusion by cortical complex cells: A system of maps, filters, and feedback networks for multiplexing distributed data. *Neural Networks*, **2**, 29-51.

Grossberg, S. & Mingolla, E. (1987). Neural dynamics of surface perception: Boundary webs, illuminants, and shape-from-shading. *Computer Vision, Graphics, and Image Processing*, **37**, 116-165.

Grossberg, S. & Todorović, D. (1988). Neural dynamics of 1-D and 2-D brightness perception: A unified model of classical and recent phenomena. *Perception & Psychophysics*, **43**, 241-277.

Neumann, H. (1993). Toward a computational architecture for unified visual contrast and brightness perception: I. Theory and model. *Proceedings of the World Congress on Neural Networks* (WCNN-93), Vol. I., 84-91. Hillsdale, NJ: Erlbaum.

Neumann, H. (1994). An outline of a neural architecture for unified visual contrast and brightness perception. Technical Report CAS/CNS-94-003, Department of Cognitive and Neural Systems, Boston University.

Pessoa, L., Mingolla, E., & Neumann, H. (1995). A multi-scale network model of brightness perception. *Vision Research*, in press.

DECISION PROCESSES IN SPATIAL VISION: DETECTION RULES

Miguel A. García-Pérez[*] & Vicente Sierra-Vázquez[°]

[*]*Dep. de Metodología and* [°]*Dep. de Psicología Basica, Facultad de Psicología*
Universidad Complutense, 28223 Madrid, Spain

Abstract

A two-dimensional space-variant multiple channel model is adopted to test four different detection rules against data on the foveal and peripheral detectability of a number of spatial gratings. The results show that none of the detection rules combined with this model can account for all of the data, although most of them account for a relatively large proportion of the data. Several amendments of the multiple channel model and/or the detection rules are discussed which may fit the data.

1. Introduction

Vision can be understood as a process involving two major stages: the creation of a visual representation of the stimulus *(early visual processing)* and the subsequent processing of that representation in order to achieve visual tasks *(late visual processing)*. An enormous variety of approaches have been taken to model early visual processing, and an also large number of late visual processing models have been proposed to account for data on visual performance in specific tasks. Quite often, models of visual processing are set up without distinguishing properly these two processing stages, thus including elements of early visual processing and a specific late processing rule that aims at accounting for the data gathered to test the joint model on the visual task of choice. As a result, failure to fit these joint models to the data does not reveal which of their elements are responsible for the outcome (and should be replaced) and which others should be kept to build an alternative model on. Similarly, success at fitting the model does not reveal which of its elements could be replaced without affecting the fit. The consequence of this situation is that there are a number of different joint models which fit the same set of data similarly despite major differences between them and, therefore, the essential visual processing characteristics which the empirical data could reveal still remain unravelled.

This situation arises because early and late visual processing models cannot be tested independently: visual representations cannot be observed directly, and data on performance in visual tasks rely on the actual visual representations obtained by the visual system. A way around this difficulty is to set up separate early visual processing models and late visual processing models for the achievement of a given task, and to test their combinations against the relevant data.

This work aims at evaluating alternative late visual processing models for pattern detection (detection models) combined with a specific early visual processing model. Section 2 describes the functional and structural characteristics of the two-dimensional (2D) space-variant multiple channel model assumed to be responsible for early visual processing. Section 3 presents the four most widespread detection models used to account for detection data. Section 4 describes the results of a test of the four detection models against data on the detectability of gratings.

2. Two-dimensional space-variant multiple channel model

2.1. Number of channels

In a 2D multiple channel model, each channel can be identified by the pair (ρ',θ') defining its spatial frequency center and preferred orientation, respectively. This work assumes 6060 distinct visual channels, resulting from the factorial combination of 101 spatial frequency centers and 60 preferred orientations. The 101 spatial frequency centers cover the 10-octave range from 2^{-5} (=0.03125) to 2^{5} (=32) c/deg, in logarithmic steps of 0.1 octaves; the 60 preferred orientations cover the range from 0 to 177 deg in steps of 3 deg. Note that, for all practical purposes (and given the bandwidths assumed in Section 2.3), this tight spacing of channels in the 2D Fourier domain is virtually equivalent to assuming an unlimited number of channels.

2.2. Point-weighting function and modulation transfer function

The point-weighting function of the (ρ',θ') channel at spatial location (x',y') will be assumed to be described by the 2D even-symmetric Gabor function

$$\psi(x,y) = \gamma(x',y',\rho',\theta')\frac{1}{2\pi s_x s_y}\exp\left[-\frac{(\hat{x}-\hat{x}')^2}{2s_x^2}\right]\exp\left[-\frac{(\hat{y}-\hat{y}')^2}{2s_y^2}\right]\cos\left(2\pi\rho'(\hat{x}-\hat{x}')\right) \qquad ,\rho'>0, \quad (1)$$

where γ is the channel gain at that location (see Section 2.4) and

$\hat{x} = x\cos\theta' + y\sin\theta'$,

$\hat{x}' = x'\cos\theta' + y'\sin\theta'$,

$\hat{y} = -x\sin\theta' + y\cos\theta'$,

$\hat{y}' = -x'\sin\theta' + y'\cos\theta'$.

The modulation transfer function of the (ρ',θ') channel at location (x',y'), Ψ, is

$$\Psi_{\rho',\theta'}(\rho,\theta) = \frac{\gamma(x',y',\rho',\theta')}{2}\left|\exp\left[-2\pi^2\left(s_x^2(\rho\cos(\theta-\theta')-\rho')^2+s_y^2\rho^2\sin^2(\theta-\theta')\right)\right] + \exp\left[-2\pi^2\left(s_x^2(\rho\cos(\theta-\theta')+\rho')^2+s_y^2\rho^2\sin^2(\theta-\theta')\right)\right]\right| . \quad (2)$$

2.3. Spatial frequency and orientation bandwidths

The spatial frequency and orientation bandwidths of even-symmetric Gabor channels depend on s_x, s_y and the tuning frequency, ρ', of the channel. In particular, the half-amplitude spatial-frequency bandwidth (in octaves), B_{oct}, can be easily shown to be

$$B_{oct} = \log_2\left[\frac{2\pi s_x\rho' + \sqrt{2\ln 2}}{2\pi s_x\rho' - \sqrt{2\ln 2}}\right], \tag{3}$$

and the half-amplitude orientation bandwidth (in degrees), B_{or}, can be shown to be

$$B_{or} = 2\arcsin\left[\frac{s_x}{s_y}\frac{2^{B_{oct}}-1}{2^{B_{oct}}+1}\right]. \tag{4}$$

These bandwidths will be assumed to be constant across channels and spatial locations, and valued at B_{oct}=1 and B_{or}=30. Therefore, from Equations (3) and (4),

$$s_x = \frac{\sqrt{2\ln 2}}{2\pi}\frac{2^{B_{oct}}+1}{2^{B_{oct}}-1}\frac{1}{\rho'} = \frac{R}{\rho'} = \frac{0.56}{\rho'}, \tag{5}$$

where R is the shape ratio of the channels and

$$s_y = \frac{s_x}{\sin(B_{or}/2)}\frac{2^{B_{oct}}-1}{2^{B_{oct}}+1} = 1.29 s_x. \tag{6}$$

Note that channels with these characteristics qualify as wavelet analyzers.[6]

2.4. Channel gains and spatial anisoplanatism

The channel gain function, γ, will be assumed to be the product of three components (see Fig. 1) each of which incorporates a different characteristic of early visual processing:

$$\gamma(x',y',\rho',\theta') = \gamma_1(\rho')\gamma_2(\theta')\gamma_3(x',y',\rho'), \tag{7}$$

where

$$\gamma_1(\rho') = \frac{42\rho'\left(20\pi^2R^2+\rho'+h(\rho')\right)\exp\left[\pi^2R^2 - \frac{\rho'\left(\rho'+h(\rho')\right)}{400\pi^2R^2} - \frac{2\rho'+h(\rho')+10}{20}\right]}{5\pi^2R^2 + 5\pi^2R^2\exp\left[-4\pi^2R^2 - \frac{\rho'+h(\rho')}{5}\right]} \tag{8}$$

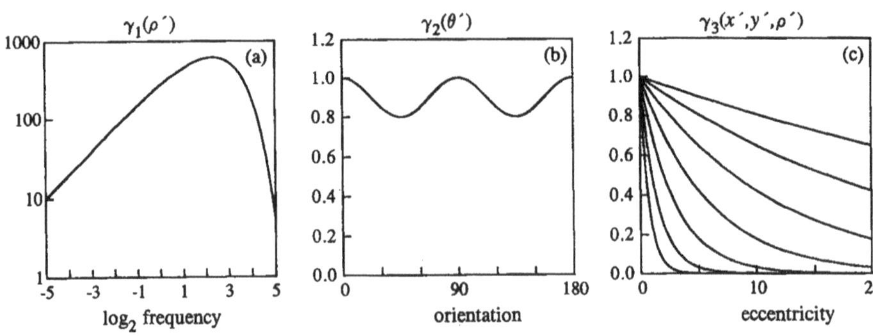

Figure 1. Components of the channel gain function. (a) Dependence on tuning frequency. (b) Dependence on preferred orientation. (c) Dependence on retinal eccentricity parameterized by tuning frequency. Top to bottom, the 7 curves in (c) pertain to channels tuned to 0.5, 1, 2, 4, 8, 16 and 32 c/deg.

with

$$h(\rho') = \sqrt{(20\pi^2 R^2 + \rho'^2)^2 - (20\pi R)^2}$$

incorporates the variation in channel sensitivity with tuning frequency,[7]

$$\gamma_2(\theta') = \frac{9 + \cos(\pi\theta'/45)}{10} \tag{9}$$

incorporates the oblique effect (lower sensitivities of channels at oblique orientations as compared to channels at vertical or horizontal orientations), and

$$\gamma_3(x',y',\rho') = 10^{-\rho' d/53.3}, \quad d = \sqrt{x'^2 + y'^2} \tag{10}$$

incorporates the decrease in sensitivity within a channel as a function of retinal eccentricity, a decrease whose gradient further depends on the channel's tuning frequency, ρ'.[13]

2.5. Spatial sampling

Units within each channel are assumed to be located at the nodes of a 2D grid with a spacing of $s_x/3.6$ deg along the x' dimension and $s_y/3.6$ deg along the y' dimension.

2.6. Channel outputs and visual representation

A channel's output to an input stimulus f is a 2D function which is a discrete version of

$$r_f(x',y';\rho',\theta') = \int_{-\infty}^{\infty}\int_{-\infty}^{\infty} f(x,y)\,\psi(x,y)\,dx dy, \tag{11}$$

obtained by sampling Equation (11) at the appropriate x' and y' locations. The collection of outputs from all channels is a discrete 4D function which can be thought of as the psychophysical representation of the input stimulus.

3. Detection rules

A detection rule specifies what condition the psychophysical representation of a pattern must meet in order for that pattern to be detected. Application of the detection rule defines a processing stage which occurs after the psychophysical representation has been built during early processing and, thus, any detection model qualifies as a late processing model. All of the detection rules ever proposed in the literature state that a pattern is detected at the minimum contrast necessary for some combination of the values in its psychophysical representation to reach a certain threshold value, t. While the various detection rules differ as to which specific combination is assumed, all of them can be derived as special cases of a single general combination rule known as Quick pooling:

$$P = \left[\sum_{\rho'}\sum_{\theta'}\left(\sum_{x'}\sum_{y'}\left|r_f(x',y';\rho',\theta')\right|^{Q_s}\right)^{Q_f/Q_s}\right]^{1/Q_f}, \tag{12}$$

where the sums over ρ' and θ' imply all channels, the sums over x' and y' imply the sampling points defined in Section 2.5, and Q_s and Q_f are the Quick exponents for space and frequency. Note that P in Equation (12) is a scalar decision variable: if P is greater than or equal to the threshold value, t, the pattern is detected; otherwise it is not.

3.1. Peak detection

The simplest detection rule states that a pattern is detected at the minimum contrast necessary for the response of any given channel at any given spatial location to reach the detection threshold. This detection rule derives from the Quick pooling equation when the Quick exponents go to infinity, yielding

$$P = \max_{\rho',\theta',x',y'} \left| r_f(x',y';\rho',\theta') \right| . \tag{13}$$

3.2. Probability summation over channels

Probability summation over channels refers to the case where each channel contributes its peak response over space to a combination involving each channel's peak response, and detection ocurring if the value of this combination exceeds the detection threshold. The precise combination rule arises from the Quick pooling equation when Q_s goes to infinity and Q_f takes on any finite and small value, usually in the range from 2 to 6. (A value Q_f=4 will be assumed for this rule in this paper.) Replacing these exponents in Equation (12) and taking limits yields

$$P = \left[\sum_{\rho'} \sum_{\theta'} \max_{x',y'} \left| r_f(x',y';\rho',\theta') \right|^{Q_f} \right]^{1/Q_f} . \tag{14}$$

3.3. Probability summation over space

Probability summation over space refers to the case where each channel produces a combination of its responses over space, and detection occurs if any channel's combination exceeds the detection threshold. The exact combination rule arises from the Quick pooling equation when Q_f goes to infinity and Q_s takes on any finite and small value, usually in the range from 2 to 6. (A value Q_s=4 will be assumed for this rule in this paper.) Replacing these exponents in Equation (12) and taking limits yields

$$P = \max_{\rho',\theta'} \left[\sum_{x'} \sum_{y'} \left| r_f(x',y';\rho',\theta') \right|^{Q_s} \right]^{1/Q_s} . \tag{15}$$

3.4. Probability summation over channels and space

Probability summation over channels and space occurs when each channel produces a combination of its responses over space and contributes this value to a second combination involving each channel's own contribution. The exact combination rule in this case is the general pooling expression in Equation (12) when both Q_f and Q_s take on any finite and small value, usually the same, and in the range from 2 to 6. (Values Q_f=Q_s=4 will be assumed for this rule in this paper.)

4. Detection experiments

4.1. Predicting sensitivity

Predicting the sensitivity of the model described in Section 2 to any given spatial pattern under a given detection rule requires obtaining the psychophysical representation of the pattern and evaluating the decision variable P for the corresponding rule. Although the internal threshold t is unknown, it can be easily shown that sensitivity, S, to a given pattern is proportional to the value of P for a unit-contrast pattern: by definition, $S=1/m$, m being the contrast necessary for an observer to detect the pattern, and detection occurs when $mP=t$; therefore, $S=P/t$. Thus, sensitivity to any pattern under any detection rule can be predicted except for a scaling factor. The following subsections compare model sensitivities arising from the four detection rules to a variety of patterns.

4.2. Sinusoidal, square and sawtooth gratings

Figure 2 shows the predicted sensitivities to sinusoidal, square and sawtooth gratings. Note that the four sine-wave contrast sensitivity functions (CSFs) differ almost exclusively by a scale factor which is related to the amount of summation in the underlying

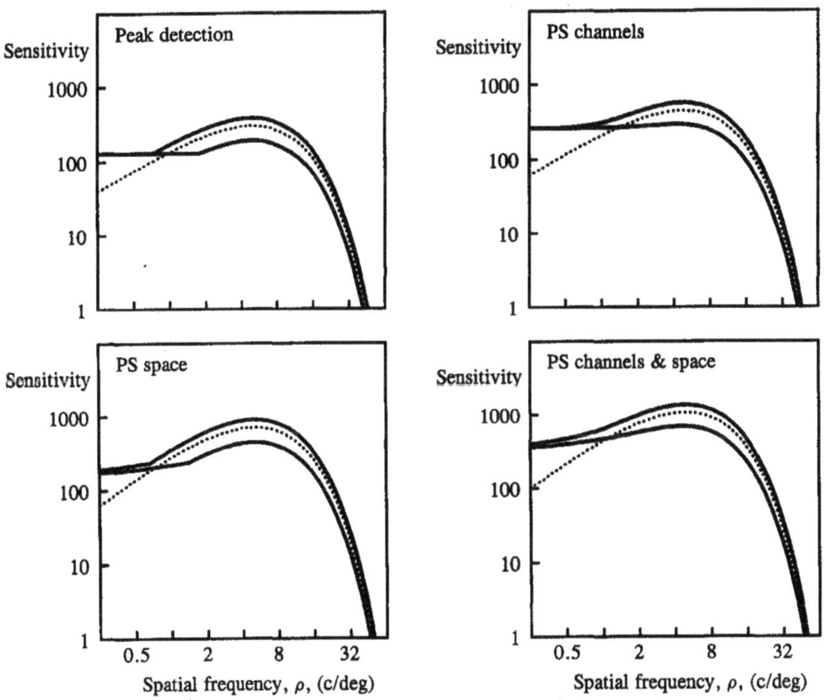

Figure 2. Predicted sensitivities to sinusoidal gratings (dotted curve), square gratings (upper continuous curve), and sawtooth gratings (lower continuous curve) as a function of grating frequency for four detection rules.

detection rule.[7] On the other hand, at relatively high frequencies, the four predicted square and sawtooth CSFs are related to the corresponding sine-wave CSF as is found empirically,[1] and at very low spatial frequencies, all four detection rules also predict equal sensitivities to square and sawtooth gratings of the same frequency.[3] The only major differences between the four detection rules as regards square (or sawtooth) CSFs occur in the shape of their low frequency ends. Empirical data show that sensitivity to low frequency square gratings is constant and independent of spatial frequency,[1-4] a result which is in agreement with the predictions from the peak detection and the probability summation over channels rules, but not with the predictions from the two other detection rules.

4.3. Rectangular gratings

Complete CSFs for rectangular gratings of various duty cycles have not been reported, but some measurements indicate that high frequency rectangular gratings of small duty cycle are detected at the contrast at which their first harmonics are detected when presented alone, and that the 11 c/deg sine/rectangular sensitivity ratio as a function of duty cycle is determined by the amplitude of the first harmonic of the rectangular grating.[1] All four detection rules predict rectangular CSFs that are in agreement with these results and, again, they only differ as to the shape of the predicted low frequency end.

4.4. Trapezoidal gratings

Empirical CSFs for trapezoidal gratings of various ramp widths[3,9] are nearly perfectly accounted for by all four detection rules, the differences among their predicted CSFs being too small to allow a decision as to which detection rule fits the data best.

4.5. Square gratings missing harmonics

Results on the detectability of square gratings missing either high or low harmonics[2] are also well accounted for by all detection rules (see Fig. 3). In particular, square gratings missing high frequency harmonics are equally as detectable as complete square gratings in the intermediate and high frequency range, while they are less detectable than complete square gratings in the very low frequency range; on the other hand, square gratings missing low frequency harmonics are equally as detectable as complete square gratings in the very low frequency range, while they are less detectable than complete square gratings in the intermediate and high frequency range. In addition, the reduced detectability in this range of frequencies is that which would be expected if the grating were detected at the contrast at which the lowest harmonic that is present reached its own contrast threshold.

4.6. Sine-wave contrast sensitivity at different eccentricities

All four detection rules predict accurately the major results found when sine-wave CSFs are measured at different retinal eccentricities. Specifically, when small grating patches of fixed retinal size are used, the sine-wave CSF moves towards lower frequencies and sensitivities with increasing eccentricity, and the family of CSFs thus obtained do not meet nor intersect at any spatial frequency.[10] On the other hand, when grating patches of a fixed number of cycles are used, the sine-wave CSF also moves towards lower frequencies and sensitivities with increasing eccentricity, but the family of CSFs thus obtained stem from the same sensitivity at some very low spatial frequency.[11]

4.7. Sine-wave contrast sensitivity as a function of grating size

Foveal measurements of the sine-wave CSF using grating patches of different sizes reveal that sensitivity increases asympotically at all spatial frequencies with increasing number of cycles in the pattern.[5,12] On the other hand, similar measurements out in the periphery reveal that sensitivity also increases at all spatial frequencies with number of cycles, but this increase is nearly linear and shows no asymptotic behavior within the range of grating sizes studied.[12] These characteristics cannot be accounted for by any detection rule in which probability summation over space is not allowed. Specifically, neither the peak detection rule nor the probability summation over channels rule predict any increase in sensitivity as the size of the grating patch increases. On the contrary, the

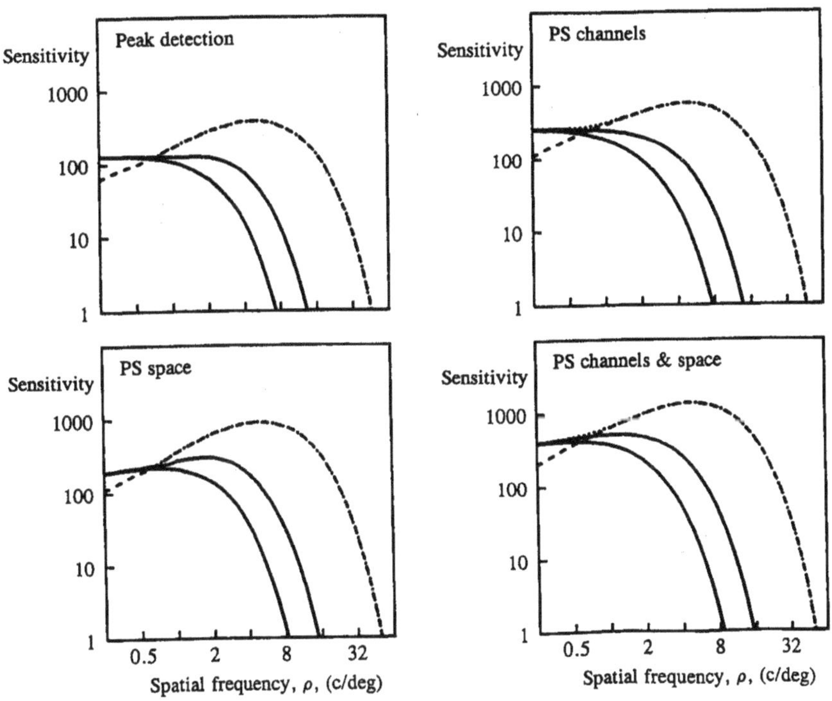

Figure 3. Predicted sensitivities to complete square gratings (dotted curve), waveforms containing only the first five harmonics of a square grating (dashed curve), and square gratings missing the first (upper continuous curve) or the three first (lower continuous curve) harmonics, as a function of grating frequency for four detection rules.

two remaining detection rules predict exactly this pattern of results (see Fig. 4).

5. Discussion

Overall, the results presented in Section 4 do not completely support any of the four detection rules considered in this paper: peak detection and probability summation over channels predict very accurately the flat sensitivity to very low frequency square and sawtooth gratings, but they fail to account for the effects of grating size on sensitivity. In contradistinction, the two other detection rules account fairly well for the latter result, but they fail to account for the former.

While more detailed contrast sensitivity measurements could be carried out in order to test the conflicting predictions from each of the four detection rules in a number of critical situations, it is important to note that the results presented in this paper cannot definitely reject any of the detection rules. It may well be that some characteristics of the predictions are directly tied to specific assumptions in the early visual processing model. Assumptions whose effects on these predictions are worth studying cover (i) the number of channels, (ii) the mathematical form of the channel point-weighting func-

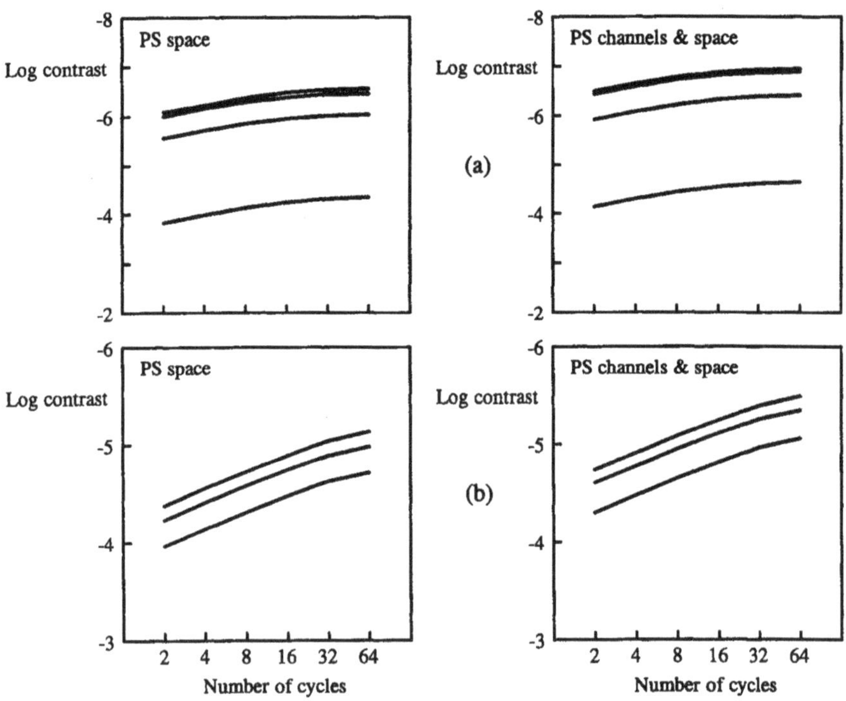

Figure 4. (a) Predicted foveal sensitivities to sinusoidal gratings of (top to bottom) 6, 3, 12, and 24 c/deg as a function of number of cycles. (b) Predicted peripheral (42 periods above fixation) sensitivities to sinusoidal gratings of (top to bottom) 3, 6, and 12 c/deg as a function of number of cycles.

tions, (iii) the spatial frequency and orientation bandwidths, and whether they are constant across channels or they change with tuning frequency, (iv) the form of the channel gain function as regards spatial anisoplanatism, and (v) the density of spatial sampling.

In addition, and besides alternative values for the Quick exponents, some amendments to the detection rules involving probability summation could also be tested against the data. Specifically, the functional area for spatial summation might be bounded so that summation does not take place over the whole extent of the grating.[8]

The results of thorough tests carried out along the directions just mentioned will surely reveal which characteristics of early visual processing must be included in a model and also which characteristics of a detection rule are necessary in order to account for extant data not only on the detectability of gratings but also on that of aperiodic patterns.

6. Acknowledgements

This work was supported by DGICYT grant PB90-0257 to the authors.

7. References

[1] F.W. Campbell and J.G. Robson (1968). Application of Fourier analysis to the visibility of gratings. *Journal of Physiology (London)*, **197**, 551-566.

[2] F.W. Campbell, E.R. Howell and J.R. Johnstone (1978). A comparison of threshold and suprathreshold appearance of gratings with components in the low and high spatial frequency range. *Journal of Physiology (London)*, **284**, 193-201.

[3] F.W. Campbell, J.R. Johnstone and J. Ross (1981). An explanation for the visibility of low frequency gratings. *Vision Research*, **21**, 723-730.

[4] C.S. Furchner, J.P. Thomas and F.W. Campbell (1977). Detection and discrimination of simple and complex patterns at low spatial frequencies. *Vision Research*, **17**, 827-836.

[5] M.A. García-Pérez (1988). Space-variant visual processing: Spatially limited visual channels. *Spatial Vision*, **3**, 129-142.

[6] M.A. García-Pérez and V. Sierra-Vázquez (1995). Visual processing in the joint spatial/spatial-frequency domain. In: *Vision Models*. E. Peli (Ed.). Hong Kong: World Scientific Publishing. In press.

[7] M.A. García-Pérez and V. Sierra-Vázquez (1993). Deriving channel gains from large-area sine-wave contrast sensitivity data. Manuscript under review.

[8] E.R. Howell and R.F. Hess (1978). The functional area for summation to threshold for sinusoidal gratings. *Vision Research*, **18**, 369-374.

[9] W. Jaschinski-Kruza and C.R. Cavonius (1984). A multiple-channel model for grating detection. *Vision Research*, **24**, 933-941.

[10] J.S. Pointer (1986). Visual representation at the cerebral cortex: Qualitative and quantitative aspects. *Ophthalmic and Physiological Optics*, **6**, 171-175.

[11] J.S. Pointer and R.F. Hess (1989). The contrast sensitivity gradient across the

human visual field: With emphasis on the low spatial frequency range. *Vision Research*, **29**, 1133-1151.

[12] J.G. Robson and N. Graham (1981). Probability summation and regional variation in contrast sensitivity across the visual field. *Vision Research*, **21**, 409-418.

[13] A.B. Watson (1982). Summation of grating patches indicates many types of detector at one retinal location. *Vision Research*, **22**, 17-25.

RECENT RESULTS IN EMERGENT VISUAL SEGMENTATION

Ennio Mingolla

Center for Adaptive Systems and
Department of Cognitive and Neural Systems
Boston University
111 Cummington Street
Boston, Massachusetts 02215 USA

Abstract

Recent progress in the understanding of the interaction of brightness perception, illusory contours, and corticogeniculate feedback through developments in a neurally based model of early visual perception are reviewed. The Boundary Contour System (BCS) of Grossberg and Mingolla (1985) finds the location and orientation of surface boundaries of a scene. The outlines of such surfaces can be defined by regions of different luminance or texture, or by illusory contours. These computations are carried out through a series of cooperative and competitive interactions, including positive feedback from cells sensitive to long-range alignments of contours to cells coding more local oriented contrast information. BCS boundaries also gate diffusive filling-in of brightness (surface property) signals within a Feature Contour System (FCS). A recently enhanced version of the BCS, along with the FCS, has been shown to offer an explanation of brightness effects often associated with illusory contours by Gove, Grossberg, and Mingolla (1994). The principal novelty of the enhanced BCS model is that top-down feedback from BCS analogs of complex cells in V1 to model lateral geniculate nucleus (LGN) cells is used to enhance contrast at line ends and other areas of boundary discontinuity. The result is an increase in the model's analog of perceived brightness just beyond a dark line's end, akin to what have been termed "brightness buttons" (Kennedy, 1979). V1 to LGN feedback is discussed in relation to the larger BCS/FCS architecture and a simulation of the Ehrenstein illusion is discussed.

The Ehrenstein (1941) illusion includes a striking juxtaposition of two distinct visual phenomena: the formation of illusory contours, and an induced brightness difference within the region enclosed by those contours. (See Figure 1a.) Decades of debate followed the introduction of such figures concerning which of these effects was "primary," and many theories sought to derive one of the phenomena (illusory contours or brightness induction) as a consequence of the other. (See Lesher, in press, and Lesher and Mingolla, in press, for reviews.) More recently, many physiological and psychophysical results have been interpreted by neurally inspired models as providing distinct, though interacting, mechanisms for brightness induction and illusory contour formation. This chapter gives a brief overview of one such model, part of a programmatic development of a neurally based theory of visual perception by Stephen Grossberg and his colleagues. Details of the work described here can be found in Gove, Grossberg, and Mingolla (1994).

The conundrum of the relation of brightness induction and illusory contours in Ehrenstein figures is illustrated in Figure 1. Why do several line segments, placed together in a certain configuration, produce a perceptual result whose existence does not even appear to be hinted at by consideration of a single segment in isolation? Kennedy (1979) refers to the formation of "brightness buttons" at the ends of dark lines, while emphasizing that perceptually subliminal effects from several such lines must apparently be pooled before the effect can be noticed. (Other terms referring to this phenomenon are "dissimilation" or "line-end contrast".) Kennedy, however, gives only a phenomenological description of candidate mechanisms; an attempt to discover possible neural substrates for brightness buttons leads to surprising results.

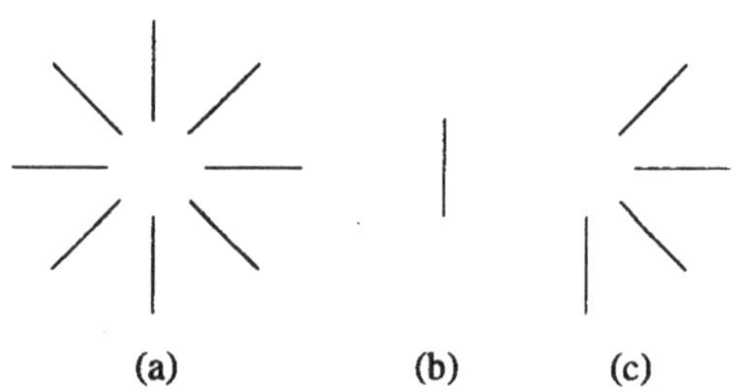

Figure 1: The Ehrenstein figure of (a) induces both a circular illusory contour and illusory brightness in the enclosed disk. An isolated line, in (b), does not induce any perceptible brightness increment at its ends. However, Kennedy (1979) spoke of thin dark lines as having "brightness buttons" at their ends, whose effects could be rendered visible by the close spatial juxtaposition of several such lines, as in (c).

Note that, as in Figure 2, brightness buttons cannot arise solely from classical mechanisms of lateral inhibition, mediated by center-surround cells. By themselves, cells with concentric (unoriented) receptive fields would be ignorant of effects arising from lines or edges, for the latter possess the attribute of *orientation*. Since area V1 of cortex is the earliest region in primates with cells of strong orientational preference, one must next look there. As shown in Figure 3, endstopped units generate maximal activity near line ends. If their activity map could be used to modulate the activity of center-surround cells -- units that presumably contribute the primary signal subserving the sensation of brightness -- then a circuit for brightness buttons could be found, as shown in Figure 4. To support this claim in a larger context requires some review of background.

Grossberg (1987, 1994) has named the theory of vision that he has developed with his colleagues FACADE theory, to emphasize the importance of the interwoven dynamics of multiple large networks of early and middle vision in simultaneously resolving Form And Color and DEpth in visual scenes. The two major subsystems of the early portion of FACADE theory are called the Boundary Contour System (BCS) and the Feature Contour System (FCS). (See Figure 5.) The theory clarifies the distinction between the recognizable boundaries that form groupings of visual

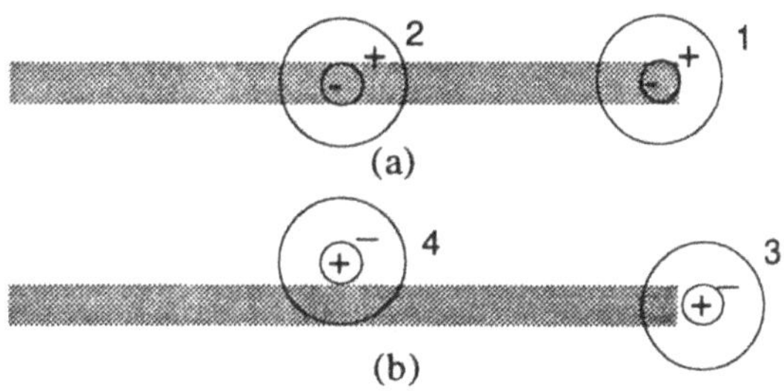

Figure 2: Lateral inhibition, resulting from center-surround processing of early cells (e.g. retinal ganglion cells or cells of the lateral geniculate nucleus) is said to enhance contrast at edges. While it might seem that it should then readily explain brightness buttons, a closer look indicates not, for the activity enhancement occurs only for cells coding the *interiors* of thin line segments. Thus the OFF-center, ON-surround cell labeled 1 in (a) *does* respond more vigorously than 2, since more light falls in its excitatory surround, but the ON-center, OFF-surround cell 3 of (b) responds *less* than 4, since more light falls in the former's inhibitory surround. Without some additional processing, the region at the end of the line ought therefore to appear, if anything, darker than the region along the sides, according to the straightforward hypothesis linking cell activity to perceptual outcome.

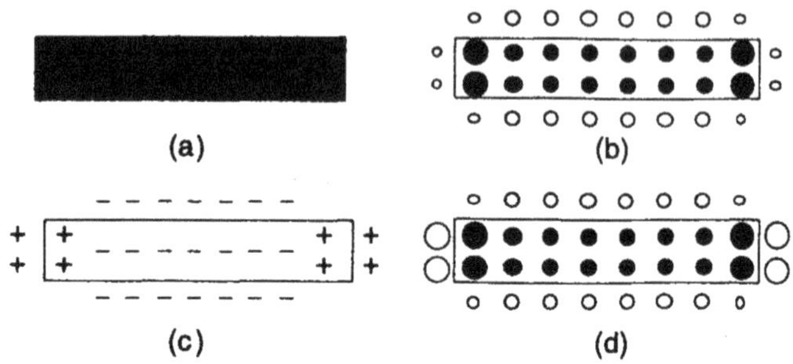

Figure 3: A schematic diagram shows that a dark bar (a) processed by center-surround cells of both contrast polarities produces a pattern of responses such as shown in (b), where circle size indicates response strength of the corresponding units (ON-center, OFF-surround for open circles, OFF-center, ON-surround for shaded circles) centered at a given place. Center-surround units are by nature ignorant of the orientation of contrast falling in their receptive fields, but endstopped (orientationally sensitive) cells in V1 produce a pattern of responses cartooned in (c), where plus signs indicate strong responses and minus signs indicate weak or absent responses. If the "weight pattern" of (c) could be used to modulate the activity pattern of (b) (e.g., by a form of gating or multiplication), then the result would appear as indicated in (d). Note that an early neural analog (or "precursor") of the phenomenon of brightness buttons occurs, whereby the results of ON-center, OFF surround processing at line ends have been dramatically boosted relative to the responses of similar cells along the sides of the lines.

surface elements and those featural qualities of surface appearance commonly described by such terms as hue, saturation, or lightness. This usage rests on the complementary dynamics of network interactions underlying the boundary/feature distinction. Long-range spatial interactions in the BCS are directed inward between two or more orientationally tuned nodes that are sensitive to amount of contrast, but not direction of contrast (light-to-dark vs. dark-to-light). A particular arrangement of short range competition and long range cooperation among orientation sensitive nodes ensures a rapid, active choice and sharpening of boundaries, including emergent boundaries. Spatial interactions in the feature system are diffusive, flowing outward from any single active node, and are sensitive to both amount and direction of contrast. Because the patterns of activity in the boundary system restrict the diffusion of activity in the feature system, regions of like color or

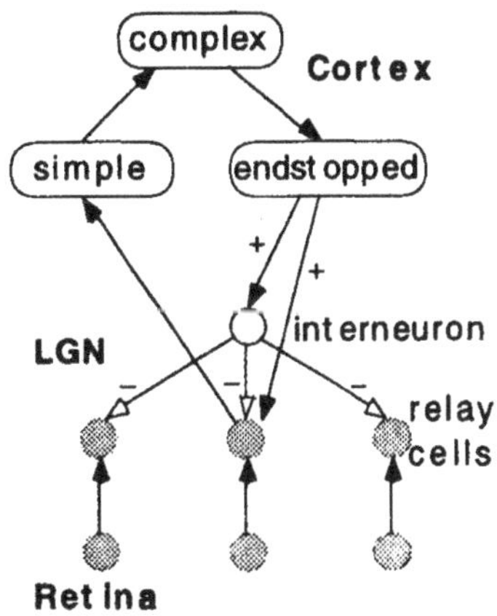

Figure 4: The V1--LGN circuit. Closed arrows stand for excitatory and open arrows for inhibitory connections. Some excitatory V1 projections go directly to LGN relay cells, while others excite interneurons that are in turn inhibitors of relay cells, in accordance with physiological data (Weber, Kalil, and Behan, 1989).

contrast typically match regions formed by boundary segmentations one-to-one. Thus the objects that we recognize in everyday environments generally have surface appearance attributes that are unitary and distinct from appearance attributes of other objects. Many laboratory curiosities, on the other hand -- including illusory contours, Glass patterns, and neon color spreading -- present the observer with a seemingly paradoxical scission between recognizable regions and phenomenal color patterns. The theory clarifies how these phenomena are inevitable consequences of strategies that the primate early visual system has evolved in order to coherently segment and group complex patterns of optical stimulation in real time while overcoming inherent uncertainties of measurement associated with the use of finite, quantized receptive fields.

More specifically, Figure 5 diagrams how early processing by ON cells and OFF cells embedded in shunting center-surround networks normalizes input dynamic range and performs local contrast enhancement. Combining ON cell and OFF cell output in the BCS to define oriented simple and complex cell filters overcomes complementary processing deficiencies of each cell type and improves sensitivity to image contours. The oriented filters send signals to stages of short-range competition that define hypercomplex cells and long-range cooperation that define bipole cells. Together these interactions segment the image into regions by cooperatively completing and regularizing the most favored boundaries while suppressing image noise and weaker boundary groupings. While not indicated in Figure 5, boundary segmentation is performed by multiple copies of the BCS at scales ranging from small to large, whose interaction distances covary with the size of the scale. Filling-in of surface representations occurs within the FCS at each scale via a diffusion process. Diffusion is activated by the normalized FCS inputs and restricted to the compartments defined within each BCS boundary segmentation.

Figure 6 offers an iconographically different view from that of Figure 5 for understanding BCS processing. Note that the early stage of monocular preprocessing is followed by a Static Oriented Contrast (SOC) Filter, until recently usually spoken of as a feedforward cascade, followed by a Cooperative-Competitive (CC) Loop. The discussion of brightness buttons has motivated a role for backprojections (labeled *"NEW!"* in Figure 6) from V1 to LGN. Figure 4 offers a detailed view of the V1 to LGN circuit employed in the model. Note that insofar as the model endstopped cells sending the backprojected signals are themselves part of the CC Loop, the theory predicts that the effects of short-range competition and long-range grouping, involving (at least) area V2, may be propagated back to reorganize signals in LGN; such is precisely the present argument with respect to brightness buttons and the Ehrenstein illusion. More generally, any closing of a feedback loop between neural layers makes suspect the distinction between "early" and "late" processing so often based on feedforward models of neural signal processing.

All machinery needed to address brightness buttons and illusory brightness induction in the Ehrenstein illusion is now in place. Consider the simulation of Figure 7. The input is a pattern of dark (low luminance) lines on a light (medium to high luminance) background. The FCS output is gotten from two types of computations. The first is center-surround filtering; the theory calls for shunting inhibition in order to discount the illuminant and preserve relative contrast (ratio) information at edges. Recall that center-surround cells occur in both ON and OFF channels; their outputs have been superimposed in the figure for simplicity. That is, pixels darker than the mean of the panel labeled FCS output indicate high activity in the OFF channel, while pixels lighter than the mean indicate high activity in the ON channel. The second, novel FCS computation is the formation of brightness buttons. Note that the description of this as an "FCS computation" is partly figurative, insofar as signals from oriented BCS units are required to generate the enhanced contrast responses at line ends in the first place. In the figure, these precursors of perceptual brightness buttons are most visible near the inward ends of the line segments.

The lower left panel of Figure 7, labeled "BCS output", shows the results of the SOC Filter and equilibrated CC Loop dynamics, resulting in an unbroken boundary segmentation of the inner

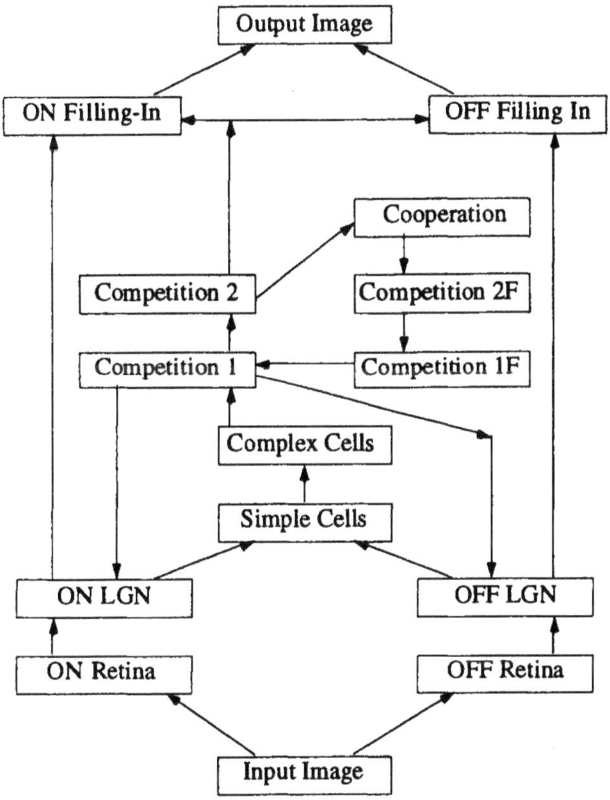

Figure 5: Macrocircuit of monocular BCS and FCS processing. In this representation, boxes at the three lowest layers (labeled image, retina, and lateral geniculate nucleus) form a stage of monocular preprocessing. The BCS stages are in the central region of the figure, while the FCS filling-in process, including a combination of the results of diffusion in an ON and OFF representation, is indicated in the upper three boxes. Note that the downward pointing paths from the box labeled Competition 1 to those labeled ON LGN and OFF LGN represent an innovation in the model circuit relative to previously published versions. See text for details.

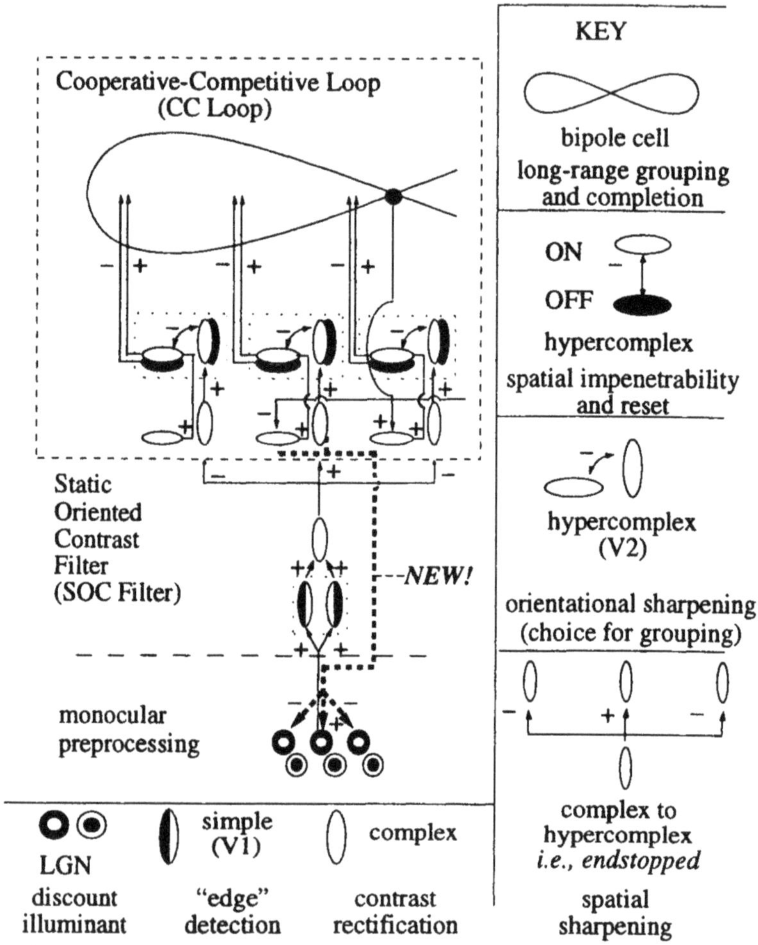

Figure 6: The stages of the Boundary Contour System appear iconographically. The key along the bottom and sides gives brief descriptors of some of the functional consequences of processing at each stage. While most of the terms used are self-explanatory, "spatial impenetrability" refers to the need to prevent the cooperative bipole cells from forming spurious or inappropriate groupings where local evidence overrules long-range coalignments, and "reset" refers to the temporal aspects of formation, persistence, and dissolution of perceptual segmentations.

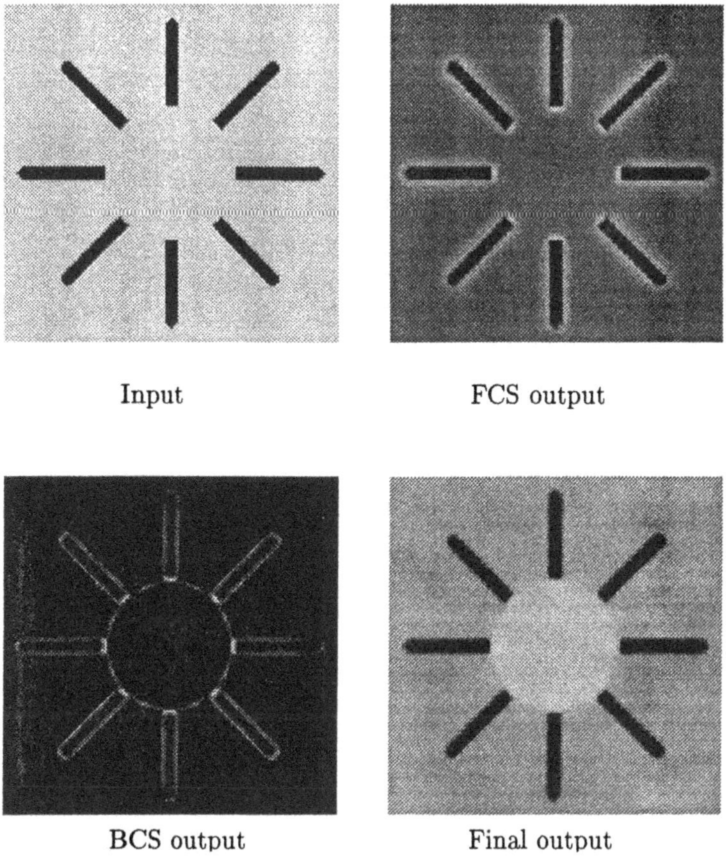

Figure 7: Simulation of Ehrenstein illusion. See text for details.

circular region, in addition to the generation of boundaries at the exteriors of each line segment. Note that the long-range cooperative bipole cells are critical to the generation of contours over distances spanning several times the diameters of inducing elements.

Finally, the result of diffusion of the FCS output within the boundary compartments formed by the BCS results in a pattern of activity that is the model analog of perceived brightness, and which, as shown in Figure 7, displays a clear brightness induction within the illusory circular figure. Note that the filling-in of brightness throughout the circle interior occurs by virtue of the FCS diffusion mechanism. Since that diffusion is nonconservative, the small enhanced values of ON signals at line ends can "fill up" the enclosed circular region without attenuation in amplitude.

The present chapter has at best merely sketched in coarse terms the outcomes of complex interactions in a model of many interacting neural mechanisms; several important considerations have, inevitably, been insufficiently addressed. Most important is the question of the functional role of the V1--LGN projections, such as employed in the present work, for it would be odd if the generation of brightness buttons were nature's *reason* for designing such a circuit! A functionally important role for corticogeniculate feedback has been proposed by Grossberg (1980), who suggested it as a critical mechanism for stabilizing codes for LGN signals that were consistent with binocularly formed scene segmentations, while suppressing those inconsistent signals. A fuller treatment of this issue and details of the model appear in Gove, Grossberg, and Mingolla (1994).

References

Ehrenstein, W. (1941). Uber abwandlundgen der L. hermannschen helliskeitserscheinung. *Zeitschrift fur psychologie*, **150**, 83-91.

Gove, A., Grossberg, S., and Mingolla, E. (1994). Brightness perception, illusory contours, and binocular corticogeniculate feedback. Technical Report CAS/CNS-TR-94-033, Boston, MA: Boston University.

Grossberg, S. (1980). How does the brain build a cognitive code? *Psychological Review*, **87**, 1-51.

Grossberg, S. (1987). Cortical dynamics of three-dimensional form, color, and brightness perception: II. Binocular theory. *Perception & Psychophysics*, **41**, 117-158.

Grossberg, S. (1994). 3-D vision and figure-ground separation by visual cortex. *Perception & Psychophysics*, **55**, 48-120.

Grossberg, S. and Mingolla, E. (1985). Neural dynamics of perceptual grouping: Textures, boundaries, and emergent segmentations. *Perception & Psychophysics*, **38** (2), 141-171.

Kennedy, J. M. (1979). Subjective contours, contrast, and assimilation In C. F. Nodine and D. F. Fisher (Eds.), *Perception and pictorial representation*. New York : Praeger.

Lesher, G. (in press). Illusory contours: Toward a neurally based perceptual theory. *Psychological Bulletin*.

Lesher, G. W. and Mingolla, E. (in press). Neural models of illusory contour perception. In M. A. Arbib (Ed.), *Handbook of brain theory and neural networks*. Cambridge, MA: MIT Press.

Weber, A. J., Kalil, R. E., and Behan, M. (1989). Synaptic connections between corticogeniculate axons and interneurons in the dorsal lateral geniculate nucleus of the cat. *Journal of Comparative Neurology*, **289**, 165-164.

Simple Mechanisms in Stereoscopic Depth Perception[1]

Hanspeter A. Mallot
Max–Planck–Institut für biologische Kybernetik
Spemannstr. 38, 72076 Tübingen, Germany
ham@mpik-tueb.mpg.de

Abstract

This paper reviews a number of psychophysical experiments designed to test popular theories on stereopsis which have been developed in machine vision. The results indicate that stereoscopic depth perception can be based on differences in image intensities which do not correspond to localized image features. In addition to this *intensity–based* stereo vision, image intensities can also modulate feature–based stereo, e.g., in the solution of the correspondence problem or in the quantification of perceived depth. Further results indicate that the underlying mechanism is based on squared contrast differences between the two halfimages. A similar mechanism has been proposed by Julesz (1971) for global stereopsis.

Introduction

Stereoscopic depth perception is based on the differences between the images of the two eyes which, in turn, are due to the differences of viewpoint (parallax) or (less commonly) specular reflectance. The mechanisms of stereoscopic depth perception have been studied intensively in Psychophysics, Neurophysiology, and, most recently, in Computer Vision. For review see Poggio & Poggio (1984), Arditi (1986), Dhond & Aggarwal (1989), Regan et al. (1990) and Blake & Wilson (1991). Most of the more recent work in all of these approaches is strongly influenced by the seminal book by Bela Julesz (1971) which treats stereo vision as a problem of information processing. Three major question are:

1. What stage of (uniocular) image processing feeds into the binocular comparison mechanism? Julesz showed that simple local image features rather that more complex object descriptions suffice, i.e., that stereopsis is a process of *early vision* (see also Julesz 1991).
2. What exactly is the binocular comparison operation? Especially in images of points in various depth positions, *correspondences* between image features depicting the same 3D object feature have to be found. Julesz proposed a cooperative, neural–network type mechanism for solving the correspondence problem.
3. What is the result of stereo processing? Julesz distinguishes two types of stereopsis: global stereopsis does not establish detailed correspondences and leads to the perception of planes or smooth surfaces (sometimes even subjective surfaces) in depth. Feature–based or local stereopsis results in a sparse disparity map with one disparity value assigned to each pair of corresponding image features.

In this paper, I will review a number of psychophysical experiments showing that (i) stereovision can take place even earlier than what is usually considered "early vision", i.e., on the level of image intensities, (ii) image intensity or local contrast is relevant in feature–based stereo as well. In the discussion, an evolutionary account of stereo processing is attempted linking levels of complexity in behaviour and neural network architecture.

Intensity–based stereo

Expt. 1 Stereo without localized image features

Stereovision is still possible with stereograms lacking intensity edges (e.g., smoothly shaded spheres, Bülthoff & Mallot 1988, see Fig. 1). We have designed binocular stimuli that lack Laplacian zero–crossings, and whose intensity extrema and centroids have identical positions (Arndt et al. *in press*). Still, the intensity distributions for the left and right eye are different (Fig. 2).

As a criterion for the use of stereo information, we tested the ability of subjects to distinguish between the two possible arrangements of two given halfimages into one stereogram. If we denote the

[1]Supported by the *Deutsche Forschungsgemeinschaft* Grants Ey8/17 3(8) and Ma1038/3–1,2. The collaboration of P. Arndt, H. Bülthoff, S. Dartsch, A. Erben, P. Unrath and W. von Seelen is gratefully acknowledged.

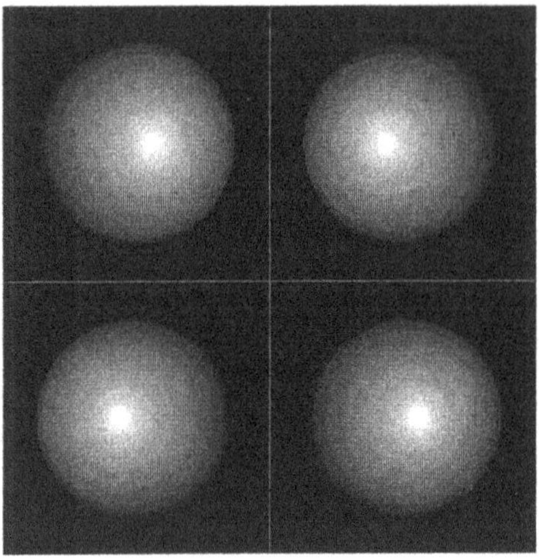

Figure 1: Stereograms of a smoothly shaded Lambertian ellipsoid with elongation 1 : 2 out of the image plane. Stereoscopic fusion can be obtained by placing a piece of cardboard vertically between the two halfimages such that the left and right eyes see only their appropriate pictures. **Top**: Orthoscopic view. **Bottom**: Pseudoscopic view (left and right halfimage exchanged). In the orthoscopic view, the object is perceived closer to the observer and with higher curvature. The difference between the two views is due to intensity-based stereopsis.

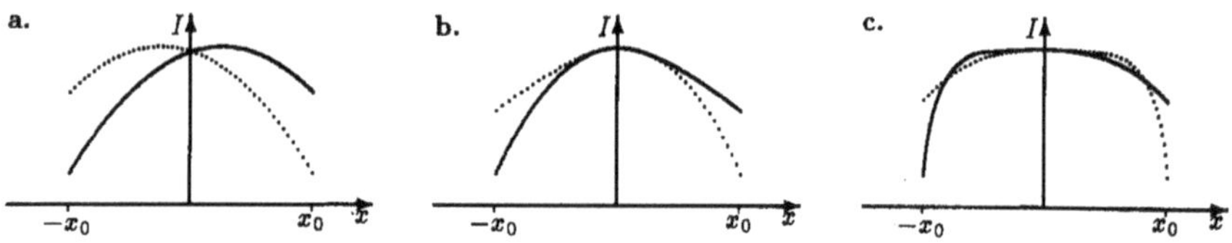

Figure 2: Intensity profiles of stimuli testing the role of localized features in stereo processing. Solid and dotted lines correspond to the two half-images of the stereograms. **a.** Parabolic greywedge: no Laplacian zerocrossings. **b.** Cubic greywedge: no Laplacian zerocrossing, intensity-peak at position $x = 0$ (i.e. without disparity information). **c.** Rational greywedge: no Laplacian zerogrossing, both intensity-peak and centroid at position $x = 0$.

intensity profiles for the two halfimages by $I_a(x)$ and $I_b(x)$ (solid and dotted lines in Fig. 2), we can form the stereograms S_{ab} (I_a left and I_b right) and S_{ba} (I_a right and I_b left). If the subject is able to tell stereogram S_{ab} from S_{ba}, we conclude that the differences between I_a and I_b have been extracted. In the experiments, we presented pairs of stereograms each of which could be either of type S_{ab} or S_{ba}. In a two alternative forced choice paradigm, subjects had to judge whether the stereograms were "equal" or "different". In addition to this discrimination task, we tested perceived depth ordering ("which one is in front?") as well (see Arndt et al., *in press*).

The data for the discrimination experiment is summarized in Table 1 indicate that binocular differences can still be perceived in the described stimuli. Subjects reported to perceive the difference between the stereograms as a depth difference. We conclude that intensity information is still available at the binocular ("cyclopean") stage of stereo computation. Localized features such as zerocrossings, intensity peaks, or centroids are not required to perceive stereoscopic depth.

Expt. 2 Mechanism of intensity-based stereo

What can be the mechanism of intensity-based stereo? For the perception of dotted regions with constant disparity, Julesz (1971, p. 122ff) proposed a simple area-based mechanism using a difference measure of the type

$$\Psi(\delta) = \int \left| I_l(x - \frac{\delta}{2}) - I_r(x + \frac{\delta}{2}) \right|^{\alpha} dx. \tag{1}$$

While Julesz proposed to choose $\alpha = 1$, we will argue in Expt. 3 that $\alpha = 2$ might be a better estimate. For an elaborated image matching algorithm based on the above difference measure, see Mallot *et al.*

Table 1: Average results of Expt. 1 for 4 subjects. "Peak condition:" intensity profiles as depicted in Fig. 2 (bright centers). "Trough condition:" inverted intensity profiles $1 - I(x)$ (dark centers). "=" and "≠" denote the "equal" and "unequal" stimulus presentations and judgements. In all cases, perception depends significantly on the stimulus presented ($p < 10^{-6}$). (Compiled from Arndt et al. *in press*)

profile	stim.	percept			
		peak		trough	
		=	≠	=	≠
parabolic	=	84	6	77	3
(Fig. 1a)	≠	11	79	0	80
cubic	=	77	13	64	16
(Fig. 1b)	≠	10	80	28	52
rational	=	48	2	40	10
(Fig. 1c)	≠	9	41	10	40

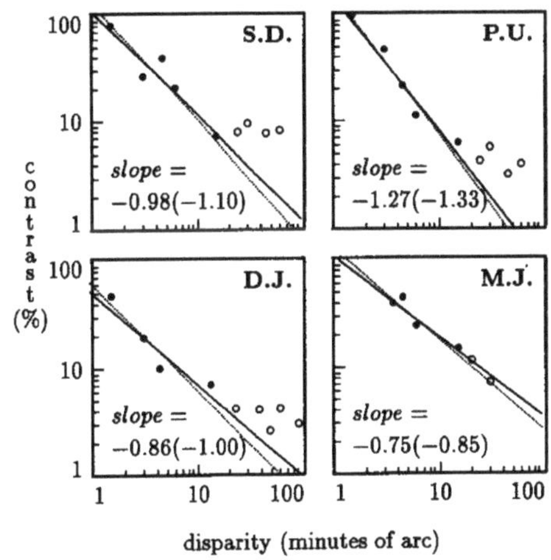

Figure 3: Results of Expt. 2. Discrimination thresholds for ortho– and pseudoscopic presentation of parabolic grey-wedges (Fig. 2a) as a function of disparity and contrast. Solid lines: regression of contrast as a function of disparity. Dotted lines: regression of disparity as a function of contrast. Since fusion becomes increasingly difficult for higher disparities (Panum's area), disparities above 15 minutes of arc have been excluded from the regression calculations (open circles).

1991 and Bohrer *et al.* 1990.

One possible test for the difference mechanism is the tradeoff of disparity and contrast at the discrimination threshold for ortho– and pseudoscopic views: if we decrease contrast by some amount, how much more disparity is needed to compensate for contrast decrease and restore visibility? Suppose that intensity–based stereo is perceived if $\Psi(0)$ exceeds some fixed value Ψ_ϑ. We assume that I_l and I_r are actually identical except for a true disparity δ_T, i.e., $I_l(x - \delta_T) = I_r(x) =: I_o + cg(x)$, where I_o is mean luminance, g a zero–mean greylevel function and c the resulting image contrast. This situation applies to the parabolic grey–wedges used in Expt. 1 (Fig. 2a). We can now expand the integrant in Eq. 1 into a Taylor–series and obtain

$$\Psi(0) \approx \int |c\delta_T g'(x)|^\alpha \, dx \propto |c\delta_T|^\alpha. \tag{2}$$

Therefore, the prediction of an intensity–difference scheme is that disparity and contrast be inversely proportional at threshold, $c \propto d^{-1}$. As has been pointed out by Legge & Gu (1989), intensity–peak matching predicts $c \propto d^{-2}$.

Fig. 3 shows contrast threshold for intensity–based stereo as a function of disparity δ_T measured with a parabolic grey wedge for four different subjects. The results show clearly that for our stimuli contrast is inversely proportional to disparity, not disparity squared. Thus, the difference mechanism is supported by this experiment.

Correspondence

The previous section summarized evidence for a stereo mechanism operating with differences of image intensities rather than with localized image features. While this result concerns largely the uniocular preprocessing for stereo vision, this section will address the binocular interaction stage, especially the correspondence problem. In psychophysical experiments, matching or correspondence can be studied by means of the *double–nail illusion*: if two vertical bars are presented to each eye, two globally consistent interpretations are possible, a frontoparallel one with the matches *left–left, right–right* and a medial (stacked) one with the matches *left–right, right–left*. The double–nail illusion was used by Mallot & Bideau (1990) to show that matches with small (relative) disparity are preferred over matches with large disparities. In the language of Computer Vision, this is evidence for an "ordering constraint" of small disparity (Marr & Poggio 1979).

A second psychophysical paradigm for studying correspondence is the so–called *wallpaper illusion* wich occurs with horizontally repetitive patterns: besides matches with the "correct" disparity, matches with an offset of one or several pattern repetitions can be perceived. Unlike the situation in pointwise matches, the correspondence problem in the wallpaper–illusion is global, i.e., one offset is selected for the entire stereogram. In Mallot et al. (1994) we present a special case of the wallpaper illusion indicating that stereo correspondences are not necessarily established in a coarse–to–fine hierarchy. Again, the difference–model of Eq. 1 accounts for our results.

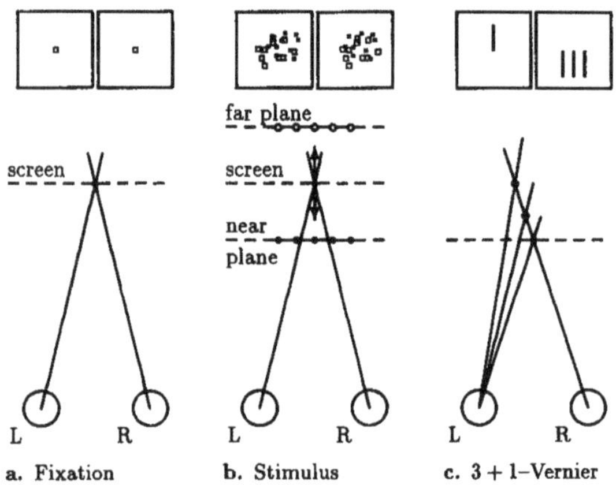

Figure 4: Procedure of Expt. 3. **a.** A fixation target is presented for 750 *ms*. **b.** A random–dot stereogram of one or two depth planes is shown for 230 *ms*. **c.** Elicited vergence is measured psychophysically with a dichoptical "3 + 1–Vernier". If fixation moved backwards, the single bar (right eye) appears to be shifted to the left relative to the three bars presented to the left eye.

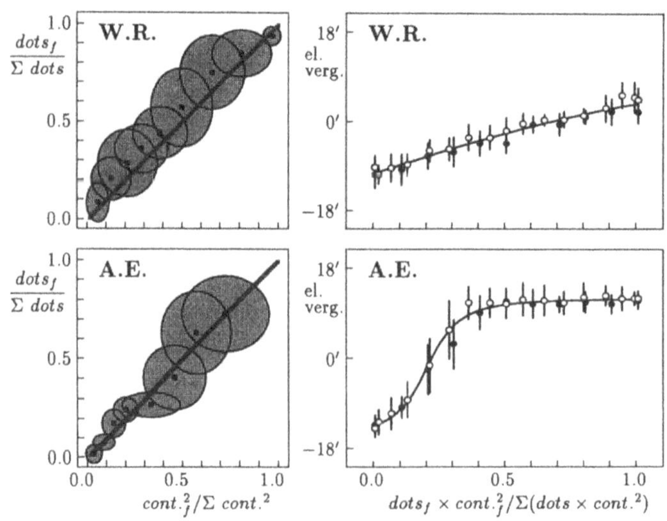

Figure 5: Results of Expt. 3. Left: Dot frequencies and contrast ratios eliciting the same vergence movements. Right: elicited vergence (Δv) as a function of relative signal energy per plane. • Variation of dot number. ∘ Variation of dot contrast. Each point is the average of 40 trials. Both curves coincide.

Quantitative perceived depth

In the experiments reported so far, we studied qualitative properties of stereoscopic depth perception, i.e. when does it occur and how do we deal with ambiguities? In this section we address the problem of quantitative perceived depth in a particularly simple case: only one depth value is assessed over the entire image. In Expt. 3 we measured the vergence eye–movements elicited by the presentation of random–dot stereograms comprising two targets at two different depth locations (Erben et al. 1993). Vergence eye–movements adjust stereoscopic vision to a 3D region of interest determined by the point of fixation and Panum's fusional area (see Collewijn & Erkelens 1990). The last experiment, Expt. 4 leads back to the initial demonstration given in Fig. 1. As already mentioned, the difference between the ortho– and the pseudoscopic view of the ellipsoid is largely a difference in perceived overall distance. This depth offset is measured in Expt. 4.

Expt. 3 Disparity–evoked vergence is directed towards average depth

Vergence eye–movements were elicited by random dot stereograms showing two transparent planes. In two series of experiments, we varied the relative number of dots per plane and the contrast of the dots in the two planes. After a fixation target, the two–plane stimulus is flashed for 230 *ms* to minimize visual feedback by the beginning vergence eye–movement. The resulting angle of vergence was measured psychophysically with a dichoptic "3+1" vernier stimulus (cf. McKee & Mitchison 1988) presented immediately after the stimulus. During the trial, subjects attended to the final presentation of the vernier stimulus (see Fig. 4).

For the brief presentation times used in our experiments, the resulting vergence is an average of the planes' depth positions. I.e., eye–movements are guided towards the empty region between the stimulus planes. If one of the planes contains more dots or dots with higher contrast, vergence approaches this plane. We propose that this depth averaging is a center of gravity process in which each plane's "pulling strength" is a function of dot number and contrast. In the left part of Fig. 5, the relative contributions of contrast and dot frequency to the assumed pulling strength are compared by plotting the values that are equally effective. The results indicate that an increase in dot number can be compensated by the same decrease in the squared contrast ratio; i.e. the combined pulling strength is a function of dot number times contrasted squared, or signal energy. On the right side elicited vergence is plotted as a function of this relative energy (correlation) per plane, $\Phi(\delta) = \int I_l(x) I_r(x+\delta)dx$, which for our stimuli amounts to Φ = (no. of dots) $\times$ (dot contrast)^2. The curves for the dependency on dot number and dot contrast coincide, indicating that signal energy is an appropriate measure of the assumed pulling strength (Erben *et al.* 1993).

To formulate the center of gravity model, we need the "pulling strength" $\Phi(\delta)$ and an "effective disparity" a_i towards which vergence is actually pulled by a stimulus presented at disparity δ_i. The effective disparities depend on the subject and can be measured independently by using just one depth plane with disparity δ_i; they are used as free parameters in the fit. We obtain the elicited vergence Δv:

$$\Delta v = \left(\frac{\sum_i a_i^p \Phi(\delta_i)}{\sum_i \Phi(\delta_i)} \right)^{\frac{1}{p}} = \left(\frac{a_{near}^p n_{near} c_{near}^2 + a_{far}^p n_{far} c_{far}^2}{n_{near} c_{near}^2 + n_{far} c_{far}^2} \right)^{\frac{1}{p}} \tag{3}$$

where p accounts for a non–linearity of the probability–summation type. The lines plotted in Fig. 5 are least–square fitts of this model. The resulting fit–parameters were $a_{near} = -10.6'$, $a_{far} = 5.6'$, $p = 1.0$ for subject W.R. and $a_{near} = -10.9'$, $a_{far} = 13.7'$, $p = 4.2$ for subject A.E. The correlation function Φ is closely related to the squared image difference as defined in Eq. 1:

$$\begin{aligned} \Psi(\delta) &= \int (I_l(x) - I_r(x+\delta))^2 dx = \int I_l^2(x)dx + \int I_r^2(x)dx - 2\int I_l(x) I_r(x+\delta)dx \\ &= \text{const} - 2\Phi(\delta); \end{aligned} \tag{4}$$

it has been proposed for image comparison in motion detection by Hassenstein & Reichardt (1956).

Vergence control is a simple type of stereo processing; it computes just one angle from the entire image. Also, stimuli that cannot even be fused are still effective to elicit vergence (Westheimer & Mitchell 1969). Our results indicate that the mechanism underlying the guidance of vergence movements is based an interocular correlation. Correlation as a mechanism for "simple" types of stereopsis has been proposed earlier for vergence control (Cormack, Stevenson & Schor 1991) and for global stereopsis (Weinshall 1991).

Expt. 4 Contrast–weighted depth averaging in intensity–based stereo

The demonstration of intensity-based stereo (Fig. 1) has shown that in the case of a pseudoscopic stereogram of an ellipsoid (i.e. a stereogram of a bowl), intensity–based stereo does not lead to a perception of a concave surface (bowl). One might conclude that intensity–based stereo cannot overcome prior assumptions such as convexity and general viewpoint. Instead, intensity-based stereo changes the perceived distance of the entire object. In the following experiment we examine this global depth effect in more detail (cf. Fig. 6).

An interactive adjustmend task was used to quantify the overall depth separation between orthoscopic and pseudoscopic stereograms. The pseudoscopic stereogram was displayed on the left half of the monitor screen with fixed disparities of ±9, ±3.6, or 0 minutes of arc in pseudorandom order. On the right half of the screen an orthoscopic version of the stereogram appeared in a randomly chosen depth position. Subjects were asked to adjust the depth position of the orthoscopic stereogram by shifting the halfimages horizontally (under computer–"mouse" control) until it appeared to be at the same depth as the pseudoscopic stereogram. The pseudoscopic image of a smoothly shaded ellipsoid is perceived as a solid albeit somewhat flatter object at an increased distance as compared to the orthoscopic presentation. The difference in perceived overall distance increases for more elongated objects, i.e. objects with a higher depth variation (Fig. 7).

If the 3D surface is reconstructed point–by–point and the depth averaging occurs after that, one would expect that the perceived offset be proportional to the displayed elongation of the ellipsoid; i.e., the expected slope in Fig. 7 would be 1.0. It, however, the image intensities are compared by the square difference mechanism, i.e., they are shifted globally such that the difference between $I_l(x)$ and $I_r(x+\delta)$ is minimized, the expected slope is 1.58 which is much closer to the values found in our experiments. As is discussed in detail by Arndt et al. (*in press*), the global image difference mechanism is equivalent to an averaging of local disparities which is weighted by local image contrast.

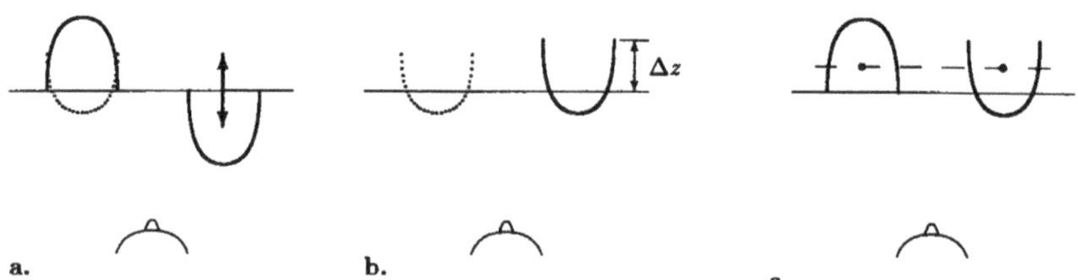

a. b. c.

Figure 6: Bird's eye view of the design of Expt. 4. **a.** Two stereograms are displayed simulating an ellipsoid (right) and a bowl. As has already been demonstrated in Fig. 1, the shaded image of a bowl will look like a less elongated ellipsoid, somewhat further away from the observer (dotted line). The subject can interactively shift the orthoscopic ellipsoid in depth. **b.** In an adjustment task, the perceived distances of both surfaces are equated. **c.** A possible interpretation is that the average depth of both surfaces (indicated by the black dot) is equated.

Discussion

Disparity map interpretation

Fig. 8 shows a useful scheme for a neural interpretation of disparity computations. While the regular arrangement of disparities in cortical space is hypothetic, disparity–tuned neurons that might built up such a map are well known (e.g., Poggio 1984). A more realistic approach using units with overlapping tuning curves has been studied by Lehky & Sejnowski (1990). In the simplest case, which actually suffices to account for our data, we might assume that the excitation of a unit at location (x, δ) depends on the product of the image intensities presented to each eye:

$$e(x,\delta) = I_l(x+\frac{\delta}{2})I_r(x-\frac{\delta}{2}) \qquad (5)$$

As was pointed out earlier (Eq. 4), this correlation scheme is computationally equivalent to a squared image difference model. The usefulness of global image differences e.g. for the perception of obstacles

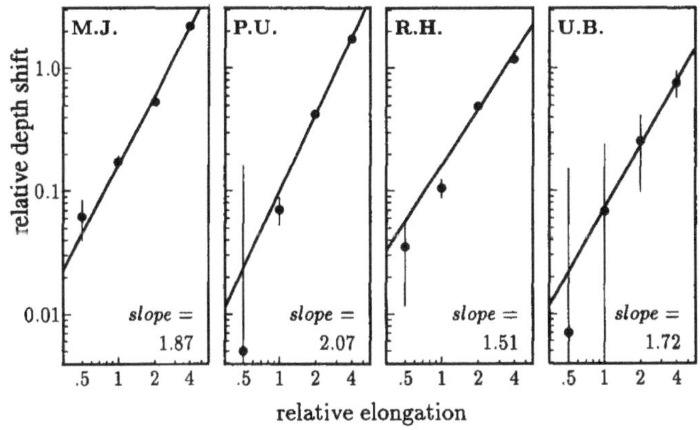

Figure 7: Relative perceived offset between the ortho– and pseudoscopic view of an ellipsoid plotted as a function of elongation (offset and elongation measured in multiples of $e_o = 5cm$). The data are averaged over the five absolute depth positions presented in Expt. 4. Error bars are standard deviations of the mean. The straight lines are regression lines minimizing the error–weighted square deviations. The slopes range from 1.51 to 2.07, i.e. way above the value 1.0 prediced from contrast–independent averaging. The mean square difference mechanism (Eq. 1 with $\alpha = 2$) predicts a straight line with a slope of 1.58.

is discussed in the accomanying paper (Mallot 1994). In contrast, most of the earlier models assume that only feature–based information is available to the binocular units (e.g., Marr & Poggio 1976).

We can now briefly review the experimental situations studied in this paper:

Intensity–based stereo: Ortho– and pseudoscopic presentation of the greywedges results in activity–peaks in the disparity–map that lie either in front or behind the zero–disparity line. To achieve this near/far distinction, only the units at the "cyclopean position" $x = 0$ in the disparity–map are actually required.

Threshold: The distinction between crossed and uncrossed disparities, i.e. between activity–peaks before and behind the zero–disparity line depends on both disparity (separation in the map) and contrast (signal strength). The quantitative result $c \propto \delta^{-1}$ allows conclusions on the detailed form of the activation function, Eq. 5.

Vergence and the double–nail illusion: The ordering constraint of small disparities might be due

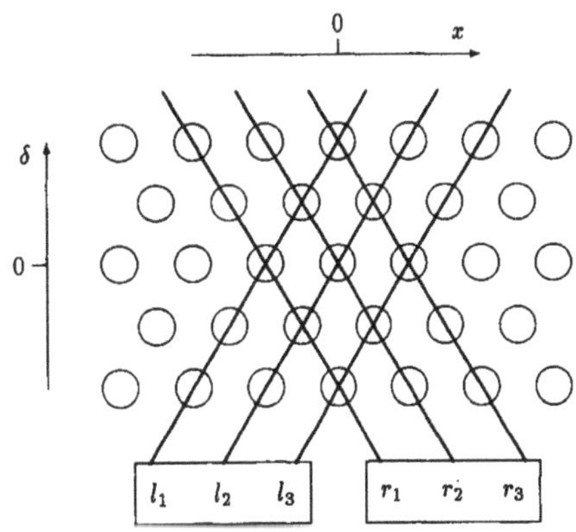

Figure 8: Disparity–map. Retinal signals from locations $l_1,..,l_3$ (left) and $r_1, ..., r_3$ (right) activate units at the locations (x, δ) satisfying $x + \delta/2 = l_i$ and $x - \delta/2 = r_j$. A survey of stereo algorithms based on the idea of a disparity map is given by Blake & Wilson (1991).

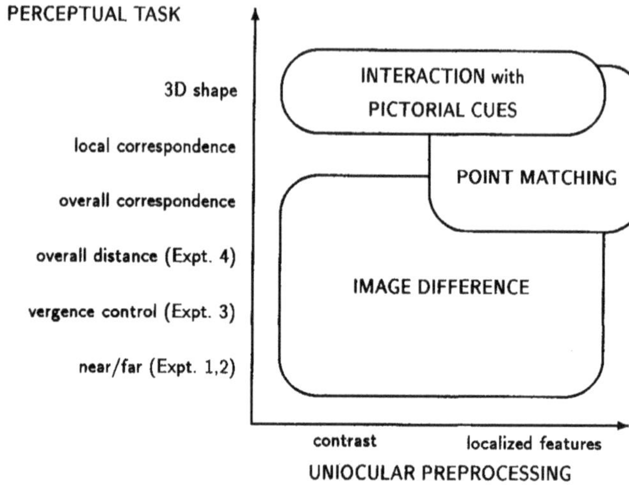

Figure 9: A classification of stereo mechanisms based on the input (uniocular preprocessing) and the perceptual tasks they subserve. For explanation see text.

to an overrepresentation of small disparities in the map. In other words, the disparity–map has a "fovea" for small disparities (see Lehky & Sejnowski 1990).

Wallpaper–illusion: If neural activity in the disparity–map is given by image correlation, i.e. basically by the inverse of the squared image difference plotted in Fig. ??c, a similarity–peak occurs at the disparities corresponding to both possible matches. One might assume that perceived disparity corresponds to the similarity (activity) peak closest to the zero–disparity line in the map. In the presence of a fovea, this might well be the strongest one, too.

Disparity–evoked vergence: The center of gravity model of evoked vergence translates into the formation of the center of gravity of the acitivity-pattern on the disparity map. The generation of motor commands by means of a center of gravity mechanism has been termed "population coding" by Georgopoulos et al. (1982). The differences between our subjects in the amount of vergence elicited by a given retinal disparity (accounted for by the "effective disparities" a_i in the fit) relate to a finding of Richards (1971): In his "pool hypothesis", he proposed that deficits in the perception of certain disparities result from a lack of the accordingly tuned neurons in the disparity map. Clearly, if the map lacks units for some disparities, signals in the concerned depth plane will not contribute to the center of gravity computation.

Average perceived distance in pseudoscopic presentation: The contrast–weighted depth averaging performed in the adjustment of the ortho– and pseudoscopic image of ellipsoids translates similarly into the alignment of the centers of gravity of the corresponding activity peaks.

In summary, all data presented in this paper have a very simple disparity–map interpretation if we assume that neural activity is a continuous function of the original image intensities. Higher mechanisms, such as the pointwise solution of the correspondence problem can be realized in disparity–maps as well (Marr & Poggio 1976). It is tempting to consider the simple center of gravity calculations discussed here as evolutionary precursors of the cooperative activation–dynamics required for these higher tasks.

Different mechanisms for stereopsis?

The data presented in this paper concern a simple stereo mechanism that performs a global computation similar to the squared image difference operation. It shares a number of properties with the mechanism termed "global stereopsis" by Julesz (1971). The main difference between intensity–based and global stereopsis seems to be that Julesz focused on feature–based stimuli.

Fig. 9 shows a highly speculative scheme for the classification of stereo mechanisms. The perceptual tasks listed on the left are increasingly more complicated in terms of the amount of information required by each of them. If more complicated preprocessing steps are included (e.g., orientation, motion, 2D shape), higher level disparities can be included along the horizontal axis. A survey of higher order disparities is given in Arndt et al. (*in press*) The mechanisms distinguished in Fig. 9 are:

1. Image–difference: Global (average) perception of sign and amount of disparity both in smoothly shaded images and arrays of localized features. Global image differences can play an important role both in animal vision (prey catching behaviour in salamanders, see Manteuffel & Roth 1993) and in mobile robots (obstacle avoidance, see Mallot 1994).

2. Point matching: This is the mechanism studied most intensively. It is relevant for the perception of isolated points in space, i.e. 3D segmentation (figure/ground separation), transparency, and high–precision manipulations such as threading a needle. Julesz (1988) proposed that a segmentation task, i.e. camouflage breaking, is the evolutionary origin of stereoscopic vision.

3. Interaction with pictorial cues: The perception of 3D shape of objects rests heavily on pictorial or monocular depth cues such as shape–from–shading (with contrast as preprocessing) or texture gradients (with feature extraction). These cues give no or little information on distance. There is only little evidence that shape–perception requires an additional module; one interesting finding in favor of this view has been presented by Blake & Bülthoff (1990), who showed that the disparity of specular highlights on curved surfaces, which differ considerably from the disparities of nearby surface markers, are still interpreted correctly. This result suggests that disparity and shading are evaluated jointly for the problem of shape perception.

Conclusion

In summary, we conclude that (i) human stereo vision does not exclusively rely on image features, (ii) it uses an ordering constraint of "small disparity", (iii) it does not always proceed coarse–to–fine and (iv) it involves a comparison stage of the correlation type. We speculate that these properties characterize an intensity–based stereo mechanism which is simple, coarse and maybe evolutionary old. It comprises only rough information as is required for vergence control and navigation tasks. Detailed depth maps at high resolution as may be useful for object recognition and manipulation tasks require a more evolved mechanism which, however, is likely to build on intensity–based precursors.

References

A. Arditi. Binocular vision. In K. R. Boff, L. Kaufmann, and J. P. Thomas, editors, *Handbook of Perception and Human Performance, Vol. 1: Sensory Processes and Perception*. John Wiley & Sons, New York, 1986.

P. A. Arndt, H. A. Mallot, and H. H. Bülthoff. Human stereovision without localized image–features. *Biological Cybernetics*, in press.

A. Blake and H. H. Bülthoff. Does the brain know the physics of specular reflection? *Nature*, 343:165 – 168, 1990.

R. Blake and H. R. Wilson. Neural models of stereoscopic vision. *Trends in Neurosciences*, 14:445 – 452, 1991.

S. Bohrer, H. H. Bülthoff, and H. A. Mallot. Motion detection by correlation and voting. In R. Eckmiller, G. Hartmann, and G. Hauske, editors, *Parallel Processing in Neural Systems and Computers*, pages 125 – 128, Amsterdam, 1990. North–Holland.

H. H. Bülthoff and H. A. Mallot. Integration of depth modules: Stereo and shading. *Journal of the Optical Society of America A*, 5:1749 – 1758, 1988.

H. Collewijn and C. J. Erkelens. Binocular eye movements and the perception of depth. In E. Kowler, editor, *Eyemovements and Their Role in Visual and Cognitive Processes*. Elsevier Science Publishers, 1990.

L. K. Cormack, S. B. Stevenson, and C. M. Schor. Interocular correlation, luminance contrast and cyclopean processing. *Vision Research*, 31:2195 – 2207, 1991.

U. R. Dhond and J. K. Aggarwal. Structure from stereo – a review. *IEEE Transactions on Systems, Man, and Cybernetics*, 19:1489 – 1510, 1989.

A. Erben, H. A. Mallot, and P. A. Arndt. How do vergence eye–movements select 3D regions of interest in complex visual scenes? In N. Elsner and M. Heisenberg, editors, *Gene – Brain – Behaviour (Proc. 21th Göttingen Neurobiol. Conf.)*, page 27, Stuttgart, 1993. G. Thieme Verlag.

A. P. Georgopoulos, J. F. Kalasak, R. Caminiti, and J. T. Massey. On the relation of the direction of twodimensional arm movements and cell discharge in primate motor cortex. *The Journal of Neuroscience*, 2:1527 – 1537, 1982.

B. Hassenstein and W. Reichardt. Reihenfolgen-Vorzeichenauswertung bei der Bewegungsperzeption des Rüsselkäfers Chlorophanus. *Zeitschrift für Naturforschung, Teil B*, 11:513 – 524, 1956.

B. Julesz. *Foundations of Cyclopean Perception.* Chicago University Press, Chicago and London, 1971.

B. Julesz. In the last minutes of evolution of life, stereoscopic depth perception captured the input layer to the visual cortex to break camouflage. *Perception*, 17:A3, 1988.

B. Julesz. Early vision and focal attention. *Reviews of Modern Physics*, 63:735 – 772, 1991.

G. E. Legge and Y. Gu. Stereopsis and contrast. *Vision Research*, 29:989 – 1004, 1989.

S. R. Lehky and T. J. Sejnowski. Neural model of stereoacuity and depth interpolation based on a distributed representation of stereo disparity. *The Journal of Neuroscience*, 10:2281 – 2299, 1990.

H. A. Mallot. View–based navigation and cognitive maps in man and machine. this volume, 1994.

H. A. Mallot and H. Bideau. Vergence eye movements influences the assignment of stereo correspondences. *Vision Research*, 30:1521 – 1523, 1990.

H. A. Mallot, H. H. Bülthoff, J. J. Little, and S. Bohrer. Inverse perspective mapping simplifies optical flow computation and obstacle detection. *Biological Cybernetics*, 64:177 – 185, 1991.

H. A. Mallot, S. Dartsch and P. A. Arndt. Is correspondence search in human stero vision a coarse–to–fine process? Technical Report 4, Max–Planck–Institut für biologische Kybernetik, Tübingen, Germany, 1994. (Available as /pub/mpi-memos/TR-4.ps.Z via anonyoums ftp from ftp@mpik-tueb.mpg.de.).

G. Manteuffel and G. Roth. A model of the saccadic sensorimotor system of salamanders. *Biological Cybernetics*, 68:431 – 440, 1993.

D. Marr and T. Poggio. Cooperative computation of stereo disparity. *Science*, 194:283 – 287, 1976.

D. Marr and T. Poggio. A computational theory of human stereo vision. *Proceedings of the Royal Society (London) B*, 204:301 – 328, 1979.

S. P. McKee and G. J. Mitchison. The role of retinal correspondance in stereoscopic matching. *Vision Research*, 28:1001 – 1012, 1988.

G. F. Poggio. Processing of stereoscopic information in primate visual cortex. In G. M. Edelman, W. E. Gall, and W. M. Cowan, editors, *Dynamic Aspects of Neocortical Function*, pages 613 – 635. John Wiley & Sons, 1984.

G. F. Poggio and T. Poggio. The analysis of stereopsis. *Annual Review of Neuroscience*, 7:379 – 412, 1984.

D. Regan, J. P. Frisby, G. F. Poggio, C. M. Schor, and C. W. Tyler. The perception of stereodepth and stereo–motion: Cortical mechanisms. In L. Spillmann and J. S. Werner, editors, *Visual Perception. The Neurophysiological Foundations.* Academic Press, San Diego etc., 1990.

W. A. Richards. Anomalous stereoscopic depth perception. *Journal of the Optical Society of America*, 61:410 – 414, 1971.

D. Weinshall. Seeing "ghost" planes in stereo vision. *Vision Research*, 31:1731 – 1748, 1991.

G. Westheimer and D. E. Mitchell. The sensory stimulus for disjunctive eye movements. *Vision Research*, 9:749 – 755, 1969.

FIGURAL COMPLETION

Walter Gerbino
Department of Psychology, University of Trieste
via dell'Università 7, 34123 Trieste (Italy)
e-mail: gerbino@univ.trieste.it

When contrasted with the fragmentary nature of the optical input, the phenomenal completeness of visual objects appears as one of the most surprising achievements of human perception. To recall a formula quite common in cognitive psychology, object perception is always a matter of "going beyond the information given".

Kanizsa [1] critically evaluated this formula, and used completion phenomena as a tool for exploring the peculiar nature of visual integrative processes involved in the unification of pieces of the sensory mosaic. Better than long verbal arguments, a couple of examples will help us understanding the problem I will deal with in my lecture.

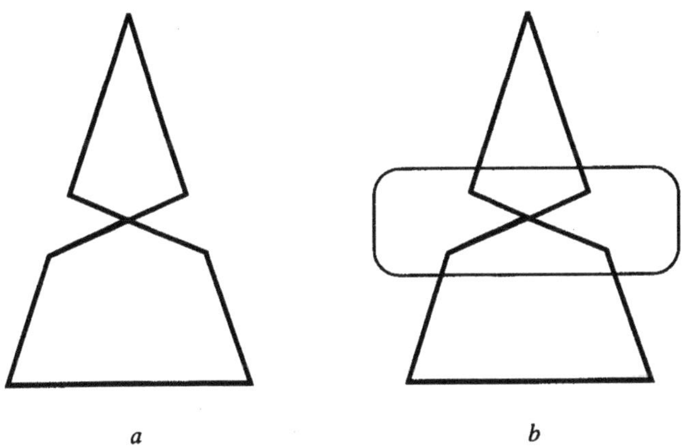

Figure 1. Look at pattern *a* and then cover its central part using an opaque occluder. Top and bottom parts should be in direct view, as illustrated in *b*. A complete triangle will appear [2].

The first is a well-known demonstration by Michotte [2, 3]. Look at Figure 1*a* and then cover its central region by your finger (or any other occluder), in such a way that only the upper and lower regions remain directly visible, as shown in Figure 1*b*. Once the proper occlusion is obtained, a complete triangle makes its appearance. Interestingly, you can repeat the same observation without affecting its salience. Remove the occluder for a while, and then cover the central crossing again. The complete triangle reappears. Knowledge of the physical state of affairs, built on memory of recent views, is irrelevant in comparison with the power of current sensory information and visual organization processes.

Michotte proposed the phrase "amodal datum" to describe this type of perceptual integration of sensory information. Such a definition captures a unique combination of features. Occluded parts are phenomenal data; i.e., they enter consciousness as properties of the external object, not as mental products. Nevertheless, occluded parts do not possess color and texture, which are the modal properties of visual objects. There are three important aspects in Michotte's demonstration.

Departure from veridicality. Observers perceive a figure different from the one they know to be on the page. Observer's surprise (which is the measure of the effectiveness of the demonstration) depends upon the discrepancy between appearance and reality.

Robustness. Repeated occlusions and disocclusions do not modify the quality and intensity of the phenomenon. Learning about physical reality might counteract the spontaneous tendency towards amodal continuation of contours; but this is not the case.

Specific shape. The literature on amodal shapes and trajectories of amodal contours is controversial. They are not easy to measure, and maybe we lack the proper concepts for evaluating how accurate is the representation of occluded parts. Amodal representation of shape is specific enough to contradict expectations based on memory of previous views, as shown by the salient aspect of the demonstration, its departure from veridicality. However, several lines of evidence suggest that amodal representation of shape is even more specific. Amodal contours are normally perceived as a continuation of modal contours (those directly in view); and amodal shapes tend to correspond to maximum order solutions, involving structural improvement.

Similar phenomena occur in the three-dimensional domain. Such observations are important, because they suggest that amodal completion is not a curiosity of pictorial perception under impoverished stimulus conditions, but a constitutive aspect of visual functioning.

Consider Figure 2, but not as a drawing. Rather, take it as a generic view of a physical object; to be precise, of the elegant little sculpture on the corner of my desk. How do these physical bodies look? Like most people, you would describe Figure 2 as a view of a complete pencil embodied into an indented block. Now suppose you are allowed to manipulate the pencil and move it away from the block. With some surprise you might discover that the pencil has an indentation exactly fitting into the intact block.

Figure 3 illustrates two solutions, both compatible with Figure 2. Figure 3*a* is consistent with the spontaneous way of perceiving pencil and block when they are in physical contact, under conditions of free visual exploration. On the other hand, Figure 3*b* shows the funny pencil and normal block I have on my desk. What happens when such a pencil contacts a common block forces us to maintain a critical attitude towards the problem of perceptual completeness.

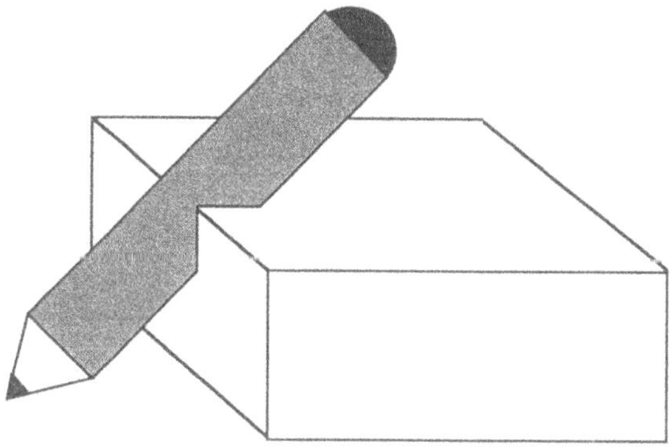

Figure 2. Amodal completion in the 3D physical world. This "piece of art" is usually perceived as a complete pencil that penetrates a block.

Please, do not dismiss the significance of the pencil-and-block demonstration by saying that complete pencils jointed with indented blocks are more likely than indented pencils jointed with complete blocks. Such a helmholtzian explanation, based on the likelihood principle, is popular; but in this particular case it is not at all clear what one should when taking into account previous experience with the world of pencils and blocks. Let us maintain an open attitude, and continue to learn from amodal completion phenomena.

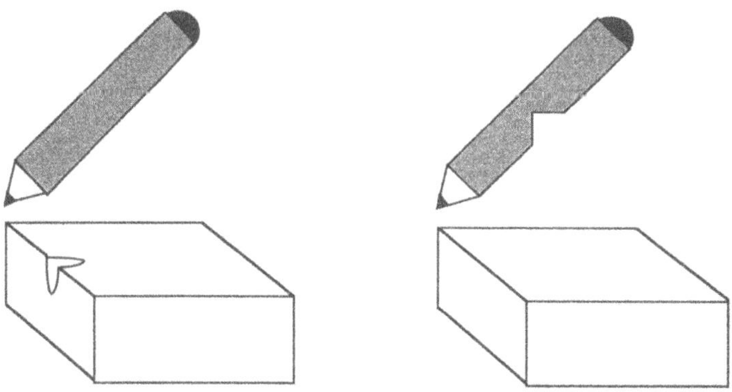

a *b*

Figure 3. In principle, the complex object shown in Figure 2 is compatible with both *a* and *b* state of affairs (among many others, of course).

Take the sculpture in Figure 2 as a reminder that tricky completions also occur in ordinary experience with three-dimensional objects, under free exploration and manipulation. Previously described features (departure from veridicality, robustness, specific shape) are present in the pencil-and-block demonstration, as well as in Michotte's graphical pattern. This is so - I suggest - because amodal completion is intrinsic to the construction of a three-dimensional model of the world out from fragmentary images. Graphical patterns that lead to the perception of overlapping layers constitute a restricted domain, quite useful for research purposes. The hypothesis that amodal completion occurs at the level of contours and surfaces, as a first transformation of the image, needs careful evaluation.

Let us develop a few theoretical considerations, now. The existence of amodal parts constitutes a problem for any sensation-based theory of perception. Properties like local color, orientation of modal contours, and texture can be explained as the result of sensory coding of the relevant proximal stimulation; but amodal parts cannot. If features that do not have a local counterpart in stimulation were phenomenally denoted as "non-objective", then visual experience would be extremely well organized. All sensory-based perceptions would be marked as objective; and all perceptual integrations would be marked as subjective. However, this is not what happens, at least in human perceptual awareness. To evaluate the function of amodal completion, consider why such an inextricable mixture of sensory-based perceptions and perceptual integrations might be desirable.

A possible answer comes from ecological ideas, which are common - I think - to both Brunswik [4] and Gibson [5, 6]. To support effective actions, perception must not take information currently available in sensory stimulation in a literal way. In some sense, its face value should be disregarded by an intelligent move; i.e., by orienting the system towards the possible origin of actual stimulation. This notion is reminiscent of the *historical* orientation of perception recently advocated by Leyton [7].

Occlusion is probably the most common feature of optic ecology, given that most physical bodies are opaque and light cannot turn around corners (at the scale of object-oriented vision, at least). In fact, it is generally true that bodies occlude background surfaces (*hetero-occlusion*) and a portion of their own surface (*self-occlusion*). Hetero-occlusion is the ecological origin of classical problems of vision, like figure/ground ambiguity. Locally, a single border cannot specify if the occluding surface is on the right or left. Motion and binocular disparity add crucial information and reduce indeterminacy usually associated with single static views [8, 9]. However, in pictorial displays, as well as in many natural circumstances, observers must rely on static monocular information alone.

This is why figural completion, which is typically studied by simple line drawings or patterns of homogeneously colored regions, is a valid tool for probing visual procedures routinely involved in ordinary vision. So, let us assume that pictorial displays constitute a reduced but valid domain for the definition of relevant aspects of perceptual completions, and look at another demonstration.

Figure 4. A pattern of slightly different rectangles. Although some regularities are apparent, no recognizable object is visible.

Look at Figures 4 and 5, that display two patterns of rectangles. Observers usually notice some regularities, but even after an explicit request to discover if there is a "hidden" object, no recognition occurs. Convince yourself and then turn the page to look at Figure 6.

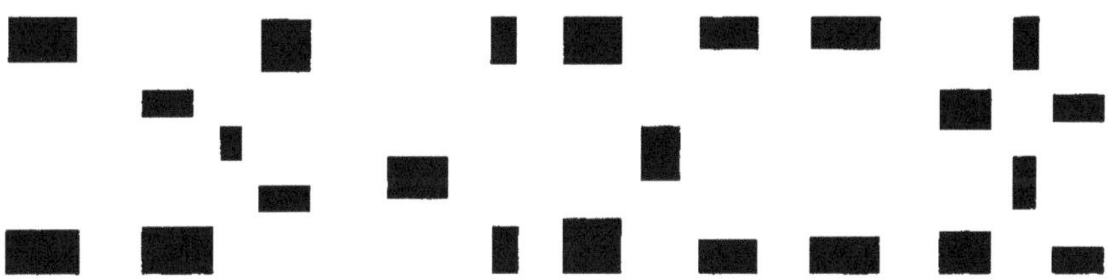

Figure 5. Another pattern of rectangles. As in Figure 4, no recognition occurs.

The word recognized in Figure 6 is very familiar, in particular to participants in this school. Therefore, it is not surprising that observers can read it. However, it is not surprising unless you compare such an achievement with the perception of patterns in Figures 4 and 5, which contain the very same black regions but provide a totally ineffective visual material for reading the familiar word.

The availability of occluders reveals the presence of a meaningful object, given the same local pieces in the optical input. Such a demonstration is a variation of well-known pictures by Kanizsa [1] and Bregman [10], and shares with them a common rhetorical structure. First, the pattern of rectangles on a homogeneous background is such that the observer perceives it in the "literal" way, to use Rock's language [11]. Then, a surprise effect is generated by adding forms that act as occluders and support figural reorganization.

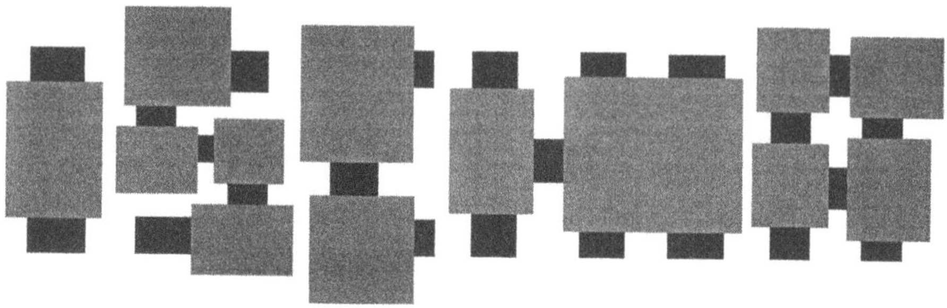

Figure 6. A word becomes visible.

The rhetorical aspect is reinforced by the expectation that the addition of occluders should *conceal* the object, and not reveal it. In fact, something deeply counterintuitive happens here. Let us analyze why.

You probably noticed the difference between Figures 4 and 5. Figure 5 contains only the pieces used in Figure 6 to build the word; whereas Figure 4 includes other irrelevant rectangles. The comparison with Figure 6 is equally effective. The impossibility of perceiving the word in Figure 4 is not due to the presence of interfering elements. In that case their concealment by grey occluders would explain the reorganization. Occluders favour the perception of the word because they set the conditions for the unification of relevant pieces into meaningful letters. These conditions can be reduced, at least in this particular case, to segmentation of T-junctions.

Consider the difference between L-junctions and T-junctions and evaluate the belongingness of contours to surfaces, the fundamental property underlying shape perception. Both segments of any L-junction in Figures 4 and 5 are perceived as borders of a black surface. So, they belong to each other and to the included shape, forming relatively isolated entities. Collinearities between segments belonging to different L-junctions (the same collinearities exploited in Figure 6) are perceptually ineffective. They can be appreciated by an act of visual attention, but such an act is not sufficient to produce a new phenomenal object. This observation plays a fundamental role in the context of experimental phenomenology, and Kanizsa [1] used it several times to stress the difference between primary and secondary processes. The operational definition of primary processes refers to perceptual objects formed out of elements which - under different but comparable conditions - remain isolated.

So, how can collinearities support unification of relevant pieces into familiar letters? In Figure 6 collinearity is involved at two levels: first, in the segmentation of T-junctions; second, in the unification of residual segments. Analyze each T-junction as the composition of three contour segments, marked according to colours of corresponding regions: B/G, W/B,

G/W. Collinearity between B/G and G/W contours supports their unification and the relative segregation from the W/B contour. According to this line of reasoning, the various W/B contours are the local residuals of the formation of grey occluders. Collinearities between some of them supports their unification and the formation of amodal shapes, eventually recognized as familiar letters.

By exploiting collinearities, the unification/segmentation process parses an image containing a collection of B/G, W/B, and G/W contours into an ordered set of unidirectional borders: B/G and G/W contours become G/- borders; W/B contours become W/- borders. As originally suggested by Rubin [12], contours tend to be perceived as borders belonging to one surface only.

Bruno & Gerbino [13] observed that the notion of image parsing provides a comprehensive framework for understanding the syntactical aspect of completion phenomena. Such an aspect has been explored systematically by Leeuwenberg and colleagues in the context of Structural Information Theory [14-16]. Recent developments of their approach are especially interesting, because of the attempt to integrate local and global determinants of completion. However, image parsing is a very formal notion, possibly inadequate to account for the production of new entities involved in visual reorganization.

Gerbino and Kanizsa [16] discussed the dichotomy between constructs and products, as metaphors for perceptual outputs. Cognitivism, as a general attitude in the study of mind, tends to obscure the intuitive difference between attentional grouping of visual objects and the emerging of new visual objects. This difference is apparent when we compare Figure 6 with Figures 4 and 5, where ISCHIA is paradoxically masked by the absence of occluders. Kanizsa and Gerbino [18] considered the production of new phenomenal data as the main functional effect of amodal completion.

The compelling and autonomous nature of productive processes involved in visual completion is revealed by some unexpected properties of figural completions. Take Figure 7, which reproduces an illusion I published several years ago [19]. Placing occluding borders on vertices of a regular polygon generates a powerful distortion, which runs against the expectation that visual organization follows a global tendency towards maximum regularity. The phenomenon has a local origin, but one might hypothesize that the structural stability of the global shape could overcome it. Instead, the violation of overall regularity just makes it more salient.

The local effect defies an easy explanation, but I think that in any case it can be taken as evidence of the dynamic nature of perceptual organization. Vision is not only a matter of putting together elements available in the proximal image; which is the aspect captured by parsing. It is something more, or possibly something different. Gestalt theorists had in mind this type of phenomena, when claimed that visual space corresponds to the distribution of organizational forces, not to the mere discovery of relationships between topographically fixed elements [20].

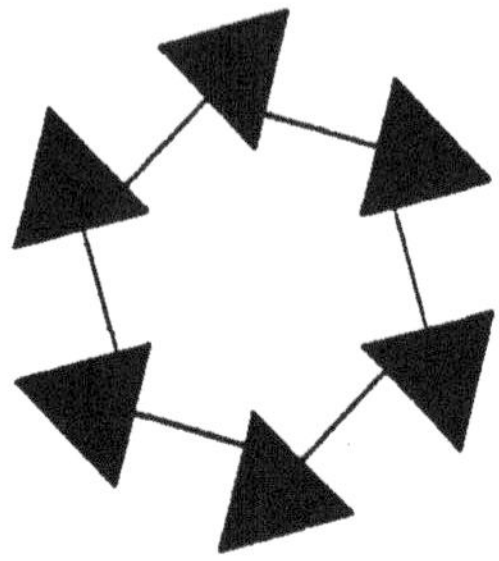

Figure 7. Distortion of a regular hexagon.

In Figure 7, amodal completion plays the following role, I think. Parsing of image contours sets the conditions for continuation of hexagon borders behind occluders. However, even a short continuation violates structural regularity, because of the coincidence between polygon vertices and occluder borders. Such a violation is perceived as a distortion of the global shape. Notions like coincidence avoidance and generic viewpoint assumption are commonly invoked to explain perceptual preferences in ambiguous stimulus situations [11, 21]. However, they do not predict the disruption of stimulus regularities. The effect in Figure 7 appears as a by-product of the tendency towards amodal continuation of interrupted borders, which normally supports the phenomenal completeness of partially occluded surfaces.

Finally, let us mention a line of research [22-26] in which amodal completion effects are evaluated by performance measures in information processing tasks. Several experiments share a common paradigm. They compare discrimination between whole and truncated targets to discrimination between whole and amodally completed targets, and measure the performance decrement in the second condition, due to whole-complete similarity.

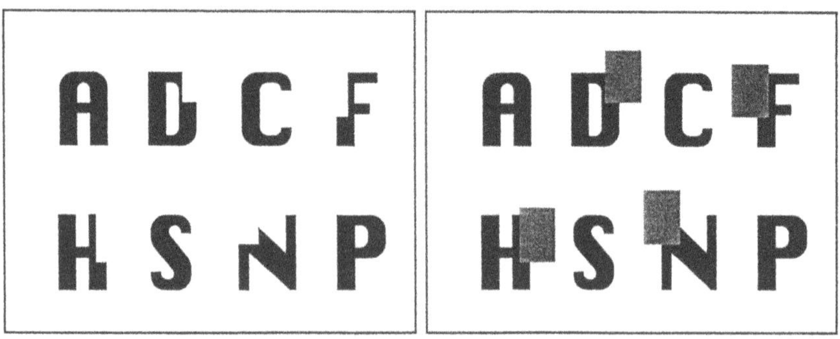

Figure 8. The matrix on the left contains whole and truncated letters. The matrix on the right contains whole and amodally completed letters.

This was also the basic idea in an old experiment of mine [27], in which subjects had to report as many letters they could after a 120 ms exposure of a matrix containing 4 whole and 4 incomplete letters. In condition 1, incomplete letters were truncated; in condition 2, they were amodally completed behind occluders, as shown in Figure 8. After the short exposure, a tone instructed the participant to report either any letter (*full report*) or a relevant subset of letters (truncated in condition 1 and completed in condition 2, *partial report*). Figure 9 describes data, scored according to the standard procedure of taking into account subset numerosity to evaluate the number of available letters.

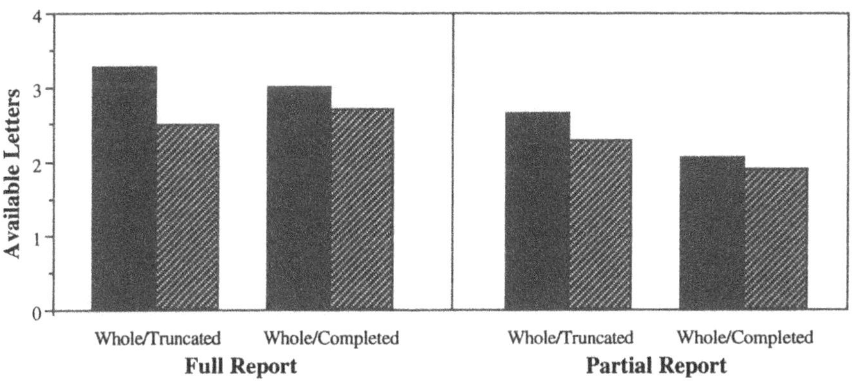

Figure 9. Average available letters under full report or partial report of the two types of matrices illustrated in Figure 8.

Two effects emerged, showing that amodally completed letters were functionally equivalent to whole letters (despite the missing parts) and different from truncated letters (despite topographical identity). On average, full report performance was identical in the two conditions. However, whole letters were reported more frequently then truncated letters; whereas this difference disappeared in condition 2, in which the matrix contained whole and amodally completed letters. This is a simple demonstration that recognition processes operate equally well on whole and completed letters, but not on truncated letters. Partial report revealed a beautiful effect. Overall performance dropped in condition 2, when the task required to discriminate between letters that are all phenomenally complete. Remember that the stimulus contained the same portions of letters in the two conditions. What made the difference was amodal completion behind occluders.

Acknowledgements. Part of this work has been supported by CNR grants CT 93.0342411 and CT 94.0065211. But I mostly indebted to Cloe Taddei Ferretti, for her constant encouragement.

References

[1] G. Kanizsa (1979) *Organization in Vision*, New York, Praeger.

[2] A. Michotte (1962) *Causalité, Permanence et Réalité Phénoménales*, Louvain, Publicationes Universitaires.

[3] A. Michotte & L. Burke (1952) Une nouvelle énigme de la psychologie de la perception: Le "donné amodal" dans l'experience sensorielle. In *Proceedings and Papers of the 13th International Congress of Psychology*, Stockholm, Bröderna Lagerström, Boktryckare.

[4] E. Brunswik (1956) *Perception and the Representative Design of Psychological Experiments*, Berkeley CA, University of California Press.

[5] J. J. Gibson (1966) *The Senses Considered as Perceptual Systems*, Boston, Houghton Mifflin Co.

[6] J. J. Gibson (1979) *The Ecological Approach to Visual Perception*, Boston, Houghton Mifflin Co.

[7] M. Leyton (1993) *Symmetry Causality Mind*, Cambridge MA, MIT Press.

[8] S. Shimojo, G. H. Silverman & K. Nakayama (1989) Occlusion and the solution to the aperture problem for motion, *Vis. Res.*, *29*, 619-626.

[9] K. Nakayama & S. Shimojo (1990) Da Vinci stereopsis: Depth and subjective occluding contours from unpaired image points, *Vis. Res.*, *30*, 1811-1825.

[10] A. S. Bregman (1982) Asking the "what for" question in auditory perception. In *Perceptual Organization*, M. Kubovy & J. R. Pomerantz (eds.), Hillsdale NJ, Erlbaum, 99-118.

[11] I. Rock (1983) *The Logic of Perception*, Cambridge MA, MIT Press, Bradford Books.

[12] E. Rubin (1921) *Synsoplevede Figurer*, Copenhagen, Gyldendalske.

[13] N. Bruno & W. Gerbino (1992) L'occlusione dinamica: implicazioni per i modelli di visione artificiale. In *I Sensi dell'Automa*, G. Adorni, W. Gerbino, & V. Roberto (eds.), Trieste, Lint, 3-20.

[14] E. Leeuwenberg & F. Boselie (1988) Against the likelihood principle in visual form perception, *Psych. Rev.*, *95*, 485-491.

[15] F. Boselie (1994) Local and global factors in visual occlusion, *Perc.*, *23*, 517-528.

[16] R. van Lier, P. van der Helm, & E. Leeuwenberg (1994) Integrating global and local aspects of visual occlusion, *Perc.*, *23*, 883-903.

[17] W. Gerbino & G. Kanizsa (1987) Can we see constructs? In *The Perception of Illusory Contours*, S. Petry & G. E. Meyer (eds.), New York, Springer-Verlag, 246-252.

[18] G. Kanizsa and W. Gerbino (1982). Amodal completion: Seeing or thinking? In *Organization and Representation in Perception*, J. Beck (ed.), Hillsdale NJ, Erlbaum, 167-190.

[19] W. Gerbino (1978) Some observations on the formation of angles in amodal completion, *Italian Journal of Psychology*, *5*, 85-100.

[20] K. Koffka (1935) *Principles of Gestalt Psychology*, New York, Harcourt Brace.

[21] D. G. Lowe (1990) Visual recognition as probabilistic inference from spatial relations. In *AI and the Eye*, A. Blake & T. Troscianko (eds.), New York, Wiley, 261-279.

[22] W. Gerbino & D. Salmaso (1987) The effect of amodal completion on visual matching, *Acta Psych.*, *65*, 25-46.

[23] N. Bruno & W. Gerbino (1987) Amodal completion and illusory figures: An information-processing analysis, In *The Perception of Illusory Contours*, S. Petry & G. E. Meyer (eds.), New York, Springer-Verlag, 220-223.

[24] W. Gerbino (1989) Form categorization and amodal completion: A reply to Wagemans and d'Ydewalle, *Acta Psychol.*, *72*, 295-300.

[25] A. B. Sekuler & S. E. Palmer (1992) Perception of partly occluded objects: A microgenetic analysis, *J. of Exp. Psychol.: Gen.*, *121*, 95-111.

[26] J. T. Enns &R. A. Rensink (1994) An object completion process in early vision. In *Visual Search III: Proceedings of the Third International Conference on Visual Search*, London, Taylor and Francis.

[27] W. Gerbino (1981) Il ruolo della completezza fenomenica nel riconoscimento tachistoscopico, *Giorn. It. di Psicol.*, *8*, 437-452.

TIME AS INFORMATION IN VISUAL PERCEPTION

L.F. GIULIO

Dipartimento di Morfofisiologia Veterinaria, Università di Torino, Viale Mattioli 25, 10125, Torino, Italy

"Vision", says David Marr [15 p. 3], "is therefore, first and foremost, an information-processing task ...". Nevertheless, he adds, "we cannot think of it just as a process". Because "the study of vision must therefore include not only the study of how to extract from images the various aspects of the world that are useful to us, but also an inquiry into the nature of the internal representations by which we capture this information and thus make it available as a basis for decisions about our thoughts and actions". "This duality", he goes on, "- the representation and the processing of information - lies at the heart of most information-processing tasks and will profoundly shape our investigation of the particular problems posed by vision".

The visible world is perceived as a unit, a "coherent percept" as Zeki and Shipp [25] call it: a whole gamut of properties of "objects", or the information from them that reaches our organs of sight and relates to colour, shape, distance, and so on, "are processed in separate, anatomically distinct regions in the visual cortex, each executing its functions with considerable autonomy" [25]. Even so, this functional specialization of the cortex, which is split into "several serial patways running in parallel, each of which ... functionally specialized", must converge to reach the "coherent percept" through integration between these physically segregated patways. It is suggested that this functional integration should be divided into two categories, "topical and confluent convergence", relating to integration "across space" and integration "between different attributes" [25]. These convergent associations between the specialized patways occur at different levels, resulting in a "multistage integration in the visual system" [25]. Provision is made in this multistage integration for temporal information as an essential feature of the process.

According to the "equivalence postulate" put forward by the psychologist Michon [16] ("equivalence of temporal and non-temporal information", "time is one of experiential dimensions of information, and should be considered equivalent to other, non-temporal, aspects of this information, such as intensity, size, etc."), time has a double role. In addition to its conventional role (as the variable "t" in dynamic equations) it is also time-as-information. In Michon's words [17]: "In my view time-as-information is a stimulus for the human organism that is essentially similar to such attributes as size, loudness or colour. What it implies is the time is explicitly represented in the mind and that investigations of psychological time should aim at establishing the properties of time as a cognitive representation ...". Yet whereas it is relatively easy to reach a common understanding as to the meaning embraced by the term "shape information" [15 p. 37], this appears to be more difficult when it comes to "time-as-information". "Almost all evidence presently available favours the view that temporal information processing is a complex cognitive activity" [17]. As an aid towards the better understanding of this "complex activity", we shall describe some experiments which show that time, or to be more precise, temporal information, is an important signal in the attainment of coherent percept in the visual field.

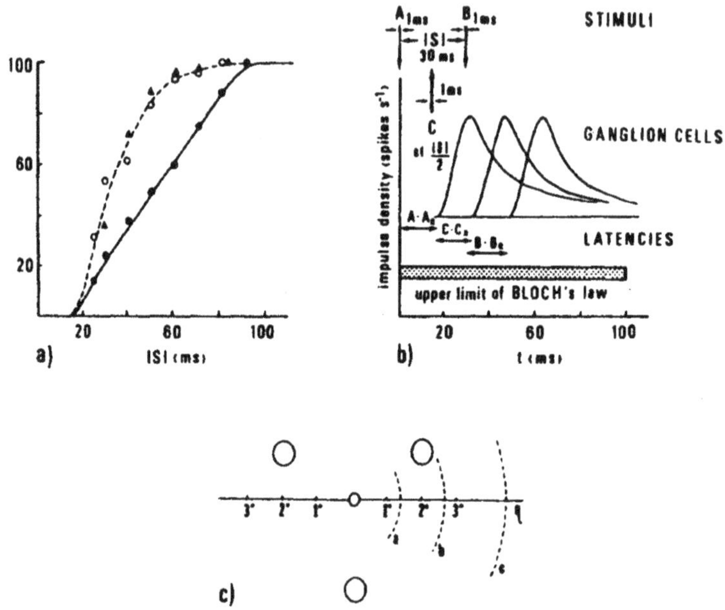

Figure 1 - **a)** Percentage of correct discrimination of temporal order, plotted vs ISI (ms) for the subject E.C. Condition α (∘): discrimination between two spots of light, delivered in succession. Condition β(•): discrimination between two spots, when a third spot, spatially separated is time interpolated at Δ t/2 (*i.e.* ISI/2). Condition γ (Δ): discrimination between two spots, when a third spot, spatially separated is present either at t = 0 or at t = Δt (= ISI), *i.e.* simultaneously with the first or the second stimulus (stimuli duration = 1 ms). **b)** Hypothetical course of the excitation pattern transmitted along the visual pathways in the experimental condition β discussed above: the graphs of time-density of impulse (spikes s^{-1}) are largely superimposed each other in time. **c)** Schematic diagram of the retinal presentation of the stimuli. F.p. = fixation point (gazed binocularly). O: projections of the spots onto retinal areas with about the same eccentricity (η). [From Ref. 6].

"Measurements of subjective responses to stimulation of localized retinal regions by double pulses of light have yielded information on both the temporal resolution and the integration characteristics of the human visual system" [18]. From a strictly neurophysiological standpoint, in fact, "the simplest stimulus which would cause a group of impulse intervals in one fiber to be followed by a similar group in another fiber would consist of a light flashed first on one region of the retina, than on another region ..." [1]. Yet the "double pulse presentation", a method used by very many reasearchers, is undoubtely an over simplification of the ordinary conditions under which visual perception of the outside world takes place. It is, in fact, enough to interpolate between the pair of optical stimuli, whose threshold one wishes to establish for the determination of simultaneity and temporal order as phenomena, a third, identical signal projected onto the retina at time Δ t/2 between the other two stimuli, and on the third vertex of an equilateral triangle not occupied by them (Figure 1c).

If this is done, the performance of temporal discrimination (between the pair of optical stimuli) becomes poorer, and the amplitude of the "temporal window" for apparent simultaneity increases (Figure 1a). Insertion of this third temporally and spatially non-coincident stimulus results in a pronounced deterioration in temporal resolution shown by a 60% increase in the threshold. It can, indeed, be regarded as a visual "noise" that decreases the channel capacity [5]. The trend of the phenomena observed can be summarized in the conclusion that "the window of perceptual simultaneity ... depends, *inter alia*, on the numerosity of the temporal signals arriving during the physical unit of time at the perceiving subject, the numerosity of purely spatial signals being uninportant" [6].

As Hirsh & Sherrick [12] have shown, the perception of simultaneity and that of successiveness (temporal order) of phenomena (in relation to the temporal interval between the stimuli) fails within a period in which the subject, altough perceiving the non-simultaneity of the stimuli, is unable correctly to evaluate their temporal successiveness (time order).

Therefore the perception, in relation to Δt between the stimuli, is articulated in the following way:

1) Simultaneity
2) Non-simultaneity, but indefinable successiveness (indefinite order)
3) Successiveness (time order).

It is possible that the perception of successiveness (time order) is more complex than the perception of simultaneity; if this hypothesis is correct, the use of more complex and more numerous stimuli, thus accentuating the incidence of the period of "undefinable successiveness", should enable us to produce a more precise and accurate study. This is what I attempted to do by using graphic marks which, when correctly projected simultaneously into the visual field, a letter with a more complex structure was formed [7] (Figure 2).

One possible explanation of the fact that segregation takes place as regards the graphic marks situated in the left visual hemifield could lead back to the order of processing in visual perception. In fact, it is well known that "letters in the left visual field are more accurately reported than letters in the right visual field, while left and right field letters have the same retinal acuity distribution and equivalent interstimulus spacings" [24]. The perception facilitated for the graphic marks projected into the left visual hemifield (in our case, the intervention of segregation which perceptively isolates the mark), may be similar to the demonstration of a left-to-right order of scanning visual inputs (of the sort proposed by Heron [cf. 21]).

Temporal information or, as Michon puts it, "what constitutes the functional stimulus or the temporal attribute of an event or an event sequence" [17], influences the perception of a particular perspective configuration of a simplified version of Necker's ambigous cube in the form of two partly superposed squares with mutually parallel sides. When this figure is transilluminated at a <20 ms interval (between the first and the second square), *i.e.* less than the corresponding temporal discrimination threshold of 29 ± 2 ms), the apparent perception of depth is such that the square presented first often seems to be further away.

This tends to give rise to the equivalence:

that which come first is distant
that which comes afterwards is close

in other words,

BEFORE = FAR
AFTER = NEAR

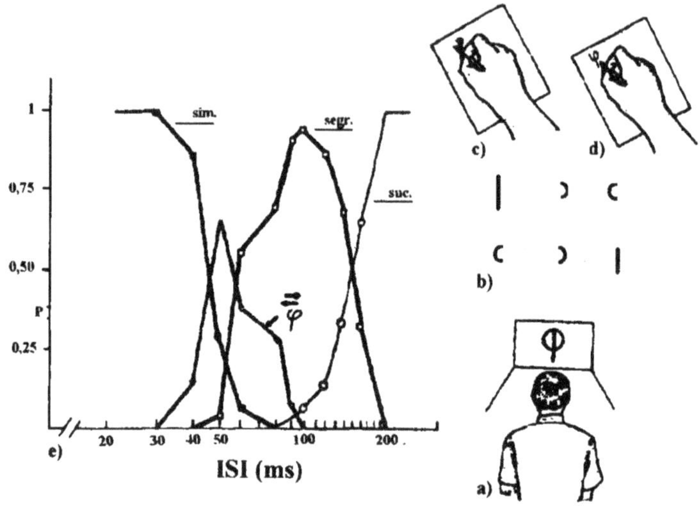

Figure 2 - The negative plates of three graphic marks (b), conveniently juxtaposed by means of a beam splitting cube, were transilluminated for a period of 1 ms ± 1.5% by means of three LEDs. The observer (a) looked binocularly at the fixation point (which coincided with the "I" graphic mark: the retinal image of the target corresponds to the rod-free fovea). The signals to the LEDs were fed from a microprocessor which delivered, for each ISI value (or Δt between the first and the third stimulus; the second one falls always at Δt/2) triplets of pulses with two temporal sequencies interspersed at random (stimuli 1, 2, 3 or 3, 2, 1 or 1+2+3 simultaneously). For ISI values <42 ms (P = 0.75 in e) the subject perceived three marks that appear juxtaposed as forming the Greek letter φ (sim = simultaneity); when the ISI value was >170 ms, the subject perceived the series of graphic marks in the order imposed by the microprocessor (suc = successiveness). In between simultaneity and successiveness, for a period between 30 and 100 ms (max 50 ms) the subject perceived the Greek letter φ as if it were being written from right to left or left to right, depending on the temporal order in which the relative graphic marks were projected (Wertheimer's φ-Phänomen). Successively (40-200 ms, mx 100 ms) the individual graphic marks appear to be partially segregated (or, which is the same, partially integrated). In other terms, the stimulus projected in correspondence with the left visual hemifield is preferentially segregated. Thus the perception becomes P, c or c, P. [From Ref. 7].

In the "distortion or displacement illusion" (to use Gregory's expression [11]), first described by the physicist Johann Christian Poggendorff in 1860 [see 2], the projection of two components, such as horizontal rectangles and obliquely orientated segments, at different times, results as following (Figure 3).

The illusion is retained even if there is a longer interval (<300 ms), provided the segments are projected first (Figure 3, a and b), whereas it becomes weaker or disappears if the rectangles come first and the time interval (between the two projections) is less than 50-100 ms (Figure 3, a' and b'). In the first situation, the phenomenal simultaneity of the components actually projected in succession appears to last very much longer than in the second case.

The explanation may be that oblique orientation may enjoy greater persistence "to compensate inferiority in oblique directions", as Ueno says [23]. Moreover, using as visual stimuli the component sets of two numbers, "1" and "7" (formed by oblique, vertical and horizontal segments), and employing variable Δt passing between the component sets, we recorded the respective visual evoked potentials (VEPs). We noticed that CII and CIII latency, corresponding to the stimulation sequence in which the oblique stimulus appeared first, showed bigger values than the sequence in which the vertical or the horizontal stimulus came before the oblique one [9].

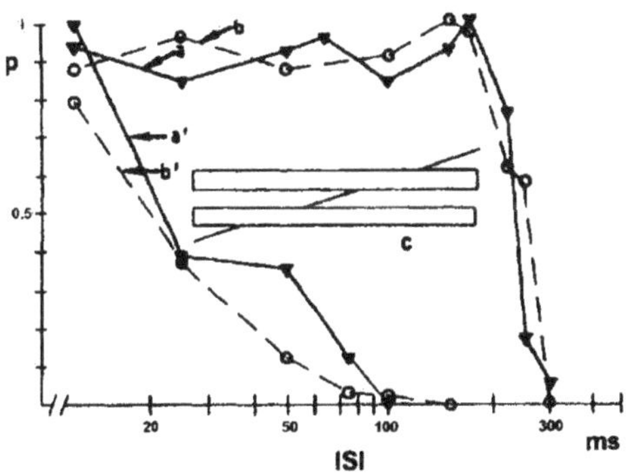

Figure 3 - Explanation in the text. ISI: interstimulus interval. c) "Poggendorffische Täuschung" [22 p. 360]. [From ref. 10].

In his "Kritik der reinen Vernunft", Kant states [13]: "times is not something that exists in itself or as an intrinsic element of things ... time is nothing other than the form of the inner sense ... it belongs to neither form, nor position, etc., but determines the relation of representations in our inner state". This was written about a hundred years after Newton"s renowned scholium in the "definitions" of his "Principia" [19]: "Absolute, true and mathematical time of itself, and from its own nature, flows equably without relation to anything external...", where time, as J.T. Fraser points out, "was independent of man". Time thus became "a mode of human understanding"; in other words, "part of our mental apparatus" [4 pp. 42-43].

"Kantian time", to use North's expression, "is the form of the intuitions of the self and of our internal state, rather than of outward phenomena. It does not subsist of itself or inhere in things as something objective. It is, in fact, a presupposition made prior to our appreciation of the coexistence and succession of things in time, and may be thought of independently of phenomena, whereas the converse is not the case" [20].

Another two centuries have passed and this way of looking at time has undergone a substantial modification. Time is no longer treated as an "a prior" necessity. According to Denbigh [3 p. 11], it is a theoretical consideration that leads to the so-called relational view of time. To be more precise, "it is the view that time is a relationship between events, and that without events there could be no concepts of time".

It is clear from the examples put forward that time (or temporal information) is a very important attribute through which to reach a coherent percept from the array of signals received by our sense organs.

As Michon states the matter, "an important question, therefore, is what constitutes the functional stimulus or the temporal attribute of an event or an event sequence" [17]. The answer he offers is that "only the order of stimuli would seem to qualify as such". As we have seen, the order in which the components of a percept are presented may have a decisive effect on the final result. Yet it is not only the order that is important. The time interval between events is also of significance. Even if "unlike for colour or loudness we seem to have no regular sense for time or duration" [17], we are none the less unable to arrange a linear series of events in order to place them in before-after relationship.

In conclusion, time appears to be an attribute of an event or an event sequence which is often essential for the attainment of a coherent percept, even if this percept is not always a faithful image of the array of stimuli received by our senses. It is not always true, as Goethe said [see 8, 14], "denn was innen, das ist aussen".

References

[1] Barlow, H.B. (1961) The coding of sensory messages, in *Current problems of animal behaviour*, W.H. Thorpe & O.L. Zangwill, eds, Cambridge: Cambridge University Press, pp. 331-360.

[2] Burmeister, E. (1896) Beitrag zur experimentellen Bestimmung geometrisch-optischen, *Z. Psychol.* 12: 335.

[3] Denbigh, K.G. (1981) *Three concepts of time*, Berlin: Springer.

[4] Fraser, J.T. (1987) *Time, the familiar stranger*, Amherst: The Massachussetts University Press.

[5] Giulio, L.F. (1983) L"ipotesi del "physiological time quantum", *Atti Accademia delle Scienze di Torino* 117: 369-376.

[6] Giulio, L.F. (1985) The hypothesis of psychophysiological time quantum, *Teorie & Modelli*, suppl. I: 113-132.

[7] Giulio, L.F. (1987) From simultaneity to successiveness: psychophysiological characteristics of the transition period, *Atti Accademia delle Scienze di Torino* 121: 104-114.

[8] Giulio, L.F. (1995) La microstruttura del presente, in: *Il Tempo nella scienza e nella filosofia*, E. Agazzi, ed., Napoli: Guida, pp. 147-166.

[9] Giulio, L.F., L. Bergamasco & S. Vighetti (1991) The oblique effect in numeral recognition: visual evoked potentials and brain mapping, (1991), *Abstracts III I.B.R.O. World Congr. of Neuroscience*, August 4-9, Montréal, CAN, 177 P 26, 13.

[10] Giulio, L.F., L. Bergamasco & M. Borgarelli (1990) Temporal information in the Poggendorf visual illusion, *Atti Accademia delle Scienze di Torino*, I, 124: 146-152.

[11] Gregory, R.L. (1974) *Concepts and Mechanisms of Perception*, London: Duckworth.

[12] Hirsh, I.J. & C.E. Sherrick jr. (1961) Perceived order in different sense modalities, *J. exp. Psychol.* 62: 423-432.

[13] Kant, I. (1968) *The Critique of Pure Reason*, transl. N.K. Smith, New York: St. Martin's Press.

[14] Köhler, W. (1920) Die physischen Gestalten in Ruhe und im stationären Zustand, Braunschweig.

[15] Marr, D. (1982) *Vision: A computational investigation into the human represensation and processing of visual information*, San Francisco: Freeman.

[16] Michon, J.A. (1972) Processing of temporal information and the cognitive theory of time experience, in: *The study of time*, J.T. Fraser, F.C. Haber & G.H. Müller, eds, Berlin: Springer, pp. 242-258.

[17] Michon, J.A. (1985) Approaches to human time experience, *Teorie & Modelli*, II, suppl. I: 29-47.

[18] Musselwhite, M.J. & D.A. Jeffreys (1983) Visual evoked potentials to doutile-pulse pattern presentation, *Vision Res.* 23: 135-143.

[19] Newton, I. (1960) *Mathematical Principles of Natural Philosophy and his System of the World*, transl. A. Motte, Berkeley: University of California Press.

[20] North, J.D. (1972) The time coordinate in Einstein"s restricted theory of relativity, in: *The study of time*, J.T. Fraser, F.C. Huber & G.H. Müller, eds, Berlin: Springer, pp. 12-32.

[21] Sekuler, R., P. Tynan & E. Levinson (1973) Visual temporal order: a new illusion, *Science* 180: 210-212.

[22] Trendelenburg, W. (1961) *Der Gesichtsinn, Grundzüge der physiologischen Optik*, Berlin: Springer.

[23] Ueno, T. (1983) Visible persistence: effects of luminance, spatial frequency and orientation, *Vision Res.* 23: 1687-1692.

[24] White, M.J. (1976) Order of processing in visual perception, *Rev. Canad. Psychol.* 30: 140-156.

[25] Zeki, S., & A. Shipp (1988) The functional logic of cortical connections, *Nature* 335: 311-317.

VISUAL ATTENTION: NEURAL AND COGNITIVE BASES

C.A. Marzi, M. Prior*, and M.C.Martini

*Dipartimento di Scienze Neurologiche e della Visione, Sez. di Fisiologia, Lab. di Psicologia, Università di Verona, Strada Le Grazie 8, 37134 Verona, and *Dipartimento di Psicologia Generale, Università di Padova, Padova, Italy*

The term attention is used in everyday life and therefore its meaning seems to be self-explanatory; it is instead rather elusive.

It is interesting from an historical point of view that the "attention" given to research on attention in the past varied according to the various school of thought in experimental psychology.

We owe to William James the great American psychologist, who was an introspectionist and therefore gave outmost importance to an analysis of the different subjective states of mind, what ~s probably still the best definition of attention.

In his two-volume treatise published in 1890 and entitled "The Principles of Psychology" [8] he defined attention in this way: "..It is the taking possession by the mind, in clear and vivid form, of one out what seem several simultaneously possible objects or trains of thought". This definition contains two important aspects of attention: selection of sensory events and selection of cognitive processing modes. In contrast to Introspectionists, Gestalt psychologists did not bother with attention because they believed that the properties of the stimuli were sufficient for perception without a need for an observer-centered specialized mechanism to make them more attractive than others. A similar dismissal of attention was typical of the behaviouristic approach albeit for entirely different reasons. Strict behaviourists did not consider attention as a respectable object of study since it is not observable. It was what we would now call a cognitive psychologist, Donald Broadbent, who in the late fifties in England resuscitated the interest of experimental psychologists for attention with his influential book "Perception and communication" [1]. Since then experimental work on attention has continued at a fast pace although many issues are still open.

Explicit and Implicit Attention

A concept that has attained in recent years great theoretical importance especially in the visual modality is the distinction between explicit and implicit attention. It is common knowledge that there are overt forms of attention that can be studied directly, such as the way we move our head and eyes to examine one object or another. It has also become increasingly clear that overt shifts of visual attention to specified spatial locations are preceded and guided by covert attentional mechanisms.
Therefore, most models of human attention include now implicit or covert as well as explicit or overt forms of attentional shifts. In this brief review we shall deal mainly with such covert attentional mechanisms and describe the main experimental paradigms used to study covert visual attention. Finally, after briefly reviewing the main theories trying to explain attentional shifts, their possible neural substrates will be outlined.
In the visual modality, covert phenomena have been particularly well documented for spatial attention although they represent a

fundamental feature of selective attention in general. It is well known from the pioneering work of Posner [12,13], that it is possible to shift visual attention to an extra-foveal location in the visual field without any overt movement. The way attention is manipulated to focus on one particular spatial location depends on the type of pre-cueing paradigm used. A first paradigm is represented by **central cues**: Typically, in a simple reaction time (RT) paradigm, a symbolic cue, such as an arrow pointing to either the right or the left hemifield, is presented immediately above the fixation point 100-500 msec prior to the presentation of a peripheral visual stimulus. On valid trials the arrow points to the hemifield of stimulus presentation whereas on invalid trials it points to the opposite field. Subjects are informed that with a probability of 80% the arrow will signal the correct hemifield of stimulus occurrence and therefore they are likely to attend covertly to the cued side. That the cue is instrumental in shifting attention is witnessed by the very fact that valid trials yield faster RTs than invalid and neutral trials, the latter condition occurring when the arrow is non-informative. In fact, in our experiments (see below) we found that central cues are effective even when they are non-informative; i.e., when there is only a 50% probability that the direction of the arrow indicates the correct hemifield of stimulus presentation.

As is well known, Posner proposed a so-called "spot-light model" to explain the effect of central cues [12,13]. According to such model the central cue induces an effortful, serial and time consuming shift of covert visual attention from the fixation point to the site of likely stimulus occurrence, see below Analog movement Theory. The successive step is an attentional engagement which allows an effective stimulus detection. Following stimulus detection attention must be disengaged from that stimulus location and moved to other sites. Importantly, Posner and his collegues have been able to show that each of these processing stages are subserved by specific neural centers. Other investigators, however, believe that focusing of spatial attention is not movement-like but represents a sort of "all-or-nothing" increase in detection speed in one hemifield at a time, see below Discrete movement theory and Hemifield activation theory.

A second paradigm is represented by the use of **peripheral cues**. As is the case for central cues, peripheral cues can be informative or non-informative. At variance with central pre-cueing, whether a peripheral cues informs on the likely site of stimulus presentation or not has substantially different consequences on the direction of subsequent covert attentional shifts.

Informative peripheral cues improve detection speed at the cued location with a mechanism that for cue-stimulus intervals of less than 150 msec is automatic or involuntary while for longer interval entails an effortful summoning of attention not unlike that of central cues. On the contrary, non-informative peripheral cues retard RTs to successive stimuli presented to the same location or hemifield. Such ipsilateral inhibition, which is known as inhibition of return, is accompanied by a facilitation of stimuli appearing in the contralateral visual hemifield. The interpretation of this remarkably consistent phenomenon is still under dispute. Posner and Cohen [13] stress the importance of such a mechanism in reducing the probability of summoning attention to an already stimulated spot while facilitating responses to new locations. Tassinari et al. [18,19] suggest that ipsilateral inhibition represents a consequence of the suppression of natural motor responses to the cue and in particular of saccadic movements for stimulus foveation.

Such inhibition would include also a slowing down of the manual response used in a RT paradigm which represents one of the various possible motor responses to the cue to be inhibited.

The third and simplest cueing paradigm is **direct voluntary allocation of visual attention** [18] which involves the measuring of simple RT to very brief light flashes appearing at one of several visual field locations. Following direct verbal instructions, subjects are tested under three attentional conditions. In one they are supposed to attend to the location where the stimulus will be presented; in another condition, they are to attend to a different location and in a third condition, they attend to all possible locations e.g., two extrafoveal sites on the horizontal meridian on each side of the fixation point. The latter is a neutral condition which allows to calculate attentional benefits and costs, i.e, decreases in RT at attended locations and increases in RT at unattended locations, respectively, both comparisons being referred to the neutral condition. With the above paradigm, it was found in our laboratory [18] that benefits are strictly limited to the attended position while costs are apparent in the entire contralateral visual hemifield and at the central location (corresponding to the fixation point) but not at the other location in the ipsilateral hemifield. This suggests a special role for the vertical meridian of the visual field in delimiting the area of costs when implicit attention is allocated to the left or the right hemifield, and this is in keeping with the so-called Hemifield-Activation Theory, see below. It is worth stressing that we believe that this paradigm has some advantages over the central cueing paradigm of Posner in that it does not require processing of the meaning of the cue and therefore it is less demanding from a cognitive point of view, see below in the description of our recent experiments.

A general conclusion that can be drawn from the ways in which the different cues induce covert attentional effects is that there are two modes of control. The first one is exogenous, of a reflex nature, and automatic; the second one is endogenous, of a voluntary nature, and controlled [9,12].

Models of Covert Attentional Shifts

As we have already mentioned above, at least three main of theories have been proposed to account for attentional shifts within the visual field. The Analog movement, the Discrete movement and the Hemifield activation theory.

Analog movement theory: According to this account, covert attention moves across the visual field in an analog manner at a fixed velocity [17,20] that can be estimated to be 8 msec per degree of visual angle. According to this theory, focusing attention to the

cued position takes more time for more distant targets and all targets along the trajectory of the "spotlight beam" benefit from it even if they were not specifically cued.
The **discrete movement** model posits that attention does not move along a trajectory from one position to the other but that attentional shifts occur discretely. This hypothesis predicts an increase in movement velocity as the distance of the target increases so that attention moves at a constant time [15]. Such a carachteristic is also shared by the overt attentional shifts subserved by saccadic eye movements.
Another aspect of this group of theories is that attention under certain conditions can be allocated to a small and well defined visual field area while under other conditions can be allocated over a wider area, even encompassing the whole visual field. This is the so-called "zoom lens metaphor" which, according to Eriksen and St. James [3], better describes attentional shifts than Posner's spotlight. Finally, the **hemifield activation** model criticizes the spotlight theory on the basis of the evidence that the benefits of attentional focusing are not restricted to one target but encompass a whole visual hemifield with the vertical and horizontal meridians serving as boundaries [6,18]. An important characteristics of this account is that processing at the unattended regions, i.e., in the opposite hemifield are broadly inhibited as witnessed by the costs associated with invalid cues. Interestingly, the stress on inhibition posed by this account is reminiscent of earlier multi-channel theories as the one originally proposed by Broadbent [1].

An Experimental Test of the Mechanisms of Covert Shifts of Spatial Attention

Before concluding with a brief review of the possible neural bases of covert attentional shifts, I shall report some recent work from our laboratory trying to test some of the above accounts of spatial attentional focusing.
We wanted to submit the analog movement and the hemifield

activation theories to a simple test by using the following paradigm: Subjects are pre-cued to attend to one hemifield only where there are three possible targets at different eccentricities along the horizontal meridian (5,25,45 degrees). The subjects' task is to press a key as soon as possible following lightening of one of the six LEDs (three for each hemifield). The order of cue and stimulus presentation is randomized and there are three possible cues: valid trials are those in which a central arrow points toward the same hemifield as that in which one of the three stimulus locations will be lit; invalid trials are those in which the arrow points to the hemifield opposite to that in which a target is lit; and finally, neutral trials are those in which a double-pointed arrow signals both hemifields simultaneously. The probability of the arrow signalling the correct hemifield of stimulus presentation is 50%, see above. There were two cue-target intervals 100 and 300 msec. An analog movement account would predict with our paradigm an interaction between cue-target interval, attentional effects and eccentricity. Covert attentional effects on speed or accuracy of response begin 50 msec after cue onset and continue to grow until they reach asymptote about 200 msec following cue offset [10]. Therefore, at the short interval (100 msec) of our paradigm, benefits should be greater at small eccentricities while at the longer interval (300 msec) there might be less differences among eccentricities (or none at all). In contrast to such a possibility, a ANOVA with Attentional Condition (Valid, Invalid, Neutral trials), Eccentricity (5,25,45 degrees) and Cue-Target Interval (100 and 300 msec) did not yield a significant second-order interaction involving the above factors, while, as expected, it showed a reliable effect of each of the three main factor in isolation. Valid trials turned out to be faster than neutral and invalid trials with no significant costs; there was a reliable speed advantage for 5 and 25 over 45 degrees of stimulus presentation but no difference between 5 and 25 degrees; and finally, the longer cue-target interval yielded an overall faster speed of response than the shorter one. These data are not in keeping with an analog-movement theory and tend to favour the hemifield activation hypothesis. Furthermore, the results of a second experiment using the same stimuli but without pre-cueing showed overall faster RT at each eccentricity and therefore substantiated the possibility outlined above that the central cue paradigm is unnecessarily complicated by the requirement of processing the cue.

Possible Neural Basis of Visual Attention

There are two enduring questions concerning the **neural basis of attention**, namely i) at what level in stimulus information processing attentional phenomena arise? and ii) is attention subserved by a unitary system or is rather the results of an interaction between attentional subsystems strictly tied up with the main sensory or motor systems? An answer to these questions is still awaited but some hints are available. I shall just briefly mention some of them.

Evidence relevant to the former question has been provided by means of electrophysiological techniques in normal human subjects. The emerging picture is that spatial selection occurs at relatively early stages and can be traced down at the level of the visual cortical areas adjacent to the primary visual cortex [11]. A consistent finding in all studies employing a spatial cueing paradigm and event-related potentials (ERPs) is that valid targets elicit an enlarged P1 component. Given that this component represents activity at the level of the visual prestriate cortex this is evidence confirming earlier psychophysical findings that pre-cueing of target location acts by facilitating sensory processing [2,4] rather than by biasing decision criteria as suggested by some authors [16].

As to the latter question, Posner and Petersen [14] recently proposed the theory that there are three neural systems involved in selective attention: an Orienting (Posterior) System, a Target Detection (Anterior) System and a Vigilance System. This proposal is in some way a compromise between models which envisage attention as subserved by one and the same system and those who consider it as a function associated to the sensory and motor systems. It takes into account brain imaging, electrophysiological (both single-unit and ERP data) as well as neuropsychological findings. The Posterior System involves areas that subserve the operations needed to bring attention to a location in space and to orient to sensory stimuli in general. Anatomically, includes parts of the parietal cortex, thalamic structures such as pulvinar and reticular nucleus and parts of the superior colliculus in the midbrain. The Anterior System involves areas that are activated every time a stimulus has to be detected (independently from the response modality). Anatomically, it involves the prefrontal cortex and in particular the anterior cyngulate gyrus and the supplementary motor area. Finally, the Vigilance System is involved in maintaining the alert state and its activation can be monitored when subjects attend to an infrequent signal in a series or during the interstimulus interval in a reaction time task. Anatomically, this system involves noradrenergic input to the cortex from the locus coeruleus [5]. Clearly, the three systems are heavily interconnected and, in particular, both the posterior and the anterior networks must be closely associated with the vigilance system in order to ensure awareness and control. An interesting neural network model that simulates aspects of the interaction between anterior and posterior attentional systems has been recently put forward by Jackson et al. [7].

References

1 D.E., Broadbent (1958) *Perception and Communication*. New York, Pergamon Press.

2 C.J., Downing (1988) Expectancy and visuo-spatial attention: Effects on perceptual quality. *J. Exp. Psychol.: Hum. Percept. & Perf.*, 14, 188-202.

3 C.W.Eriksen and J.D. St. James (1986) Visual attention within and around the field of focal attention: a zoom lens model. *Percept. Psychophys.*, 25, 249-263.

4 H.L., Hawkins, S.A., Hillyard, S.J, Luck, M., Mouloua, C.J. Downing and D.P., Woodward (1990) Visual attention modulates signal detectability. *J. Exp. Psychol.: Hum. Percept. & Perf.*, 16, 802-811.

5 C.W. Harley (1987) A role for norepinephrine in arousal, emotion and learning? *Progr. Neuropsychopharm. Biol. Psychiat.*, 11, 419-

458.
6 H.C., Hughes and L.D., Zimba (1987) Natural boundaries for the spatial spread of directed visual attention. *Neuropsychologia*, 25, 5-18.
7 S.R., Jackson, R., Marrocco, and M.I., Posner (1994) Networks of Anatomical Areas Controlling Visuospatial Attention. *Neural Networks*, 7, 925-944.
8 W. James (1890) *The Principles of Psychology*. London, MacMillan and Co.,Limited, pp.403-404, 1890.
9 J. Jonides (1981) Voluntary versus automatic control over the mind's eye. In *Attention & Performance IX*, J. Long & A. Baddeley (Eds.),Hillsdale, N.J., L. Erlbaum, pp. 187-203.
10 R.A., Kinchla (1992) Attention. *Ann. Rev. Psychol.*, 43, 711-742.
11 G.R., Mangun, S.A. Hillyard and S.J. Luck (1992) Eletrocortical Substrates of Visual Selective Attention. In *"Attention and Performance XIV"*, D.E. Meyer and S. Kornblum (Eds.), Cambridge, Mass., MIT Press, pp. 219-243,
12 M.I., Posner (1980) Orienting of attention. *Quart. J.Exp. Psychol.*, 32, 3-25.
13 M.I. Posner and Y. Cohen (1984) Components of visual orienting. In *"Attention and Performance X"*, H.Bouma and D.G. Bowhuis (Eds.). Hillsdale, NJ, L. Erlbaum, pp. 137-157.
14 M.I. Posner and S.E. Petersen (1990) The attention system of the human brain. *Ann. Rev. Neurosci.*, 13, 25-42.
15 R., Remington and L., Pierce (1984) Moving attention:evidence for time-invariant shifts of visual selective attention. *Percept. Psychophys.*, 35, 393-399.
16 M.L., Shaw (1984) Division of attention among spatial location: a fundamental difference between detection of letters and detection of luminance increments. In *"Attention and Performance X"*, H. Bouma and D.G. Bouwhuis (Eds.), Hillsdale, New Jersey, L. Erlbaum, pp. 109-121.
17 G.L., Shulman, R., Remington, J.P., MacLean (1979) Moving attention through visual space. *J. Exp. Psychol.: Hum. Percept. & Perf.*, 5, 522-526.
18 G., Tassinari, S., Aglioti, L., Chelazzi, C.A., Marzi, and G. Berlucchi (1987) Distribution in the visual field of the costs of voluntary allocated attention and of the inhibitory after-effects of covert orienting. *Neuropsychologia*, 25, 55-71.
19 G., Tassinari, M., Biscaldi, C.A., Marzi, and G. Berlucchi (1989) Ipsilateral inhibition and contralateral facilitation of simple reaction time to non-foveal visual targets from non-informative visual cues. *Acta Psychol.*, 70, 267-291.
20 Y. Tsal (1983) Movements of attention across the visual field. *J.Exp.Psychol.:Hum. Percept. & Perform.*, 9, 523-530.

VISUAL NEGLECT

LUIGI PIZZAMIGLIO

Dipartimento di Psicologia, Università "La Sapienza", Roma, IRCSS S. Lucia, Roma, Italy

This paper deals with a description of the main clinical features of visual neglect in humans. Examples will be given of perceptual, motor neglect and of imagery neglect within personal and extrapersonal space. Similar impairments can also be found in other sensory modality, such as acoustic and tactile. Clinical development of the syndrome and the most frequent anatomical correlates will be discussed.

Two main interpretation have been proposed for this complex variety of disorders. The first one points to a disturbance in the capacity to direct the attention to the surrounding or to the personal space. The second one considers the disorder in exploring the space as a consequence of asymmetrical representation of the space.

In this way, I shall present two groups of findings which support the representational interpretation of the neglect.

The first part deals with stimulations which produce a transient reduction of the neglect disorders. The second part will describe a new finding concerning a dissociation between perceptual and imagery neglect.

A description of a number of facts, collected in relatively similar domains, will be followed by a possible interpretation.

1. REDUCTION OF NEGLECT DISORDERS

A new, large group of findings was stimulated by an early observation of Silberpfening (1941) and more recently applied to a sample of neglect patients by Rubens (1985).

The stimulation of the left ear with cold water or warm water in the right ear produces a nystagmus with a slow movement toward the left side and a quick movement toward the right side. This stimulation also produces a substantial and, in some cases, spectacular reduction of neglect as measured by a segment cancellation test and a test of sentence reading. The effect lasts from 5 to 10' and then the patient's performance returns to how it was prior to the stimulus.

This improvement has been replicated by Cappa, Sterzi, Vallar and Bisiach (1987) in four cases, with the important addition that two of these patients also had reduced anosognosia for almost half an hour.

In 1990 and then in 1993, Vallar and colleagues showed that vestibular stimulation could transiently reduce hemianesthesia on the left side of the body. In two separate studies, the same group of Authors argued that in neglect patients the hemianestesic disturbance may in part be independent from a lack of primary sensory information, since the same patients showed a skin conductance response and normal somatosensory evoked potential to left tactile stimuli, which they failed to perceive consciously (1991a, 1991b).

Also, vestibular stimulation produced a positive although transitory reduction of somatoparaphrenic delusion associated with neglect in an elderly woman (Bisiach *et al.*, 1991).

Finally Rode and Perenin (1994) showed improvement in describing items on the left side of the mental image following heat stimulation.

Similarly, but using an optokinetic stimulus moving at a critical speed in both directions, Pizzamiglio *et al.* (1990) produced a reduction of the left bias in neglect patients when the stimuli moved leftward and a worsening of their performance in a line bisection task when they moved rightward.

A recent research by Vallar *et al.* (1993, 1995) reports a positive effect on the horizontal and vertical position sense in neglect patients for leftward OKS and a worsening of this response to stimuli moving toward the side ipsilateral to the lesion.

Finally, following Biguer and colleagues' (1988) observation on the effects of vibration applied to the neck muscles, Karnath, Christ and Hartje (1993) were able to produce transient remission of neglect disorders.

Their *rationale* was that (quote) "both vibration of muscle or muscle tendons produces kinesthetic illusion. The illusion involves the perception of a constant movement as well as a spatial displacement of limb position, due to an altered somatosensory representation and the relation between the body and the environment."

In particular, the Authors claim that either the lengthening of the neck muscle produced by the rotation of the trunk or the vibration of the neck muscles on one side results in a displacement of the egocentric midline of the body-centered representation of visual space.

Three patients with right hemisphere lesions and left-side neglect showed a reduction of the neglect symptom when either one of the proprioceptive stimulations were used (see also Karnath *et al.*, 1991; Karnath, 1994).

All of these data can be briefly summarized and a general explanation proposed.

The first point is that vestibular, optokinetic and vibratory stimulation can reduce neglect disorders. This implies that neglect is a reversible phenomenon under certain conditions.

Second, all of these stimulations are important in normal individuals for generating a body-centered system of coordinates, necessary for organizing the perception of space and relations with the body.

Third, these stimulations proved to be effective in reducing neglect disorders, and can also decrease the disorders of tactile, proprioceptive, and visual perception. Their effect is so robust that even imaginative and delusional symptoms can be transiently reduced.

To try to explain this group of findings, we might refer to a model which can account for the conscious perception of the spatial position of the body.

Two stages may be necessary: one in which the processing of the primary sensory information by each hemisphere produces a somatotopic representation; the other, a perceptual processing which maps the sensory information into an egocentric representation of the body.

Vestibular, visual and proprioceptive information is particularly relevant to the localization of body parts with reference to the egocentric coordinates.

It is also necessary to hypothesize that both hemispheres are competent in processing contralateral information, while the ipsilateral representation of the left side of the body in the left hemisphere is comparatively weaker than the right one.

Unilateral cerebral lesions, and particularly right-sided ones for the latter assumption, may disrupt the bilateral cerebral activity mediating the egocentric, body-centered representations of the body and of the external space.

The right-sided lesion inroduces imbalance into the egocentric representation, resulting in a rightward bias of body and space perception.

The displacement produced by the lesion may be temporarily reduced by stimulations producing leftward bias, such as the ones we have described above.

It is also important to note that since the lesion produces a displacement of the whole egocentric representation, the effect of the vestibular and particularly of the OKS are direction specific; that is, they can increase or decrease spatial exploration as well as tactile or proprioceptive impairment, depending on the direction of the stimulation. This is important since it makes any interpretation of these effects in term of general hemispheric activation very unlikely.

Another point is that a partial compensation of egocentric displacement can be obtained by using one or the other of these stimulations. Although these are only a few points, they provide interesting evidence for the possible summation of two of these stimuli (Karnath, 1994).

2. DISSOCIATION BETWEEN PERCEPTUAL AND IMAGERY NEGLECT

The main characteristic observed of the dissociation between perceptual and imagery neglect is a disorder in reporting a well know visual image (squares geographic maps, apartment map, etc.) from a certain vantage point (Bisiach & Luzzatti, 1978).

It was never systematically investigated how frequent is the association between the impaired imaging disorder and a perceptual one.

From a clinical experience it is however clear that, in spite of the difficulty in testing imagery, often a perceptual or motor neglect is accompained by some degree of difficulty in representing spatial arrangement from a long term memory. In some other cases one can see perceptual, but not imagery impairment.

This common observation is also in agreement with several evidences that perception and imagery share the same neural substrate.

A case, never described so far, where imagery was impaired with intact visuo-perceptual processes, is here described (Guariglia *et al.*, 1993).

The report which follows documents such an exceptional patient.

A 59-year-old right-handed man was examined 16 months after an ischaemic stroke, which produced a left hemiparesis. A computerized tomography (CT) scan showed a large lesion involving the right frontal and anterior temporal lobe. A SPECT study showed a hypoperfusion in the lesioned frontal area, but not a perfusional deficit remote from the infarct, except in the contralateral cerebellum. A neuropsychological examination showed no mental deterioration or visual field defects, some difficulty in abstract reasoning and face recognition, and a mild impairment in selective attention and in constructional apraxia.

The patient's performances on tests for visuo-spatial (Line and Letter Cancellation, Wundt-Jastrow Area Illusion and Sentence Reading) and personal neglect were normal. In contrast, when he had to describe different but very familiar piazzas from a given vantage point, he consistently reported several elements on the right side and none or very few on the left side. When requested to describe these piazzas from the opposite perspective, he again reported more elements on the right side of his mental image. The patient's performances were very similar when retested at one- and two-month intervals.

Interestingly, the patient misplaced some semantically relevant details from the left to the right side of one piazza; that is, the ticket office where he had taken a bus to work for 20 years was described as being on the non-neglected side.

Another set of tests was done to identify possible selected forms of neglect, or asymmetry in perceiving targets in a far or near space. On tasks involving the generation and manipulation of mental images, he performed like normal controls in comparing mental representations of both colour and weight of objects, and performed normally in a test of mental unfolding.

Through mental imagery, the patient was able to identify shapes differing on either the left or right side, with no asymmetry for either side. He could also read and spell a word orally without omitting its left side.

Finally, the patient was taken into an unifamiliar room with different objects and told to observe carefully all details in order to describe them afterwards. When asked to describe the room from two opposite perspectives, the patient consistently omitted half of the elements on the left and none on the right side. Conversely, he made no errors in identifying the observed details by multiple choice (choosing between two different armchairs).

The reported observations describe an impressive dissociation between a long-standing deficit in the mental representation of the hemispace contralateral to the lesion without any sign of impairment in exploring and manupulating the corresponding perceptual space and in representing the ipsilateral hemispace. All previously reported representational neglect was observed in patients with personal and/or extrapersonal neglect. This patient was not affected by a general mental imagery disorder; when comparing mental representations within working memory he performed quite normally.

These discrepancies of impaired mental imagery could then rely on good access to the immediate memory versus a difficulty in retrieving visuo-spatial information from a long-term store. This interpretation does not entirely correspond with our data as the patient was very competent in tasks relying extensively on long-term acquisitions (comparing colours and weights). The patient seems able to generate the global image of an object or some of its characteristics, but this is insufficient for correctly describing a familiar piazza or a well-inspected room from memory. In this case, the subject has to generate a "skeletal image" of the square in order to: (1) superimpose a system of egocentric coordinates to organize that space; (2) move an 'attentional window' to select a number of items that belong to the image; and (3) locate the specific items in their respective positions. The latter two operations seem defective in this patient.

The second important finding concerns the predominantly right frontal localization of the lesion. Previous studies have proposed that the neural structures involved in mental imagery were coincident or adjacent to the circuitries involved in visual perception (parieto or temporo-occipital cortex); unlike the present case, no frontal lobe involvement was described.

The tasks used in the psychology literature, however, dealt with the generation and transformation of single objects, symbols or abstract figures, whereas the selective impairment found in the present case only emerged in organizing a complex material from a long-term store into an egocentric frame. Our patient, although showing a good capacity to retrive many items in the piazzas and in the unifamiliar room from long-term memory, failed when he had to attend to details and position them with respect to the internal spatial coordinates. This matching between a representation of external events and egocentric space may require a new module which shares attentional properties.

Both groups of findings on unilateral stimulations and the imaginal neglect can be looked at as a consequence of a disruption of a very basic step in organizing our spatial cognition. I used the label "egocentric representation" to indicate a frame of reference based on a variety of body and external information which are necessary to organize both spatial perception as well as a particular kind of visual imagery.

However, it may eventually be necessary to use a more articulated model in order to explain how we organize our perception of the outside world *versus* the organization of our orientation within the external world.

REFERENCE

[1] Biguer, B., Donaldson, I.M.L., Hein, A. and Jeannerod, M. (1989) Neck muscle vibration modifies the representation of visual motion and direction in man. *Brain* 111, 1405-1424.

[2] Bisiach, E., Rusconi, M.L. and Vallar, G. (1991) Remission of somatoparaphrenic delusion through vestibular stimulatio. *Neuropsychologia* 29, 1029-1031.

[3] Cappa, S., Sterzi, R., Vallar, G. and Bisiach, E. (1988) Remission of hemineglect and anosognosia during vestibular stimulation. *Neuropsychologia* 25, 775-782.

[4] Guariglia, C., Padovani, A., Pantano, P. and Pizzamiglio, L. (1993) Unilateral neglect restricted to visual imagery. *Nature* 364, 235-237.

[5] Karnath, H.O. (1994) Subjective body orientation in neglect and the interactive contribution of neck muscle proprioreception and vestibular stimulation. *Brain* 117, 1001-1012.

[6] Karnath, H-O., Christ, K. and Hartje, W. (1993) Decrease of controlateral neglect by neck vibration and spatial orientation of trunk midline. *Brain* 116, 383-396.

[7] Karnath, H-O., Schenkel, P. and Fischer, B. (1991) Trunk orientation as a determining factor of the controlateral deficit in the neglect syndrome and as the physical anchor of the internal representation of body orientation in space. *Brain* 114, 1997-2014.

[8] Pizzamiglio, L., Frasca, R., Guariglia, C., Incoccia, C. and Antonucci, G. (1990) Effect of optokinetic stimulation in patients with visual neglect. *Cortex* 26, 535-540.

[9] Rode, G. and Perenin, M.T. (1994) Temporary remission of representational hemineglect through vestibular stimulation. *Neuroreport* 5, 8, 869-872.

[10] Rubens, A.B. (1985) Caloric stimulation and unilateral neglect. *Neurology* 35, 1019-1024.

[11] Silberpfenning, J. (1941) Contribution to the problem of eye movements. III Disturbances of ocular movements with pseudo hamianopia in frontal lobe tumors. *Confinia Neurologica* 4, 1-13.

[12] Vallar, G., Sterzi, R., Bottini, G., Cappa, S., Rusconi, M.L. (1990) Temporary remission of left hemianestesia after vestibualr stimulation. A sensory neglect phenomenon. *Cortex* 26, 123-131.

[13] Vallar, G. Bottini, G., Sterzi, R., Passerini, D. and Rusconi, M.L. (1991a) Hemianestesia, sensory neglect and defective access to conscious experience. *Neurology* (Cliveland) 41, 650-652.

[14] Vallar, G., Sandroni, P., Rusconi, M.L. and Barbieri, S. (1991b) Hemianopsia, hemianestesia and spatial neglect: a study with evoked potentials. *Neurology* (Cleveland) 41, 1918-1922.

[15] Vallar, G., Antonucci, G., Guariglia, C. and Pizzamiglio, L. (1993) Deficits of position sense, unilateral neglect, and optokinetic stimulation. *Neuropsychologia* 31, 1191-1200.

TEMPORAL ASPECTS IN VISUAL PERCEPTION AND COGNITION

JIRINA RADILOVA'

Academy of Sciences of the Czech Republic, Prague, and Department of Psychology, Helsinki University, Finland

This short review article describes some of our experiments concerning temporal aspects in visual perception and cognition.

1. Spontaneous Illusory Figure Reversals

In the first group of experiments temporal aspects of spontaneous illusory figure reversals have been analysed. The basic assumption was that in case of ambiguous two-dimensional representations of three-dimensional objects the temporal intervals between illusory changes of their interpretations did reflect probably the duration and mumber of sequential processing steps underlying cognition. Reversible figures (the Necker cube, Nc, and its modifications, including hexagonal and quasi-hexagonal patterns, and a modification of the Schroeder staircase, as well as a special "double reversible figure" designed by the author for the purpose in 1974) were presented to healthy paid subjects, usually University students and the staff of the Institute. Time intervals between consecutive figure reversals were recorded and analysed statistically using different computer programs [2].

First, the permanently illuminated Nc pattern, drawn with black lines on white background (Fig. 1) was presented to Ss sitting in a darkened acustically (and electrically) shielded chambre. They had to signalize by a two-positional switch the moment of the perceived figure reversals. Statistical evaluations showed that both types of subjective interpretations, which were the results of two different types of perceptual processes, were equivalent in duration and temporal development, as well as in mutual dependencies of adjacent time intervals corresponding to both types of interpretations. Thus it could be assumed that both types of cognitive interpretations were equivalent as well [2,14].

When Nc was flashed repetitively with regular intervals 3.6, 2.4, or 1.2 s to the Ss, the dependence of the actual interpretation from the preceding ones was much more expressed in comparison with the condition when Nc was permanently illuminated. The sequential occurence of different interpretations could be described as a Markovian process of the fourth or higher order. This phenomenon pointed to the significance of perceptual-memory mechanisms in determining the actual cognitive interpretation of the pattern [13].

In other experiments EEG Event Related Potentials (ERPs) were registered during repetitive expositions of the Nc by flashes (the regular interstimulus intervals being 3 or 5 s) from the occipital scalp. ERPs were classified into two groups according to the actual interpretation of Nc signaled by the Ss by pushing the corresponding button. In some subjects a difference in averaged ERPs during the first 150 ms after the stimulus could be observed (Fig. 2). That happened, however, only in cases, when slow oscillatory waves at alpha frequency, which might have been related to stimulus expectation, appeared during the 500 ms period preceding the stimuli [8].

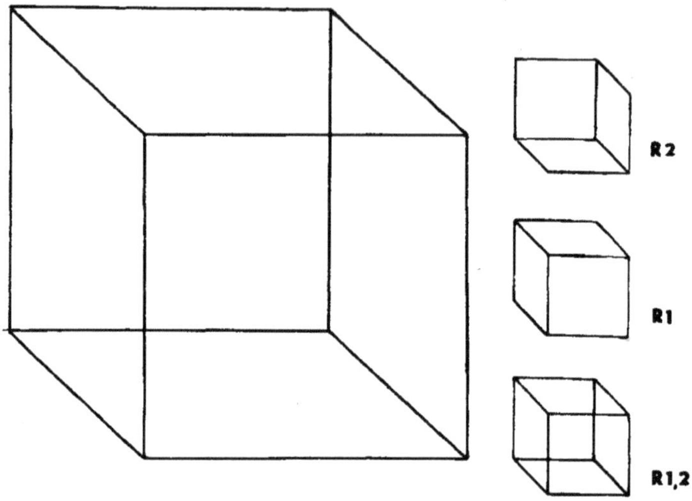

Figure 1 - Pattern of the Necker cube.

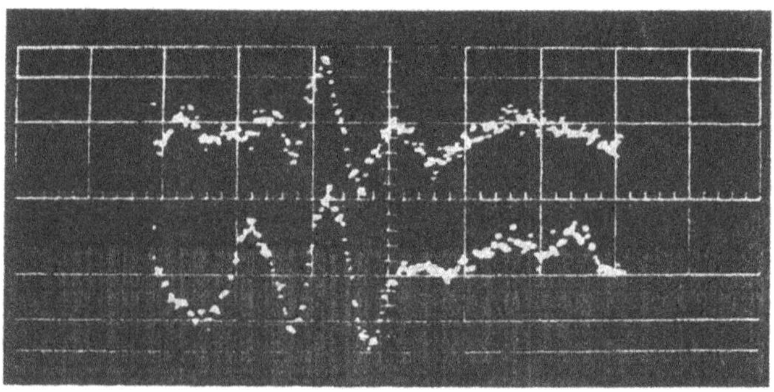

Figure 2 - Averaged ERPs corresponding to the two perceptual interpretation of the repetitively flashed pattern of the Necker cube (64 realisations each, duration of the traces 512 ms).

In similar experiments Nc was flashed repetitively with regular 512 ms intervals and EEG activity recorded by means of non-polarizable Ag/AgC1 electrodes, and averaged. A global negative potential shift, lasting about 2.5 s, increasing in a linear way and reaching a final amplitude of about 11 mV, developed gradually mainly in Fpz, Cz, Pz and Oz before the stimulus onset (Fig 3). A positive correlation was found between the duration and amplitude of these DC shifts. This phenomenon, which could be considered a version of the so called CNV wave, and which was probably related to the activity of glial cells, could also reflect cortical processes related to expectation of regularly repeated perceptualy relevant stimuli [15].

When dealing with reversible figures we considered also the possible role of cognitive programs related to pattern conceptualization. We found that the reversal rate could be influenced by specific instructions emphasizing certain details in the figure. For instance, the reversal rate of Nc could be decreased dramatically by emphasizing a "candy like structure" in the pattern by means of an appropriate psychological instruction (Fig. 4) [9].When the permanently illuminated Schroeder staircase pattern (Fig.5) was studied, Ss were supposed to see either "stairs leading up" or "stairs hanging down from the ceiling". Some of the Ss had difficulties with percieving the second variant. However, when they received a more detailed instruction on the possible geometrical interpretations involved, reversal rates doubled, approximately. Perceptual intervals corresponding to the more usual interpretation from the cognitive point of view: "stairs leading up", were always longer than the unusal ones. Thus it was possible both to decrease and increase the rate of figure reversals by means of appropriate instructions addressing the cognitive program underlying interpretation of the patterns [10].

In additional experiments the pattern of the Schroder staircase was drawn either in the usual way by black lines on white background, or *vice versa*. An acceleration of the reversal rate was observed in the latter case when the figure was presented in a negative form, *i.e.* drawn with white lines on black background (Fig.6). That was probably caused by the circumstance, that the white-light lines did carry both the sensory information and the cognitive meaning of the pattern. It seemed to be probable therefore, that the information content of the pattern does influence the rate of figure reversals [6].

We confirmed that a suitable spatial projection of the Nc could be seen either as a 3-D object (the cube), or a 2-D hexagonal pattern (Fig.7). We found that the figure was reversible under both conditions (in the latter case as an alternating "figure-background" pattern). The reversal rate, however, was considerably slower in the former case, when due to the higher complexity of perceptual-cognitive processes required for stimulus manipulation in 3-D, more individual steps were needed probably in comparison with the less demanding, from the informational point of view, 2-D variant [18].

The relationship between 2-D and 3-D subjective reversals in physically identical patterns, proceding in accordance with the cognitive instruction given, has been further studied using a special "double reversible figure" (Fig. 8). It can be seen in 2-D mental space either as a vertical "spool", or a horizontal "butterfly", and in 3-D either as a concave "tunnel", or a convex "form". The basic finding was that reversal is slower in 3-D space than in 2-D space, perceptual intervals being about two times longer in the former case [11, 17]. When increasing the amount of additional semantic items in this pattern, emphasizing its 3-D nature (and increasing the amount of information to be processed) reversal rate was reduced further (Fig. 9). However, as reversal rate in this case became slower for 2-D interpretation as well, the dimensional specificity of the phenomenon remained questionable [16, 19]. In complementary pilot experiments incomplete representations of the "double reversible figure" were adopted. The rate of

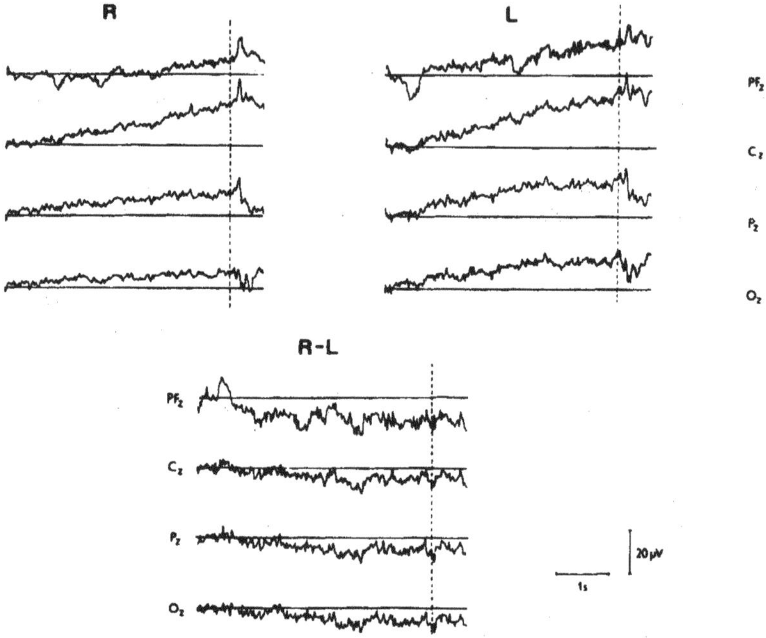

Figure 3 - Example of averaged EEG and DC recordings, 4608 s before and 0.512 s after presentation (vertical line) of the Necker cube at both perceptual interpretations (R, L), and the subtracted curves (R-L).

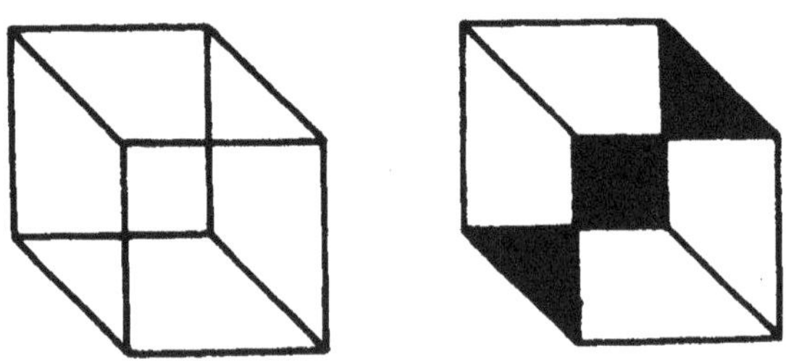

Figure 4 - Left: the Necker cube pattern used. Right: schematic representation of the 2-D spatial perceptual cue emphasized by the experimental instruction.

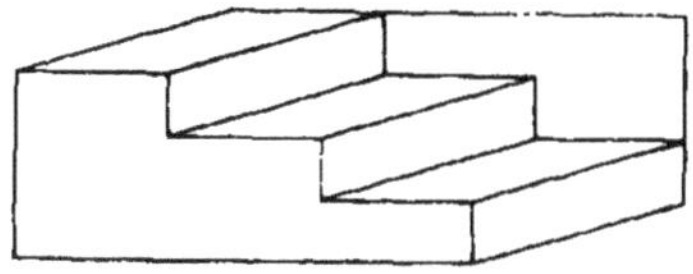

Figure 5 - The adaptation of the Schroeder staircase pattern used.

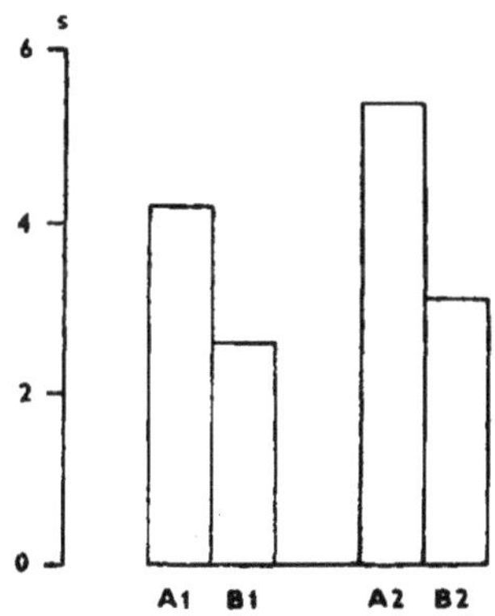

Figure 6 - Mean percepual intervals for white on black (A1, B1), and black on white (A2, B2) variants of the Schroeder staircase pattern. A: "stairs leading up. B:"stairs hanging down from the ceiling".

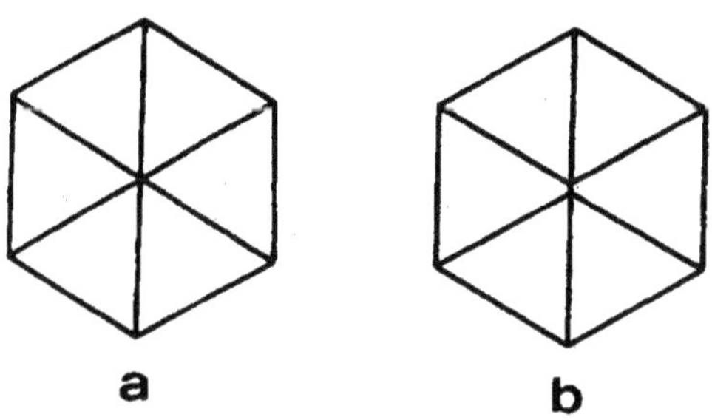

Figure 7 - The hexagonal and quasi-hexagonal pattern interpreted either as a Necker cube viewed from a certain angle (a reversible 3-D figure), or a figure background pattern (reversible in 2-D).

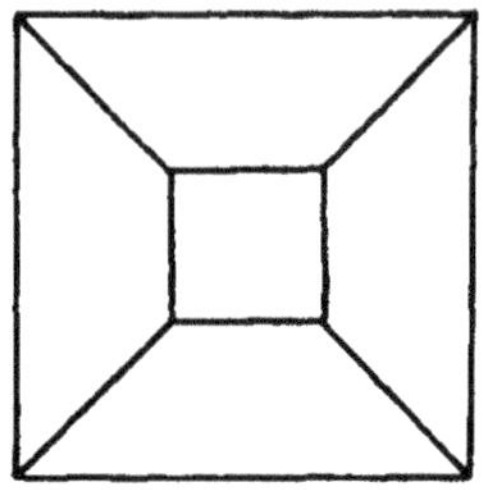

Figure 8 - The Double reversible figure.

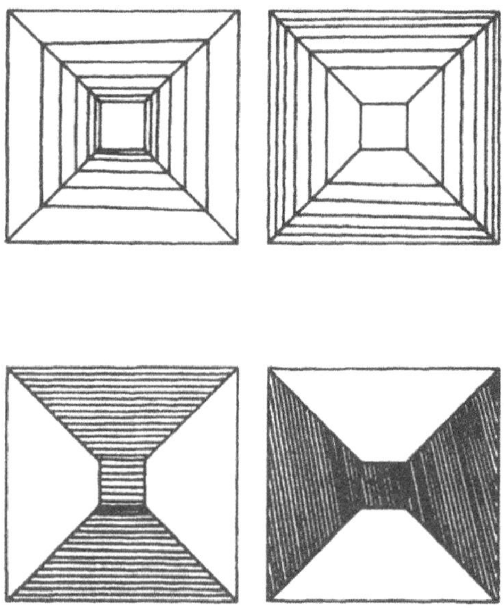

Figure 9 - The Double reversible figure with additional cues.

figure reversals decreased under this condition both for the 3-D and 2-D interpretations in a way indirectly proportional to the amount of cognitively relevant information in the figure. Thus it seems to be probable that both an increase and decrease of specific information in the pattern, to be processed additionally, or to be substituted in order to complete or reconstruct the pattern respectively, requires additional operations needing time.

2. Late Positive Component (the P300) of EEG ERPs

The second group of our experiments deals mostly with the late positive component (the P300) of EEG ERPs (recorded at Oz) induced by repetitive tachistoscopic presentation of visual stimuli, each consisting of a certain number (6-9) of white light squares randomly distributed over a dark screen, appearing for 200 or 300 ms in different experiments. The task of the Ss in this experiment (derived from Millers 7±2 paradigm) was to report on the number of items presented simultaneously (and/or on the subjective certainty/uncertainty of their judgement) in every trial. The ERPs were classified into two groups according to those binary cognitive criteria, averaged separately, and compared mutually.

In the first group of experiments dealing with jugment correctness/incorrectness, and subjective certainty/ uncertainty during judgements, the amplitude of the P300 in the average ERP was higher for correct recognition of the number of items in the successive individual stimuli in comparison with incorrect recognition. The P300 was of higher amplitude as well for subjective certainty during recognition, when compared with uncertainty (Fig. 10). Special experiments revealed that both binary cognitive factors were independent, although interrelated. Thus it followed from these results, that cognitive factors seem to codetermine in a still unknown way the amplitude of the P300 component of ERPs. That means that simple cognitive processes of the above type took place already after about 300 ms from stimulus onset. It was also possible to assume, that the underlying neural operations were related to "chunking" the information percieved (a concept which became popular in cognitive psychology) in subunits suitable for further processing [7].

The following experiments aimed to analyse the P300 during intensionally false versus correct statements regarding the number of light squares in the stimuli presented successively always for 200 ms, in order to find out whether the P300 could be used as a "lie detector". In cases of "small" and "bigger" lies generated by the Ss the averaged P300 was higher for correct answers. However, when "very big" lies were elaborated, the average P300 amplitude was practically identical when compared with correct answers (Fig. 11). Thus it seemed to be probable that in the latter case the amount of unitary neural processes contributing to the P300 amplitude, and underlying the mental operations of judging correctly or lying preponderantly, could have been similar. It would be certainly useful we would have a "truth verifier", which could determine the truthfulness of any statement. Our findings, however, revealed differences only between "small" and "bigger" lies and true answers, and not between true answers and "very big" lies, which should be detectable by any reliable "lie detector" [1].

The aim of additional experiments was to analyse whether the emotional (negative, stress-inducing) impact of visual stimuli would also be reflected in the ERP P300 amplitude. Ss watched series of slides with the light squares mentioned, presented always for 300 ms, alternating with horrifying pictures, like murdered people, corps after car accidents, etc.

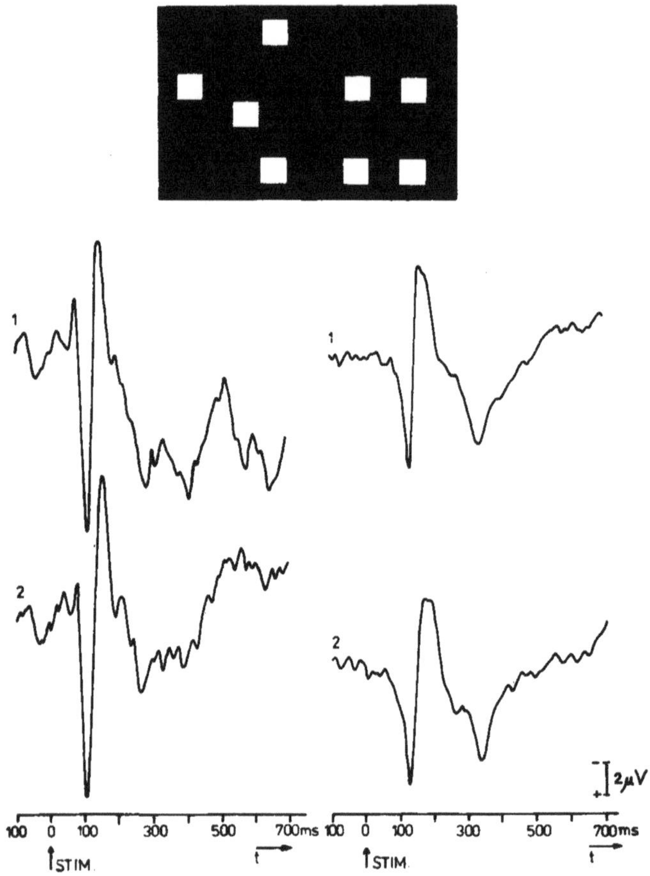

Figure 10 - Examples of pairs of averaged ERPs recorded from Oz corresponding: on the left to correct (1) and incorrect (2), on the right to certain (1) and uncertain (2) recognition of the number of light squares in the stimulus.

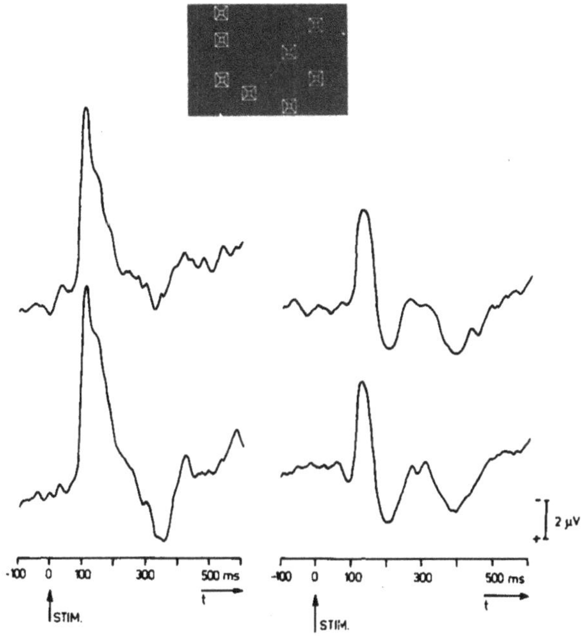

Figure 11 - Examples of averaged ERPs from Oz corresponding to intensionally false statements (upper left "small lies", lower left "bigger lies", upper right "biggest lies"), and correct answers (lower right).

Only ERPs induced by the slides containing the light squares were recorded. The Ss had to tell again after every stimulus presentation the number of squares detected, and whether or not they were certain about their statements. ERPs induced by the slides containing the light squares were recorded, automatically classified, this time into four groups: according to subject's certainty or uncertainty about the number of squares in the trial in question, as well as according to their instantaneous heart rate, automatically monitored and classified as high or low on the basis of a criterion set individually, using an instrument constructed for this purpose. The average P300 was found to be of higher amplitude when the subject was certain in his/her judgement and when his/her heart-rate was elevated (Fig. 12).

This experiment demonstrated that also vegetative-emotional factors did influence the amplitude of P300, and showed possible interactions between those factors and perceptual and cognitive processes taking place several ms after the onset of stimuli [5].

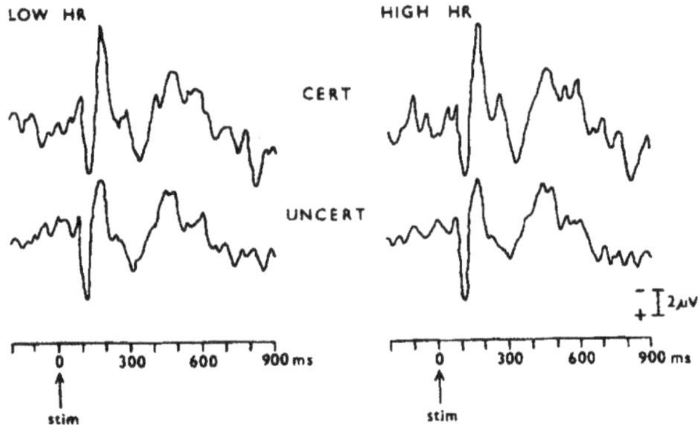

Figure 12 - Examples of averaged ERPs classified according to the criteria: certain (CERT), uncertain (UNCERT) judgment, and low and high heart rate (HR).

In additional experiments only the horrifying pictures alternating with photographs of flowers and beautiful architecture (all being of similar global luminousity) were presented tachistoscopically, and averaged ERPs evoked by both types of stimuli compared. The amplitude of the P300 component was higher in the former case (Fig. 13).

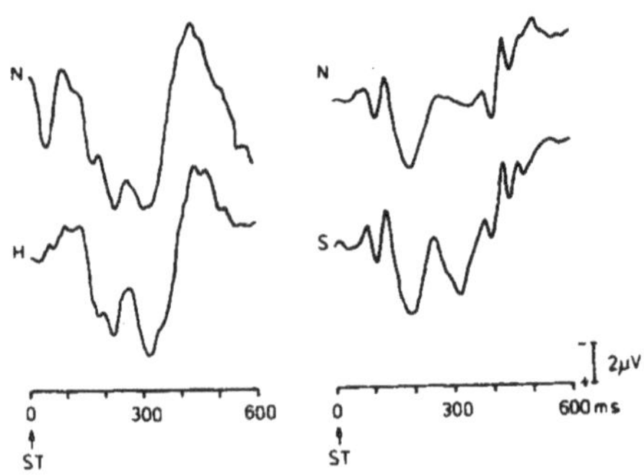

Figure 13 - Averaged ERPs corresponding on the left to horrifying (H) and irrelevant (N), on the right to erotic (S) and irrelevant (N) pictures.

These experiments proved that visual stimuli emotionally relevant in the negative sense, *i.e.* carrying the corresponding message stored in the memory, did activate more neural resurces contributing to the events taking place about 300 ms after stimulus presentation, in comparison with irrelevant ones [3]. The emotional-vegetative state of young male subjects was altered in a possibly positive way in another study, inducing sexual arousal, detected by means of penis plethysmography, by presenting erotic slides. During penile erection the average P300 wave induced by the slides with the light squares, the number of which had to be judged by the Ss, decreased markedly (Fig. 14).

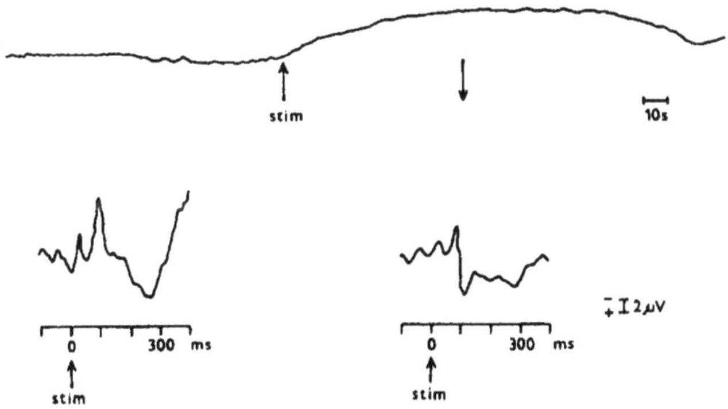

Figure 14 - Plethysmographic recording of penile erection induced by erotic slides (upper curve), and ERPs evoked by the light squares (lower traces).

However, the erotic slides themselves, when presented tachistoscopically in experiments similar to those with the horrifying pictures, did evoke high amplitude ERPs in comparison with the indiferent ones (Fig. 13). These experiments demonstrated that stimuli related to an emotional-vegetative state, which could be considered positive, did activate the processes underlying the P300 component as well. Thus the intensity of emotional involment and not its orientation (positive or negative) seemed to matter in this case. Besides, the mechanism in question was apparently state dependent. During sexual arousal the allocation of processes involved became apparently shifted from the cognitive ones toward those linked to sexual activity [4,12].

In the last group of experiments recognition correctness, analysed in the described way, and performance on Raven's Progressive Matrices were correlated in 8, 9, and 10 years old school-children in a game-like situation, in this case without recording ERPs. A negative correlation between the performance on the Raven Matrices and errors in the 7±2 paradigm, was found. Thus the specific cognitive operations involved in correct recognition in the experimental paradigm adopted both in this case and in the experiments with recording the P300 under various conditions mentioned, seemed to be related to more general mental abilities reflected by the individual scores on Raven's Matrices [4].

The experiments listed might shed some light upon the temporal aspects of simple cognitive processes taking place during several hundreds of ms after the onset of visual stimuli under the conditions studied.

References

[1] Radilova' J., Right and intensionally false statements concerning visual stimuli influence the late positive component of evoked potentials. *Activ. Nerv. Sup.* 24: 147-148, 1982.

[2] Radilova' J., Reversible Figures. (in Czech), Academia Publisher 1983, 97 pp.

[3] Radilova' J., Emotions and late positive component of evoked potentials. *Activ. Nerv. Sup.* 30: 281-282, 1988.

[4] Radilova' J., Detection of simultaneous visual stimuli by school-children. *Percept. Mot. Skills* 70: 601-602, 1990.

[5] Radilova' J., Toward a state-dependent psychophysiology of visual perception and cognition. In: Geissler H.G., Mueller M.H., Prinz W. (Eds.) *Psychophysical Explorations of Mental Structures.* Hogrefe and Huber, Toronto, 1990, pp. 333-338.

[6] Radilova' J., Poeppel E., The perception of figure reversal as a function of contrast reversal exemplified with the Schroeder staircase. *Acta Neurobiol. Exp.* 50: 37-40, 1990.

[7] Radilova' J., Radil-Weiss T., Some EEG correlates of perception and cognition. In: Sinz R., Rosenzweig M.R. (Eds.) *Psychophysiology* 1980, Gustav Fischer Verlag, Jena, Elsevier Biomedical Press, Amsterdam, 1982, pp. 349-353.

[8] Radilova' J., Radil T., Evoked responses during different subjective perception of reversible figures. (In Russian) *Fiziologia Cheloveka* 9: 211-216, 1983.

[9] Radilova' J., Radil T., Necker cube reversal reduced by two dimensional spatial perteptual cue. *Acta Neurobiol. Exp.* 43: 45-49, 1983.

[10] Radilova' J., Radil T., Influence of geometric conceptualisation on the reversibility of the Schroeder staicase figure. *Studia Psychol.* 25: 141-149, 1983.

[11] Radilova' J., Radil-Weiss T., Subjective figure reversal in two and three dimensional perceptual space. *Int. J. Psychophysiol.* 2: 59-62, 1984.

[12] Radilova' J., Figar S., Radil T., La composante tardif des potentials evoquees visuel au cours de l'excitacion sexuelle chez l'Homme. *J. de Physiol.* 78: 5, 1982.

[13] Radilova' J., Radil T., Havranek T., Sequential dependencies of perceptual interpretations of a repetitively illuminated reversible figure. *Int. J. Psychophysiol.* 2: 45-50, 1984.

[14] Radilova' J., Radil-Weiss T., Krekule I., Sequential relationships of perceptual interpretations of a reversible figure. *Activ. Nerv. Sup.* 15: 44-46, 1973.

[15] Radilova' J., Pocock V.P., Cooper R., Radil T., DC potential shifts preceding repetitive perceptually relevant stimuli and its possible origin. *Int. J. Psychophysiol.* 4: 221-225, 1986.

[16] Radilova' J., Radil-Weiss T., Riani M., Tuccio M.T., Subjective reversal of figures of increasing complexity in three and two dimensional perceptual space. *Atti del V Congresso Nazionale di Cibernetica e Biofisica* , Pisa (I), 1979, pp. 299-306.

[17] Radilova' J., Radil-Weiss T., Spunda J., Indra M., Visual pattern interpretable as reversible or figure background. *Activ. Nerv. Sup.* 17: 54-55, 1975.

[18] Radilova' J., Riani M., Tuccio M.T., Radil T., Borsellino A., Reversible figures percieved in two- and three-dimensional subjective space. *Studia Psychol.* 25: 76-79, 1983.

[19] Riani M., Tuccio M.T., Borsellino A., Radilova' J., Radil T., Perceptual ambiguity and stability of reversible figures. *Percept. Mot. Skills* 63: 191-205, 1986.

THE POSSIBLE ROLES OF NOISE IN THE PROCESSING OF VISUAL SIGNALS

MASSIMO RIANI and ENRICO SIMONOTTO

Istituto Nazionale per la Fisica della Materia
Dipartimento di Fisica, Università di Genova
via Dodecaneso 33, 16146 Genova, Italy
email: riani@genova. infn. it

Abstract

The problem of information processing in the presence of noise is addressed within the context of artificial neural networks modelling multistability in the perception of ambiguous patterns. While, in general, noise is an element to be overridden and neglected during processing, from the point of view of the phenomenon of ***stochastic resonance***, the presence of noise of suitable intensity can be utilized by a nonlinear system to obtain a higher signal-to-noise ratio, hence to facilitate information processing. We present computer simulations of an autoassociative network, modelling the perceptual alternation phenomenon. In the presence of a weak periodic perturbation, such a network exhibits the characteristic dependence of the signal-to-noise ratio, at increasing noise intensity, that is typical of stochastic resonance.

1 Introduction

The presence of noise can be easily detected at all system levels of the nervous system of a biological organism: from receptors and sensory neurons to the neurons of all the sensory, associative, and motor areas of the central nervous system (CNS) [1, 2]. Nevertheless, the nervous system is very efficient in overcoming the troubles linked to the presence of noise. Very weak stimuli are detected and elaborated by receptors and sensory neurons at the first stages of neural transmission and elaboration, even in very simple animals. At a higher level of neural elaboration, the CNS is extremely effective in extracting the structured and significant parts of an incoming message (i.e., a signal or *a figure*) from the unstructured and insignificant ones (i.e., noise or *the ground*). Therefore, it appears evident that natural neural systems are very effective in information processing, even in, or despite, the presence of a notable amount of noise, of both *eternal* origin and *internal* origin (i.e., due to the spontaneous activity of neural cells). On the other hand, the presence of *internal* noise of significant intensity causes the question to get raised as to whether such a noise source, apparently undesirable, may play some useful role in biological organisms. A possible positive reply to this question has been supported by the discovery of the phenomenon of *stochastic resonance* for nonlinear systems [3, 4].

This phenomenon has pointed out a new way of considering noise, both in the most studied bistable systems (i.e., those with two possible steady states (*potential wells*) between which such systems can oscillate during their dynamical evolutions) and in a class of deterministically resettable systems (i.e., those characterized by a single potential well, that is, a stable fixed point, and by a single barrier, that is, an unstable fixed point [5]). In all such systems, the effect of a weak periodic perturbation on the rate of transition between the two stable states (for the former systems), or on the rate of recovery to the stable fixed point, after the barrier has been surmounted (for the latter systems), is notably increased by the presence of noise of optimal intensity. In fact, if the periodic perturbation is too weak to surmount the potential barrier, the dynamic system is confined to the occupied potential well. In the presence of a suitable amount of noise, the jumps of the system over the barrier are merely random jumps; the combined presence of a weak modulation and noise results in a correlation effect between them that facilitates the overcoming of the barrier at the rate of the periodic modulation. Obviously, such correlation is not effective at both low and high noise intensities, whereas, over the range of intermediate noise values, the degree of correlation increases up to a maximum value, at a specific noise intensity, and then decreases toward zero. This phenomenon is called *stochastic resonance.*

Therefore, in such a phenomenon, noise is of positive significance to the evolution of a system and is not a disturbing element to be eliminated in any possible way.

In the past few years, there has been a growing interest in studying the phenomenon of stochastic resonance associated with nonlinear systems [6]. The possibility that an increase in the noise affecting a system may result in an increase in the signal–to–noise ratio (SNR) in the system's output (signal) has aroused the interest of researchers in different areas of natural science, both biological and non-biological, where nonlinear systems are currently being extensively investigated (see [7] for a collection of recent research papers). In the framework of biological systems, the interest of researchers focused mainly on sensory neurons, as they are essentially intrinsically noisy two-state systems (i.e. firing or not firing). Some evidence of noise-induced switching in periodically stimulated neurons was obtained by Longtin et al. [8] in 1991. Conclusive experimental evidence that an appropriate noise level can improve information transmission in sensory neurons was provided in 1993 in the paper by Moss and colleagues [9] on the mechanoreceptor hair cells of the crayfish *Procambarus Clarkii.*

All these findings have stimulated a renewal of the studies of bistable phenomena, also at higher cognitive levels of neural integration in the brain [10, 11, 12].

A well-known example of bistability or multistability in a cognitive process [13, 14] is given by the perceptual alternation of two (or more) competitive interpretations *percepts* of an ambiguous figure.

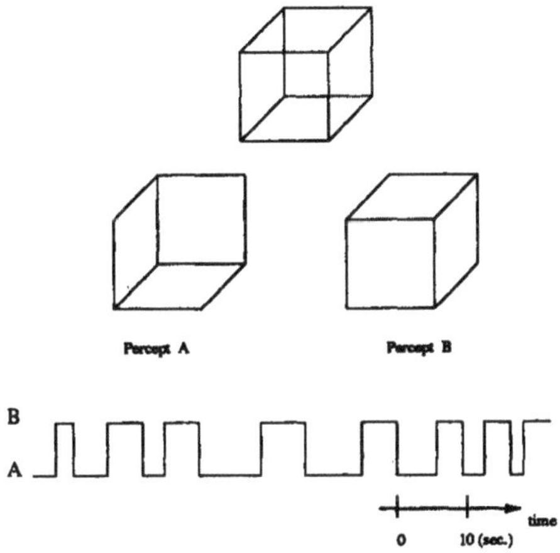

Figure 1: Necker cube with its two alternative interpretations and a sample of the time series of switching behavior recorded during the experiment

In such a figure (see an example in Fig. 1), the same visual input can elicit two (or more) different interpretations, giving rise to a cyclic perceptual alternation of the competitive percepts, during a prolonged observation of the figure. This phenomenon is then characterized by a sequence of time intervals over which one of the two different interpretations of the pattern becomes alternatively the dominant one, (i.e. the two interpretations are never seen simultaneously). Numerous experimental works [15, 17] have described the time course of such perceptual changes and pointed out its stochastic nature. In particular, it has been shown that the stochastic distribution of such reversals is generally well fitted by a Gamma distribution [17, 16, 18]. Therefore, during the observation of one of these patterns (e.g., the Necker cube), a perceptual bistable process with a stochastic time course occurs.

It is then interesting to analyze the dynamic behaviour of a mathematical model of this phenomenon in the presence of a weak periodic perturbation. The analysis begins with the study of the dynamic behaviour of an artificial system that models some properties of the aforesaid perceptual phenomenon; this makes it possible to control the parameters of the system at a non-negligible level.

In the past few years, different models of this phenomenon have been proposed that were based on artificial neural networks [19, 20, 21] or dynamic systems [22, 23, 24]. The neural network model described in this work was developed in our laboratory [21] in order to reproduce the experimental results obtained for ambiguous figures with non-equivalent alternative interpretations [25, 26]. To this end, we used an autoassociative network with two stable states, corresponding to the two different interpretations of a pattern, previously stored in the learning phase of the phenomenon. By changing the parameters of the network, we were able to vary the energies of the two minima associated with the two stable states of the network so that each of the two alternative percepts might have an equal or different probability of being the perceived interpretation. Therefore, it was possible to modify the output of the bistable dynamic system under study so that it might oscillate between equivalent states in a stochastic but symmetric fashion, in the case of equal energy minima, or spend different mean times around the two attractors, in the opposite asymmetric case. The network behaviour was also studied in the presence of a weak periodic perturbation [12], and all the main characteristic signatures of the phenomenon of stochastic resonance were found in the computer simulations of the network dynamics.

2 The neural network model of perceptual alternation

As previously stated in the brief description of the phenomenon of perceptual alternation during the observation of an ambiguous pattern, a stimulus can elicit two alternative interpretations, which, however, are never perceived simultaneously. Then the model has to be based on a neural network divided into two subunits, each consisting of *neurons* associated with the perceptual characteristics of one of the two interpretations, A and B, of the ambiguous stimulus. The excitation of the neurons of the subunit (or population) A must inhibit the subunit B, and vice versa.

The neurons of the network have discrete activation values, 0 and 1, and the network is completely interconnected, that is, each neuron has synaptic connections with all the other network neurons. The connection matrix made up of these synaptic connections is obtained by using a Hebbian rule during the learning phase, i.e., by reinforcing the connection between two neurons (the pre-synaptic and the post-synaptic ones) simultaneously activated during the process of learning the two competitive interpretations of the ambiguous pattern. In this phase, two vectors code the network activity in response to each interpretation of the stimulus. The first of these vectors has positive components for a population, or subunit, (A), and negative values for the opposite population (B); the second vector has positive and negative components that are opposite to the components of the first vector.

As a consequence, the connection matrix consists of four blocks: two square blocks, E_{AA} and E_{BB}, containing the autocorrelations of the related populations, and two blocks, E_{AB} and E_{BA}, containing the inhibitory connections between the two populations. Then, the connection matrix is symmetric and has zero elements on its diagonal, that is, there are no autoconnections.

For the sake of simplicity, we decided to set all non-zero connections in each block to the same value. The resulting connection matrix is:

$$W = \begin{pmatrix} 0 & p & p & \cdot & \cdot & \cdot & \cdot & -r & -r \\ p & 0 & p & \cdot & \cdot & \cdot & \cdot & -r & -r \\ p & p & 0 & \cdot & \cdot & \cdot & \cdot & -r & -r \\ & \cdot & & & & & & \cdot & \\ & \cdot & & & & & & \cdot & \\ -r & -r & -r & \cdot & \cdot & \cdot & \cdot & 0 & q \\ -r & -r & -r & \cdot & \cdot & \cdot & \cdot & q & 0 \end{pmatrix} \tag{1}$$

where p (q) is the intensity of the excitatory connections among the neurons of the population A (B) and $-r$ is the intensity of the inhibitory connections between the neurons of the two different populations.

As shown by Hopfield [27], for a neural network of this type, one can consider an energy function:

$$E = -\frac{1}{2}\sum_{i=1}^{N}\sum_{j=1}^{N} S_i W_{ij} S_j \tag{2}$$

where S_i denotes the activation level of the i-th network neuron and $N = n_A + n_B$ denotes the number of neurons of the network, divided into subunits, or populations, of dimensions n_A and n_B, respectively. Note the close analogy to spin systems: the status of the system elements is described by a two-value single variable; the energy for the neural network has the form of a Hamiltonian for an Ising spin system, where exchange interactions are described by the elements of the matrix W.

The peculiar form of the matrix W defined in (1) allows us to simplify the energy function. One can find that all the network configurations with the same number of neurons a (b) active in the population A (B) are isoenergetic. The related energy values can be written as:

$$E = -\frac{1}{2}a(a-1)p - \frac{1}{2}b(b-1)q + abr. \tag{3}$$

Energy can be considered both as a function of the network state and as a function of the couple of integer numbers (a, b) indicating the active neurons in the two populations.

Let us now introduce into the model a dynamic evolution of the network according to Hopfield's rule. This rule minimizes the energy value by allowing only the transitions that decrease the energy value and in which the activation value of only one neuron at a time can be changed. Hopfield proposed a rule of this type: at the step $t+1$, a network neuron i is randomly drawn, the sum $\sum_j W_{ij}S_j(t)$ is computed, and the new activation value of the neuron i is set to $S_i = 1$, if this sum is positive, otherwise, S_i is set equal to 0. It is very simple to show that, at each step, $\Delta E \leq 0$; consequently, after a finite number of steps, the system stops when a minimum (possibly local) value of the energy function is reached.

In order that the network may model the perceptual alternation phenomenon, one must fulfil the following two constraints:

- only the two configurations of the network that are linked to the two alternative interpretations of the ambiguous figure must be stored as energy minima of the network during the learning phase. Each of these two configurations is made up of a completely active population of one subunit, linked to the perceived interpretation of the pattern, and of a completely inhibited population of the other subunit, linked to the alternative interpretation. Such configurations are $(n_A, 0)$ for the interpretation A and $(0, n_B)$ for the alternative percept B.

- the dynamic evolution of the network must allow the transitions between the two interpretations, i.e., the two energy minima; therefore, Hopfield's rule has to be slightly modified.

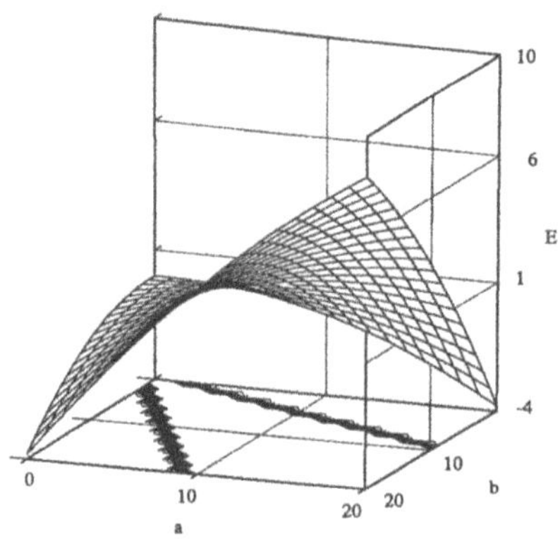

Figure 2: The energy landscape. The basins at $T = 0$, discussed in the text, are bounded by the dark lines lying in the a b plane.

For the first constraint, it is easy to verify [21] that the configurations $(n_A, 0)$ and $(0, n_B)$ are local minima, in the non trivial case of $n_A, n_B > 1$, if $p, q, r > 0$. Moreover, since the energy minima of the network cannot coincide with a generic state of the network [21], it follows that, to ascertain that these configurations are really the only two local minima for the system, we must consider only the other two vertices of the energy domain, $(0,0)$ and (n_A, n_B). Whereas the vertex $(0,0)$ has always the same energy as the first two near stable states (all with energy $E = 0$), the vertex (n_A, n_B) is not an energy minimum if:

$$r \geq \min(\frac{n_A - 1}{n_B} p, \frac{n_B - 1}{n_A} q) \tag{4}$$

Therefore, for any choices of the parameters p, q and r, given the dimensions (n_A and n_B) of the two populations, it is sufficient to check that inequality (4) is verified in order to ensure that only the two vertices considered (i.e., $(n_A, 0)$ and $(0, n_B)$) are energy minima. Once this condition on the parameters has been fulfilled, every network trajectory will end in one of the two configurations stored.

The network configurations can then be divided into three regions: the first includes all the configurations that evolve dynamically toward the minimum $(n_A, 0)$, the second includes all the configurations that evolve toward the minimum $(0, n_B)$, and the third is made up of the configurations that, in every simulation, can evolve randomly toward one or the other of the two minima.

An example of an energy function is given in Fig. 2, which shows a network with symmetric populations, i.e., $n_A = n_B$ and $p = q$. In the figure, the boundary between the two attraction basins of the system is emphasized.

Concerning the second constraint imposed on the network by the phenomenology of perceptual alternation, it is possible to consider the aforesaid partition of the phase space into three regions linked to the stochastic dynamics of the system when it performs transitions from an excited state to one of its fundamental states, at zero absolute temperature ($T = 0$). This dynamic rule can be extended, in a very natural way, to the study of the same system at temperature $T \neq 0$.

The new dynamic rule is based on the algorithm developed by Metropolis [28]: a network neuron is randomly drawn, its activation is inverted (from $0 \to 1$, or vice versa) and the new energy value is computed. The transition is accepted with probability 1, if $\Delta E \leq 0$, and with probability $P = \exp(-\Delta E/kT)$, otherwise.

Obviously, at low temperature values, the system remains close to one of the two energy minima for a long time. It performs transitions between the two basins, with an appreciable frequency, only at intermediate and higher temperatures. Therefore, the effects of the dynamics at non-zero temperature, allow the system to cross over the barrier between the two minima and hence to change the perceptual interpretation of the ambiguous pattern. This probabilistic effect can be regarded as being due to the addition of an amount of *noise* to the neural network; the noise increases with increasing temperature.

This characteristic can be exploited to model perception multistability. The interpretation A of a stimulus is perceived by the system when its state is in the attraction basin of the interpretation A, or when its state is included in the set of states that do not belong to any attractor, and comes directly from the basin A. The interpretation B is perceived in the same way. The transitions between the two perceptions are governed by a two–threshold mechanism similar to the one associated with the transitions of a Schmitt trigger.

Computer simulations of the network evolution were performed to study perception fluctuations. Each simulation included two phases. In the first phase, we let the system *thermalyze* for a certain number of steps (ranging from eight million to sixteen million, depending on temperature). In the second phase (from sixteen million to thirty-two million steps), the perception durations (i.e., the time taken by each perceptive state) were recorded to obtain the histogram of the durations of each population, or the power spectrum of the network output. The duration histograms showed Gamma-like distributions (see Fig. 6 (left) for an example), whereas the power spectra had a Lorentzian shape.

2.1 Stochastic resonance in the model

To verify the occurrence of the phenomenon of stochastic resonance, the network behaviour was studied even in the presence of a periodic perturbation. A suitable way to introduce a periodic signal into the system is to add to the unperturbed energy a sinusoidal term:

$$E = -\frac{1}{2}\sum_{i=1}^{N}\sum_{j=1}^{N} W_{ij}S_iS_j + H_0\sin(2\pi\nu_0 t)\left(\sum_{i\in A} S_i - \sum_{i\in B} S_i\right) \tag{5}$$

where the first term is the unperturbed energy and, in the second term, H_o and ν_0 denote the amplitude and the frequency (measured in $(n.steps)^{-1}$) of the periodic perturbation, respectively; the summations are extended to all neurons of the two populations.

In every half-cycle, the added term favours the activation of a population and inhibits the activation of the competitive population. It follows that, in a half-cycle, the transitions toward one of the two stable vertices are favoured, whereas, in the next half-cycle, the transitions towards the opposite stable vertex are favoured. Therefore, transition probabilities are modulated over time and hence perceptual fluctuations are characterized by a certain degree of consistency with the added signal.

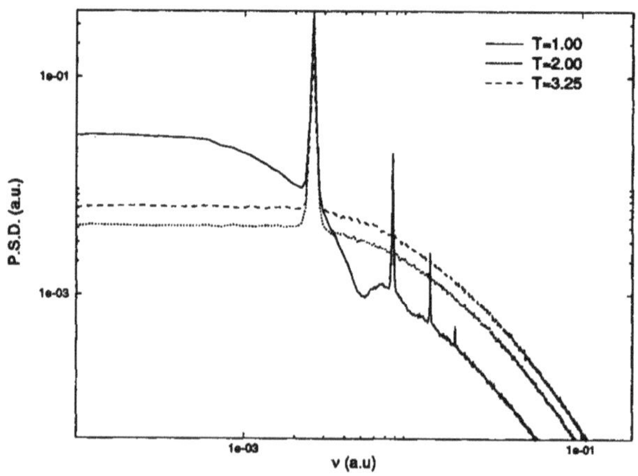

Figure 3: Simulation results for a network made up of two symmetric populations. Power spectra of the time series representing sequential perceptive interpretations for three temperatures: solid, $T = 1.00$; dotted, $T = 2.00$ and dashed, $T = 3.25$ (scaled for $k_B \equiv 1$).

AS such consistency depends on temperature, the effect of the perturbation on the network dynamics was evaluated by analyzing the power spectra of the temporal series of alternations of perceptions at different temperatures. We considered both symmetric populations (i.e., with an equal number of neurons ($n_A = n_B$) and with equal strength of the autoconnections ($p = q$)) and asymmetric populations.

Power spectra generally consist of two components: one refers to the signal and the other to noise. The noise-related component is a wide-band component of Lorentzian shape. The signal-related component includes one or more marked peaks, for the signal frequency and the signal harmonics. The power spectra allow us to derive the signal-to-noise ratios as a function of temperature.

For networks made up of symmetric populations, and for not too low and intermediate temperature values (Fig. 3), the power spectra show, a pronounced peak at the modulation frequency, and other peaks at the odd harmonics. The peak amplitudes depend on temperature: at increasingly high temperatures, the odd-harmonic peaks disappear and only the peak at the fundamental frequency remains.

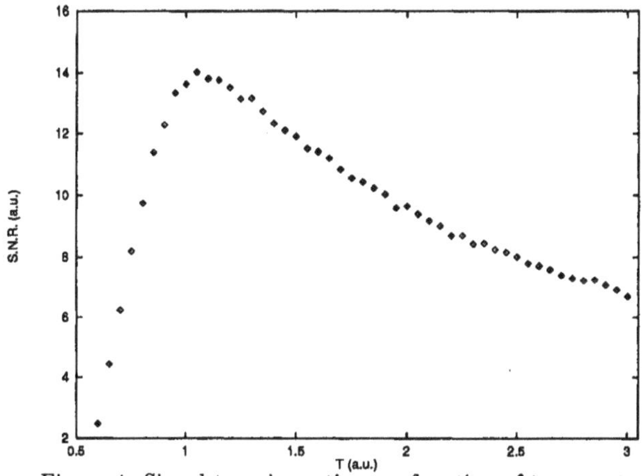

Figure 4: Signal-to-noise ratio as a function of temperature.

The signal-to-noise ratio (SNR), too, is temperature-dependent and shows a clear signature of the phenomenon of stochastic resonance. At low temperatures, the SNR approaches zero; at rising temperatures, it increases up to a maximum value; then, at higher temperatures, the SNR decreases progressively (see Fig.4).

The power spectra related to asymmetric populations and intermediate temperature values show peaks at all the modulation harmonics. At low enough temperatures and at a large modulation amplitude, interesting effects on the spectra were observed. Under such conditions, the effects of autocorrelation zeros predicted by Shneidman et al. [29] appear as large pronounced dips at the even modulation harmonics. At higher temperatures, such dips are less pronounced, and the usual peaks reappear at these frequencies. The power spectrum in Fig. 5 shows both features (i.e., dips and usual peaks).

The use of residence-time histograms constitutes another approach to the study of

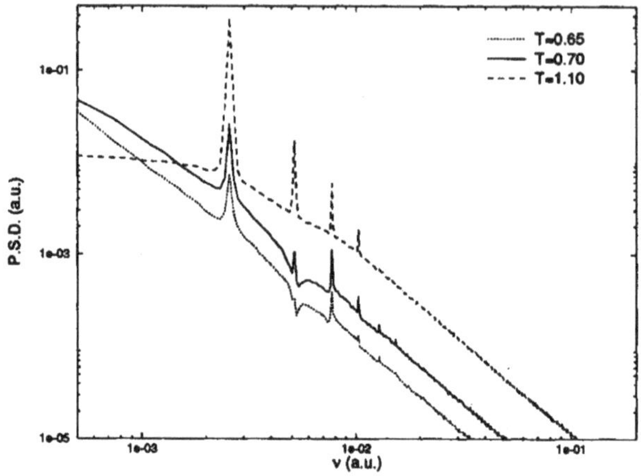

Figure 5: Asymmetric populations. Power spectra for three temperatures values: dotted, T=0.65; solid, T=0.70; dashed, T=1.10.

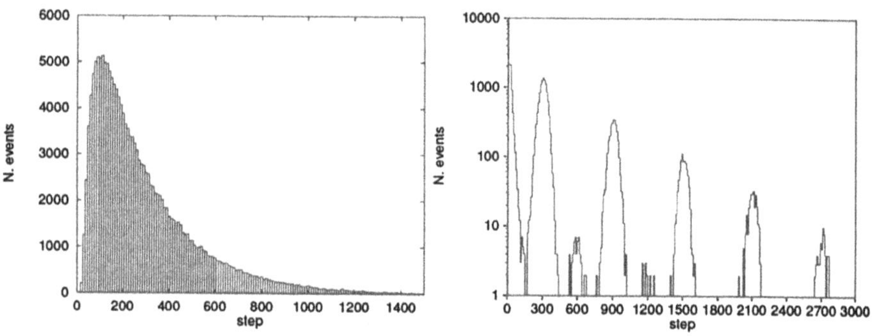

Figure 6: Residence time histograms for the neural network with (right) or without (left) the periodic modulation.

stochastic resonance. Without a periodic modulation, these histograms show duration-time distributions similar to Gamma distributions. With periodic modulation, the same histograms are multimodal and show peaks centred at the odd multiples of the modulation half cycle. The peak amplitudes decrease exponentially, as shown by the semi-log plot reported in Fig. 6 (right).

3 Conclusions

The nature of the role played by noise in perceptual processes is not yet well known, even though this role has long been acknowledged. In particular, it is not clear whether noise may be useful or not. Experimental results on the auditory nerve of the squirrel-monkey[8] and on the crayfish mechanoreceptors[9] prove that noise-induced fluctuations are basic to sensorial-information coding. As pointed out by the results reported in the previous section, it is very important that the methodologies usually adopted to study nonlinear systems affected simultaneously by noise and periodic modulations should be extended to the study of neural systems even at the cognitive levels. Moreover, the occurrence of stochastic resonance has been observed by us in a simple, yet realistic, model of the phenomenon of perceptual alternation of ambiguous figures. This fact points out the possible usefulness of noise also at the higher cognitive levels.

References

[1] G. Gerstein, and B. Mandelbrot, Random walk models for the spike activity of a single neuron, *Biophys. J.*, **4**, 41, (1964).

[2] J. Taylor, Spontaneous Behaviour in Neural Networks, *J. Theor. Biol.*, **36**, 513, (1972).

[3] R. Benzi, S. Sutera, and A. Vulpiani, The mechanism of stochastic resonance, *J. Phys. A*, **14**, *L*453, (1981).

[4] F. Moss, in *Contemporary Problems in Statistical Physics*, ed. G. H. Weiss, SIAM, 205 (1994).

[5] K. Wiesenfeld, D. Pierson, E. Pantazelou, C. Dames, and F. Moss, Stochastic resonance on a circle, *Phys. Rev. Lett.*, **72**, 2125, (1994).

[6] F. Moss, D. Pierson, and D. O'Gorman, Stochastic resonance: tutorial and update, *Int. J. of Bifurc. and Chaos*, **4**,1383, (1994).

[7] *Proc. NATO Advanced Research Workshop on Stochastic Resonance in Physics and Biology*, edited by F. Moss, A. Bulsara and M. F. Shlesinger, J. Stat. Phys. **70**, 1 (1993).

[8] A. Longtin, A. Bulsara, and F. Moss, Time-Interval sequences in bistable systems and the noise-induce d trasmission of information by sensory neurons, *Phys. Rev. Lett.*, **67**, 656, (1991).

[9] J. K. Douglass, L. Wilkens, E. Pantazelou, and F. Moss, Noise enhancement of information transfer in crayfish mechanorec eptors by stochastic resonance, *Nature*, **365**, 337 (1993)

[10] D. R. Chialvo, and A. V. Apkarian, Modulated noisy biological dynamics: three examples, J. Stat. Phys., **70**, 375, (1993).

[11] J. Kelso, P. Case, T. Holroyd, E. Horvath, J. Raczaszek, B. Tuller, and M. Ding, in *Multistability in Cognition*, eds. P. Krouse, and M. Stadler, Springer Series in Synergetics, Springer.

[12] M. Riani, and E. Simonotto, Stochastic resonance in the perceptual interpretation of ambiguous figures: a neural network model *Phys. Rev. Lett.*, **72**, 3120, (1994).

[13] F. Attneave, Multistabilty in Perception, *Scientific American*, **225**, 62 (1971).

[14] K. T. Brown, Rate of apparent change in a dynamic ambiguos figure as a funcio n of observation time, *Amer. J. Psychology*, **68**, 358 (1955).

[15] A. Borsellino, A. De Marco, A. Allazetta, S. Rinesi, and B. Bartolini, Reversal Timc diotribution in perception of visual ambiguos stimuli, *Kibernetik*, **10**, 139 (1972).

[16] A. De Marco, P. Penengo, A Trabucco, A. Borsellino, F. Carlini, M. Riani, and M.T. Tuccio, Stochastic Models and Fluctuations in Reversal Time of Reversible Figures *Perception*, **6**, 645 (1977).

[17] A. Borsellino, F. Carlini, M. Riani, M. T. Tuccio, A. De Marco,P. Penengo, and A. Trabucco, Effects of Visual Angle on Perspective Reversal for Ambiguos Patterns, *Perception*, **11**, 263 (1982).

[18] M. Riani, G. A. Oliva, G. Selis, G. Ciurlo, and P. Rossi, Effect of luminance on perceptual alternation of ambiguos patter ns *Percept. & Mot. Skills*, **58**, 267, (1984).

[19] A. H. Kawamoto, and J. A. Anderson, A Neural Network Model of multistable Perception, *Acta Psychologica*, **59**, 35 (1985).

[20] M. Riani, F. Masulli, and E. Simonotto, Perceptual alternation of ambiguos patterns: a model based on an artificial neural network, in *Proc. SPIE*, **1469**, 166 (1991).

[21] F. Masulli, M. Riani, E. Simonotto, and F. Vannucci, Boltzmann ditributions and neural networks: models of unbalanced interpretations of reversible patterns, in *Proc. SPIE*, **1710**, 267 (1992).

[22] T. Ditzinger and H. Hacken, Oscillations in the perception of Ambiguos Patterns, *Biol. Cybern.*, **61**, 279 (1989).

[23] T. Ditzinger and H. Hacken, The impact of fluctuations on the recognition of ambiguos patterns *Biol. Cybern.*, **63**, 453 (1990).

[24] H. Gang, T. Ditzinger, C. Z. Ning, and H. Haken, Stochastic resonance without external periodic force, *Phys. Rev. Lett.*, **71**, 807 (1993).

[25] M. Riani, M. T. Tuccio, A. Borsellino, J. Radilova, and T. Radil, Perceptual ambiguity and stability of reversible figures, *Percept. & Mot. Skills*, **63**, 191 (1986).

[26] F. Masulli and M. Riani, Ambiguity and Structural Information in the perception of reversible figures, *Perception and Psychophysics*, **45**, 501 (1989).

[27] J. J. Hopfield, Neural networks and physical systems with emergent collective computation abilities, *Proc. Nat. Acad. Sci. USA*, **79**, 2554 (1982) and**81**, 3088 (1984).

[28] N. Metropolis, A. Rosenbluth, M. Rosenbluth, A. Teller, and E. Teller, Equation of state calculations by fast computing machines, *J. Chem. Phys.*, **21**, 1087 (1953).

[29] V. A. Shneidman, P. Jung, and P. Hänggi, The weak-noise limit of stochastic resonance, *Phys. Rev. Lett.*, **72**, 2682 (1994).

BOTTOM-UP AND TOP-DOWN INTERACTIONS IN MULTISTABLE PATTERN PERCEPTION

CLOE TADDEI-FERRETTI[1], CARLO MUSIO[1], SILVIA SANTILLO[1], ROMANO F. COLUCCI[2] & ANTONIO COTUGNO[1]

[1]*Istituto di Cibernetica, CNR, 80072 Arco Felice (NA), Italy*
[2]*Datitalia Processing S.p.A., Napoli, Italy*

Introduction

The mapping between an object and its retinal image is always indirect and ambiguous, because it is always many-to-one. An inference process is needed to obtain a perception. Such inference leading to the perception event utilizes both the information flowing in the bottom-up direction from the sensory input and that flowing in the top-down direction from the association systems of the brain.

The information flowing in the bottom-up direction is used for the generation of 3D representations of objects, which develop in an automatic, preattentive, rapid, unavoidable way, by utilizing feature processing mechanisms [17-18, 33].

It is assumed that the 3D representation is object-centered in order to avoid continuous re-description of the object shape with the viewpoint change. However, recent results speak in favour of multiple viewpoint specific representations from which 3D object structure is inferred by view interpolations [3, 24].

The information flowing in the top-down direction is used in a non-automatic, attentive, slow way, which is flexible in accordance to different circumstance requirements and is utilized for the object identification.

There has been no general consensus regarding the relative contribution of the bottom-up and top-down information and the kind of their interaction. While some authors [9-11] stressed the prevailing importance of the bottom-up information, other authors, since Bruner, 1957, [2, 13-14, 19] have given evidence of the effects due to attention, memory, imagery, expectation, previous knowledge, and intention on the perception mechanism.

In the case of the multistable patterns, two (in some cases, more than two) main interpretations may be equally inferred to be consistent with the sense data, giving rise to two equally possible perceptions. None of the possible interpretations is preferential in the light of past experience and the two interpretations continuously alternate.

There are several types of multistable patterns and the influence of many factors on the alternance mechanism as been ascertained [8]. Experimental and theoretical researches on the perception of perspective-reversible figures are well known [1, 4, 25-29].

We have taken into account the influence of only few factors on the perception alternance of perspective-reversible figures: 1) a bottom-up factor, consisting in the fact that a subliminal presentation of a biased version of a multistable pattern affects the unbiased version perception [12, 20]; 2) a particular top-down factor, that does not intervene in usual perception events and that consists in the fact that a voluntary control over the perception reversal is possible [16, 21-23]. In particular, we investigated the relative contribution of such bottom-up and top-down factors. We focused on the already well studied Necker cube. Our preliminary results are published elsewhere [5, 30-32].

Materials and Methods

A special purpose, specifically designed and implemented programming language was used for the multistable pattern and subliminal stimuli presentation, for the monitoring of the subject reactions and for the organization of the output data [6].

We used a modification of the Necker cube, in which the two internal vertices coincide: in such a way it is avoided the increased probability of perception reversal induced by the fixation of one or the other of such vertices, which can be consedered "focal areas" [8] for selective interpretations of the pattern [7, 15].

This pattern gives rise to the perception either of a flat figure, or of a cube oriented either to one directio or to the opposite one (either to the left or the right, if the main axis of the pattern is aligned horizontally; either above or below, if it is aligned vertically), in spite of the fact that both perceived cubes are perspectively impossible (and, at a more rational inspection, they have to be interpreted as two-dimensional sketches of truncated pyramids).

The pattern was presented in front of each of five subjects, at the centre of a computer screen, at a distance of 50 cm, under a view angle of 15°. The beginning of each type of perception was manually signalled by the subject through a mouse and the space bar of a computer keyboard.

We used different experimental protocols, each one composed by a variable number of 2 min duration phases of pattern presentation.

Each experiment was structured in 3 main types: it included either phases of subliminal presentation of biased versions of the pattern, or phases of continuous application of a sustained will effort to obtain and/or maintain one particular interpretation of the pattern, or phases with the application of both types of factors, always in addition to control phases with presentation of the pattern in undisturbed conditions. The requested fixation point was always the centre of the pattern.

In the course of a phase of subliminal stimulation, single biased versions of the pattern replaced, each time during 14 ms, the normal pattern continuously administered during that phase. It was previously ascertained for each subject that 14 ms is a period not sufficient for the generation of a conscious perception of the presented figure. The biased versions of the pattern were such that the thickness doubling of some lines either at one or the other side of the pattern itself suggested either one o the other of the two possible perspective orientations, thus acting each as a seconding stimulus for one type of interpretation of the pattern and a contrasting stimulus for the opposite interpretation.

Results

1) General observations

No difference was observed between left and right interpretation preference with a horizontally oriented pattern, while a slight above-interpretation preference with respect to the below one was noted with a vertically oriented pattern. For this reason, in the following experiments performed in order to investigate the effect of several factors on the reversal mechanism, only the horizontal orientation of the pattern was used.

In all cases, the perception of the pattern as a flat figure alternated with the various types of perception as a solid figure, from a minimum of 1 time to a maximum of 23 times for each phase. The mean number of perception of the flat figure during one phase was 8,887.

The perception of the pattern as a flat figure was less favoured than that as a solid one. It occurred either as a station between two opposite perceptions as solid figures, or as a station between a solid figure perception and the same perception at a subsequent time.

A higher number of perception reversal was observed in the final control phase with undisturbed condition if compared with the first control phase with undisturbed condition.

2) Bottom-up influence

The mean of the reaction time at the first pattern presentation was 3,266 s. Lower values were detected after subliminal stimulations.

A difference was observed between: i) the number of perception alternations when each single perception of the pattern as a cube (either oriented to one direction, or the opposite one) was contrasted by a subliminal pattern presentation 1000 ms after the signalling of the beginning of such perception, and ii) the number of perception alternations when a seconding subliminal pattern was administered in the same conditions. The mean of the ratios between the first and the second situation was 1,644.

The above effect of the subliminal stimulus decreased with the increase (3000 ms, 5000 ms) of the time interval between the perceptual signalling and the stimulus presentation.

When a continuous alternance of presentation of the two types of subliminal patterns occurred every 2000 ms (total period 4000 ms), which suggested alternately one and the other of the two possible choices of the solid pattern interpretation, a matching between the type of each signalled perception and of the immediately preceding suggested pattern was observed. The ratio between the mean number of the matching and not matching cases was 4,745.

By increasing the alternance period between the presentation of the two above subliminal stimulus (4000 ms, 6000 ms; total period 8000 ms, 12000 ms), the effect was less marked.

3) Top-down influence

A sustained willing effort was continuously applied during some phases of an experiment in order to obtain and/or maintain the preception of the pattern as a solid oriented to one direction (either to the left, or to the right) and during other phases in order to obtain the opposite orientaion.

Positive values of the ratios between the number of cases of matching between one type of perception and the same type of willing interference and the number of matching between the same perception and the opposite type of willing interference were obtained. The mean of the ratios was 12.772.

When presenting two neighbour identical patterns (or two neighbour classical Necker cubes), it was also possible, by volitional effort, to perceive them with different orientation one with respect to the other.

The task was easier if the effort was not in order to obtain two different orientations, but to maintain for one pattern the orientation already obtained for both patterns and to change the orientation for the other one.

4) Reciprocal interference of the two types of influence

Subliminal stimuli, which either second (during some experimental phases) both the perception of the pattern as a solid oriented to the left and that to the right, or contrast (during different experimental phases) both the perception to the left and that to the right, were applied 1000 ms after each signalling of a perception of the pattern as a solid oriented either to the left or to the right, in addition to the continuous application of a sustained willing effort (during the same phases) to obtain and/or maintain the perception of the orientation either to the left or to the right.

Positive values of the ratios between the number of cases of one type of orientation perception under one type of interference combination (in which the will effort was matched by the perception) and the number of the same type of orientation perception under the different type of interference combination (in which the opposite will effort was matched by the perception) were obtained, both in the case of seconding subliminal stimuli and in the case of contrasting ones.

Better results were obtained with contrasting subliminal stimulus than with seconding ones. The mean of the above ratios was 1.793 in the case of stimuli seconding both types of perception, and 12.727 in the case of cotrasting stimuli.

Discussion

The analysis of the results lead us to formulate the following comments.

First, the flat-figure interpretation of the pattern as a station between two identical solid-figure interpretations suggests that the flat interpretation be a zeroing of the mechanism of the solid-figure interpretation reversal; after such zeroing, the perception reversal starts again and the last previous interpretation may be either resumed or not. It is possible to interfere to the perception alternance mechanism by means of subliminal either contrasting or seconding stimuli that, if presented alone, are not sufficient to produce a conscious perception. It is possible to entrain the perception alternance mechanism by means of repetitive subliminal stimuli. All that speaks in favour of the existence of two systems triggering the two interpretations of the sensory input and producing the perception alternance at an unconscious level by mutually inhibiting each other.

Second, a cultural bias of the sensory input interpretation cannot be excluded for the preference of the pattern interpretation as a cube oriented above with respect to the below interpretation in the case of a vertically oriented pattern. This cultural factor is of the same kind of other factors, such as expectation, previous knowledge, that are known to interfere with the interpretation process that gives rise to a perception. However, in the case of multistable patterns, the ambiguity of which cannot be definitely resolved, also the willing factor, that does not intervene in usual perception events, may play a role in a conscious way on the perception generation acting at the level of the interpretation alternance mechanism.

Third, the fact that, when subliminal stimuli unconsciously received were added to a willing effort consciously applied, better results were obtained with subliminal stimuli contrasting both the willed and non willed perception than with subliminal stimuli seconding both the willed and non willed perception is an indication that the effect of contrasting and seconding stimuli do not simply add algebrically. This means that non-linear interactions intervene among top-down and bottom-up influences of the same and different sign, as well as that the influences characterized by same type and different sign upon the two types of solid-figure perception have possible different weight. Further experiments with combinations of subliminal stimuli and willing interference, in which the two types of interference have always the same sign, should clarify this point.

In summary:

i) by using subliminal stimuli that, without reaching the conscious level, affect the perception reversal rate of a multistable pattern, either by seconding or contrasting a previous interpretation of the pattern, we have confirmed the contribution of the bottom-up flow of sensory information to the constitution of the possible perception of such multistable pattern;

ii) by means of a sustained willing effort, that is selectively directed toward one of the two possible perceptions of the pattern as a solid and that affects the relative duration of the two types of perception, we have ascertained the contribution of the top-down flow of mental information;

iii) with a simultaneous application of bottom-up and top-down inputs, variously combined, the concurred contribution of both bottom-up and top-down factors on the perception generation has been observed; non linear interactions between bottom-up and top-down inputs have been evidentiated.

Acknowledgements

We are indebted to the late E.R. Caianiello for fruitful comments and suggestions.

References

1. Borsellino, A., A. De Marco, A. Allazetta, S. Rinesi & S. Bartolini, (1972), Reversal time distribution in the perception of visual ambiguous stimuli, *Kybernetik*, 10:139-144.
2. Bruner, J.S., (1957), On perceptual readiness, *Psychol. Review*, 64:123-152.
3. Bülthoff, H.H., (1994), Image based object recognition: Psychophysics, *Proc. 17th ENA Meeting*, Wien, 40.05.
4. Caglioti, G. & E.R. Caianiello, (1978), A model for non-resolvable ambiguities, *Biol. Cybern.*, 31:205-208.
5. Colucci, R.F., A. Cotugno, M. Martino, C. Musio & C. Taddei Ferretti, (1992), Influence of subliminal stimulation on multistable visual pattern perception, *Neurosci. Lett.*, 43:S28.
6. Colucci, R.F., C. Musio & C. Taddei-Ferretti, (1994), EXPLAN - A programming language for complex visual stimuli presentation, *Int. J. Bio-Medical Computing*, 37:29-39.
7. Ellis, S.R. & L. Stark, (1978), Eye movements during the viewing of Necker cubes, *Perception*, 7:575-581.
8. García Pérez, M.A., (1992), Eye movement and perceptual multistability. *The Role of Eye Movements in Perceptual Processes*, C. Chekaluk & R.K. Llewellyn, eds., Amsterdam, London, New York, Elsevier, 73-109.
9. Gibson, J.J., (1950), *The Perception of the Visual World*, Houghton Mifflin, Boston.
10. Gibson, J.J., (1966), *The Senses Considered as Perceptual Systems*, Houghton Mifflin, Boston.

11. Gibson, J.J., (1979), *The Ecological Approach to Visual Perception*, Houghton Mifflin, Boston.
12. Goolkasian, P., (1987), Ambiguous figures: Role of context and critical feature, *J. Gen. Psychol.*, 114:217-228.
13. Gregory, R.L., (1970), *The Intelligent Eeye*, McGraw-Hill, New York.
14. Gregory, R.L., (1972), Seeing as thinking, *Times Literary Suppl.*, June 23.
15. Kawabata, N., K. Yamagami & M. Noaki, (1978), Visual fixation points and depth perception, *Vision Res.*, 18:853-854.
16. Liebert, R.M. & B. Burk, (1985), Voluntary control of reversible figures, *Percept. & Mot. Skills*, 61:1307-1310.
17. Marr, D., (1976), Early processing of visual information, *Phil. Trans. Roy. Soc. London B*, 275:483-524.
18. Marr, D. & H.K. Nishihara, (1978), Representation and recognition of the spatial organization of three-dimensional shapes, *Proc. Roy. Soc. B*, 200:269-291.
19. Neisser, U., (1967), *Cognitive Psychology*, Appleton-Century-Crofts, New York.
20. Owen, L.A., (1985), Dichoptic priming effects of ambiguous picture processing, *British J. Psychol.*, 76:437-477.
21. Pelton, L.H. & C.M. Solley, (1968), Acceleration of reversals of a Necker cube, *Amer. J. Psychol.*, 81:585-588.
22. Peterson, M.A. & I. Hochberg, (1983), Opposed-set measurement procedure: A quantitative analysis of the role of local cues and intention in form perception, *J. Exp. Psychol.: Human Percept. & Performance*, 9:183-193.
23. Phillipson, O.T. & J.P. Harris, (1984), Effects of chlorpromazine and promazine on the perception of some multi-stable visual figures, *Quart. J. Exp. Psychol.*, 36A:291-308.
24. Poggio, T. & S. Edelman, (1990), A network that learns to recognize three-dimensional objects, *Nature*, 343:263-266.
25. Radilova', J. & E. Pöppel, (1990), The perception of figure reversal as a function of contrast reversal exemplified with the Schröder staircase, *Acta Neurobiol. Exp.*, 50:37-40.
26. Radilova', J. & T. Radil-Weiss, (1984), Subjective figure reversal in two- and three-dimensional perceptual space, *Int. J. Psychophysiol.*, 2:59-62.
27. Riani, M., F. Masulli & E. Simonotto, (1991), Perceptual alternation of ambiguous patterns: a model based on an artificial neural network, *SPIE Application of Artificial Neural Networks II*, 1469, 166-177.
28. Riani, M., G.A. Oliva, G. Selis, G. Ciurlo & P. Rossi, (1984), Effect of luminance on perceptual alternation, *Percept. & Mot. Skills*, 58:267-274.
29. Riani, M., M.T. Tuccio, A. Borsellino, J. Radilova' & T. Radil, (1986), Perception ambiguity and stability of reversible figures, *Precept. & Mot. Skills*, 63:191-205.

30. Taddei-Ferretti, C., C. Musio & R.F. Colucci, (1994), Top-down intrerference in visual perception, *ICANN '94*, M. Marinaro & P.G. Morasso, eds, Berlin, Heidelberg, New York, Springer, 30-33.
31. Taddei-Ferretti, C., C. Musio & S. Santillo, (1995), Non linearity of the interactions between bottom-up and top-down signals in multistable visual perception, *Il Nuovo Cimento D*, 17:941-947.
32. Taddei-Ferretti, C., C. Musio, R.F. Colucci & M. Martino, (1993), Multistable ambiguous pattern perception, *WCNN-INNS '93*. vol. II, Lawrence Erlbaum Ass., Hillsdale, N.J., vol. II, 155-158.
33. Ullmann, S., (1984), Visual routines, *Cognition*, 18:97-159.

Toward a Theory of Visual Abductive Thinking

Lorenzo Magnani
Guido Previde Massara
Computational Philosophy Laboratory
Department of Philosophy
University of Pavia
I–27100 Pavia - Italy
lorenzo@philos.unipv.it, guido@philos.unipv.it

Abstract

The general objective is to consider how the use of visual mental imagery in thinking may be relevant to hypothesis generation. There has been little research into the possibility of visual and imagery representations of hypotheses, despite abundant reports that imaging is crucial to scientific discovery. Glasgow and Papadias have suggested an interesting technique for combining image–like representations and processing with linguistic information. The system they describe can make inferences using such cognitive representations as visual mental imagery ones: it achieves conclusions without laborious chains of inferences. We plan to explore whether this kind of hybrid imagery/linguistic representation can be modified and used to model image–based hypothesis generation, that is to delineate the first cognitive and computational features of what we call *visual abduction.*

1 Introduction

Many psychological and physiological studies have been carried out to describe the multiple functions of mental imagery processes: there exists a visual memory that is superior in recall; humans typically use mental imagery for spatial reasoning [1]; images can be rebuilt in creative ways [3]; they preserve the spatial relationships, relative sizes, and relative distances of real physical objects [9]; for a more complete list, see Tye [28].

Kosslyn introduces visual cognition as follows:

> Many people report that they often think by visualizing objects and events [...] we will explore the nature of visual cognition, which is the use of visual mental imagery in thinking. Visual mental imagery is accompanied by the experience of seeing, even though the object or event is not actually being viewed. To get an idea of what we mean by visual mental imagery, try to answer the following questions: [...] How many windows are there in your living room? If an uppercase version of the letter *n* were rotated 90° clockwise, would it be another letter?" ([11] p. 128).

We can build visual images on the basis of visual memories but we can also use the recalled visual image to form a new image we have never actually seen. Certainly, imagery is used in everyday life, as illustrated by the previous simple answers, nevertheless imagery has to be considered as a major medium of thought, as a mechanism of thinking relevant to hypothesis generation. Some hypotheses naturally take a pictorial form: the hypothesis that the earth has a molten core might be better represented by a picture that shows solid material surrounding the core. There has been little research on the possibility of visual imagery representations of hypotheses, despite abundant reports (e.g. Einstein and Faraday) that imaging is crucial to scientific discovery. It is well-known that the German chemist Kekulé used spontaneous imagery to discover the structure of benzene; Watson and Crick have reported the use of mental imagery in the interpretation of diffraction data and in the determination of the structure of the DNA molecule [7,13,15,16,26,27].

2 Computational Scheme and Abduction

The central theme of the recent imagery debate in the cognitive science has concerned the problem of representation.

How can we represent images? Are mental images represented depictively in a picture or like sentences of descriptions in a syntactic language?

According to Kosslyn's *depictionist* or *pictorialist* view [10], mental images are *quasi-pictures* represented in a specific medium called the *visual buffer* in the mind. Kosslyn builds the analogy that visual information stored in memory can be displayed as an image on a CRT screen. Kosslyn's model of mental imagery proposes three classes of processes that manage images in the visual buffer: the *generation* process forms an image exploiting visual information stored in long-term memory, the *transformation* process (for example, rotation, translation, reduction in size, etc.) modifies the depictive image or views it from different perspectives, the *inspection* process explores patterns of cells to retrieve information such as shape and spatial configuration. According to Pylyshyn's *descriptionist* view [21] mental imagery can be explained by the tacit knowledge used by humans when they simulate events rather than by a pictorialist view related to the presence of a distinctive mental image processor.

As shown by certain experimental results, Hinton proposes that imagery involves viewer-centered information appended to object-centered structural descriptions of images [6]. Finke [2] proposes five unifying principles of mental imagery to summarize some of the properties of imagery that a computational model should represent. Finally, Tye has recently suggested a theory, related to some of the aspects of Marr and Nishihara's theory of visual perception [14], in which mental images are represented as interpreted, symbol-filled arrays [28].

In Glasgow and Papadias' computational model [4,5] the imagery processes are simplified by a depictive knowledge representation scheme that involves inferencing techniques related to generation, inspection and transformation of the representation. According to Kosslyn's cognitive model, the knowledge representation scheme of mental imagery is composed of two different levels of reasoning, *visual* and *spatial*, the former concerned with what an image looks like, the latter depending on where an object is located relative to other objects. The different representations of these ways of reasoning exist at the level of working-memory and are generated (according to Kosslyn's theory of imagery [9]) from a *descriptive* representation of an image stored in long-term memory in a hierarchical organization. The separate long-term memory contains visual and spatial information as descriptions.

The spatial representation outlines the image constituents in a multi-dimensional, symbolic array, that maintains spatial and topological properties and avoids distracting details, in accordance with the features of the image domain and the questions involved;

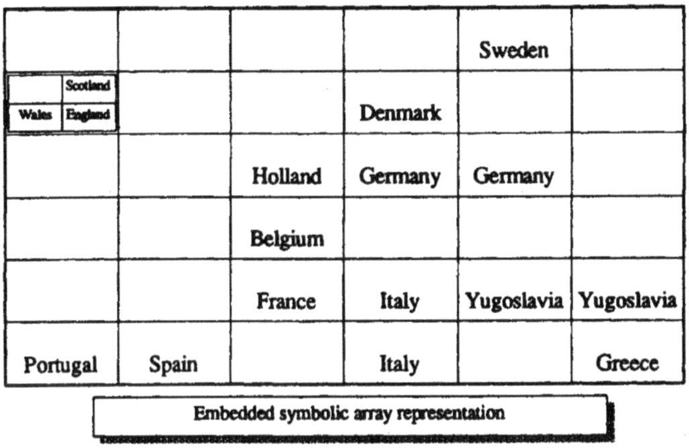

Figure 1: Embedded symbolic array representation [5]

moreover, the arrays are *nested*, to express multiple levels of the structural hierarchy of images (Fig. 1).

The spatial representation is processed using primitive functions that *transform* and *inspect* the symbolic array. The functions implemented express simple operations and reasoning techniques based on the transformation and inspection of image representations.

Moreover, a multi–dimensional array is able to depict spatial (e.g. *left–of*, *north–of*) and topological (e.g. *adjacent–to*, *inside*) relations among image parts and can also denote non–spatial dimensions such as speed. Sometimes, a part occupies more than one element in the array, to capture all the spatial relationships involved. Hence, an image frame is related to an individual image or to a class of images sharing the same features. Images can be represented in a viewer– or an object–centered frame of reference: unlike Kosslyn's approach, in Glasgow and Papadias' computational model [4] image representations are typically three–dimensional and object–centered (viewer–independent).

The "uninterpreted" visual representation depicts the space occupied by an image as an *occupancy array*. The related primitive functions concern operations for manipulating and retrieving geometric information, such as *volume*, *shape* and *relative distance* (rotate, translate and zoom, which change the orientation, location or size of a visual image representation). In a nested symbolic array each symbol in the array corresponds to a frame in long–term memory: we can extract a subimage and express it by its symbolic array parts.

The main objective is now to consider how the use of visual mental imagery in thinking may be relevant to hypotheses generation. What is abduction? After having illustrated the concept of abduction we propose an image–based hypothesis generation.

Abduction is becoming an increasingly popular term in AI [18,19,20,23,24,25] especially in the field of medical KBSs [8,12,22]. A hundred years ago, Peirce [17] interpreted abduction essentially as an inferential *creative process* of generating a new hypothesis.

There are two main epistemological meanings of the word abduction: 1) abduction that only generates plausible hypotheses (*selective* or *creative*)[1] and 2) abduction considered as *inference to the best explanation*, that also evaluates hypotheses. All we can expect of our "selective" abduction, is that it tends to produce hypotheses that have some chance of turning out to be the best explanation. Selective abduction will always produce hypotheses that give at least a partial explanation and therefore have a small amount of initial plausibility. In this respect abduction is more efficacious than the blind generation of hypotheses.

We should remember, as Peirce noted, that abduction plays a role even in relatively simple *visual phenomena*. *Visual abduction*, a special form of abduction, occurs when hypotheses are instantly derived from a stored series of previous similar experiences. It covers a mental procedure that tapers into a non–inferential one, and falls into the category called "perception". Philosophically, *perception* is viewed by Peirce as a fast and uncontrolled knowledge–production procedure. Perception, in fact, is a vehicle for the instantaneous retrieval of knowledge that was previously structured in our mind through inferential processes. By perception, knowledge constructions are so instantly reorganized that they become habitual and diffuse and do not need any further testing. Many visual stimuli are ambiguous, yet people are adept at imposing order on them: "We readily form such hypotheses as that an obscurely seen face belongs to a friend of ours, because we can thereby explain what has been observed" ([24] p. 53). This kind of image–based hypothesis formation can be considered as a form of *visual abduction*.

3 Visual Abduction

3.1 Image–Based Explanation

We will now discuss, from a computational perspective, a *visual abductive problem solving strategy*. To provide manageable bounds to our very general objective, i.e. to analyze the role of visual hypothesis generation, which is so crucial to scientific discovery, we have initially limited ourselves to the subtask of illustrating some structurally similar examples from the field of common sense reasoning, where it is very easy to find many cases dealing with what we have just called visual abductive problem solving. Moreover, we have limited ourselves to the spatial representation.

The spatial representation does not add information that cannot be expressed by propositions; notwithstanding this, the spatial representation is not computationally equivalent to a descriptive one. In several imagery–related tasks (e.g. inspecting) spatial representation may reduce the computational complexity of the solution.

Although there is considerable agreement concerning the existence of a high–level visual and spatial medium of thought as a mechanism relevant to abductive (selective and creative) hypothesis generation, the underlying cognitive processes involved are still not well understood. Notwithstanding this, we will attempt to work around this gap in our understanding: although describing a computational model able to "imitate" the real ways the human brain works when it makes visual abductions would be best, our

[1] To illustrate from the field of medical knowledge, the discovery of a new disease and the manifestations it causes can be considered as the result of a creative abductive inference. Therefore, *creative* abduction deals with the whole field of the growth of scientific knowledge. This is irrelevant in medical diagnosis where instead the task is to *select* from an encyclopedia of pre–stored diagnostic entities [22].

primary concern is its expressiveness, efficiency and inferential adequacy, rather than its explanatory and predictive power as regards psychological research.

Let us consider the example above concerning the ambiguity of an obscurely seen face: more generally, we can face an *initial* (eventually)[2] observed image in which we recognize a problem to solve. For example, given a visual or imagery datum, we may have to explain: 1) the absence of an object; 2) why an object is in a particular position; or 3) how an object can achieve a given task moving itself and/or interacting with the remaining objects in the scene/image. How can "visual" reasoning perform this explanation? To answer this question it is necessary to show how visual abduction may be relevant to hypothesis generation, that is, how an *image-based explanation* is able to solve the problem given in the initial image.

Faced with the initial image, in which we have previously recognized a problem to solve, as stated above, we have to work out an *imagery hypothesis* that can explain the problem-data. Thus, the formed image will acquire a hypothetical status in the inferential abductive process at hand.

The generation of a "new" imagery hypothesis can be considered the result of the *creative abductive* inference previously described; in this respect we can consider how the imagery representations of new hypotheses lead to scientific discovery. The selection of an imagery hypothesis from a set of pre-enumerated imagery hypotheses, stored in long-term memory, also involves abductive steps, but its creativity is much weaker: this type of visual abduction can be called *selective* [12] (see Sect. 2).

3.2 Imagery Hypotheses

As stated above, to illustrate the role of visual hypotheses generation we have initially limited ourselves to the subtask of illustrating some structurally similar examples from the field of common sense reasoning. It is not difficult to mention a few cases in which we recognize a visual or imagery datum as containing a problem to solve. As we have seen in the previous section, we could have to explain: 1) the absence of an object; 2) why an object is in a particular position or 3) how an object can achieve a given task moving itself and/or interacting with the remaining objects of the scene/image.

We shall clarify the first case with the following example, which is based on common sense reasoning: we see a broken horizontal glass on the floor, near the table. On the floor there are also some leaves and we see that the window is open. If we retrieve from long-term memory another *visual* (imagery) description still containing the glass (intact), the table, and the window, and we recognize this new representation as being a slightly different version of the previous one, we have to explain the presence of the leaves and broken glass in the initial image. They constitute an *anomaly* that needs to be solved (explained). If we can link the leaves to the presence, say, of wind, we are in turn transported to a new imagery explanatory hypothesis.

The second case deals with the capacity to justify the absence of a given object in a scene. Let us consider the following example: one of our friends is accustomed to travel the same route every day. The road passes near to a little bridge, under which ducks can usually be *seen* swimming. On a particularly cold day our friend does not *see* the ducks. He asks himself where the ducks could be, but, since he has never seen any ducks in a different setting, while he is able to detect the anomaly he is unable to explain it. The

[2]Of course the initial spatial image can be the representation of a real case.

imagery explanatory reasoning is impossible: therefore, it is stopped. On the contrary, if our friend had previously seen the ducks, say, under the roof of a farmhouse, once he has detected the absence of the ducks he can retrieve from long-term memory the image of the ducks sleeping under the roof. The imagery explanatory hypothesis is immediately achieved.

The third case deals with the well-known monkey-banana problem. In a room there is a banana, a box, and a monkey. The monkey cannot reach the banana because it is on the ceiling, but it can push the box to a point below the banana, climb on top of it and so reach the banana. Every visual representation of the effect of a sequence of actions the monkey can perform may be considered as an hypothesis generation. Such an hypothesis, if successful, is viewed as the one that gives a solution of the problem.

3.3 Visual Abduction System [VASt] Structure

We are developing a system (VASt) that focuses on spatial reasoning tasks and is ontologically organized into a *universe* of nested *spatial worlds*. At the computational level, the universe is composed of a collection of *array representations* that describe each spatial world, and of a *navigation device*. Thus, the different worlds can communicate with each other by means of the navigation device. Indeed, each spatial world is represented, according to Glasgow and Papadias' computational model, by a nested symbolic array and by a collection of rules identifying it (e.g. Newtonian mechanics). Imagery objects [3] are of two kinds: *active* and *passive*. Both kinds of objects "know" the rules of the world where they are included; moreover, they are individually characterized by a family of functions that imply some definite kinds of action they are able to perform. Finally, 1) each active object may interact with the remaining objects in a world; and 2) each passive object in a world may only act after interacting with an active object.

To represent interesting cognitive relations explicitly among the objects of the universe of spatial worlds we have to enrich Glasgow and Papadias' knowledge representation scheme by means of topological and combinatorial considerations upon the hierarchy of spatial worlds. Since the universe of spatial worlds can be mathematically represented as a tree, we can consider its adjacency matrix. A well known theorem in graph theory allows us to manage easily paths of any length and therefore to consider a lot of relations beyond the mere inclusion (e.g. "to be next to but not included", "o be elsewhere but not far from here" and so on). Such a capability induces the possibility to match spatial worlds consistently with these new relations, in order to provide a detailed computational problem solving strategy for solving visual abduction problems described above (that is the explanation of the absence or of the position of an object).

We are developing a set of functions with several capabilities, such as object manipulation, array retrieval, mapping and so on. The principal interest is to design functions of cognitive interest in array theory. The complete set of such functions covers the functionality of the *navigation device*. Here are some examples. Consider a hierarchy of nested arrays (Fig. 2). We are trying to give a precise meaning to "cognitive" sentences like "Array A is more interesting than array C", "There is a stronger link between G and F than between G and I" or "The object represented by array A is better known that any other". Someone could wonder why we think array A is more interesting than C, seeing

[3] By object in this situation we mean nothing else that a cell in the array, to which corresponds in turn another array at the lower level of representation

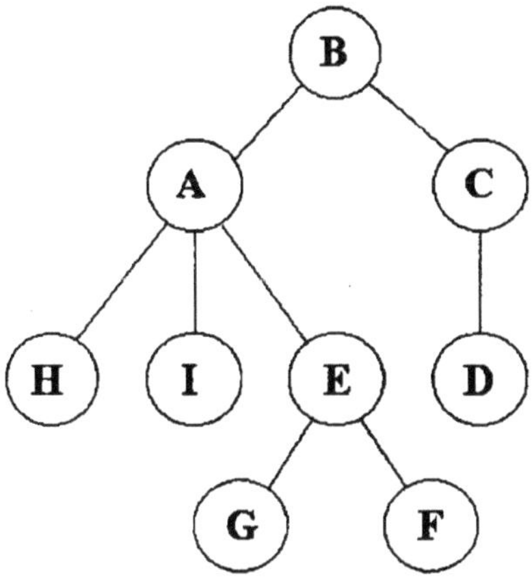

Figure 2: Hierarchy of nested arrays

that we do not know what the arrays A and C are representing. We think that such considerations can be simply suggested by the topological structure of the hierarchy of the arrays (that is, actually, a tree).

The basic idea is to take into account not only paths of length one between the nodes (e.g. the path E–F in the above illustration), but also paths of length more than one (e.g. the two–lenght path G–E–F). Obviously, from a cognitive point of view, a path of length one is more "important" than a path of length two. In order to consider this situation we give different weights to n–length paths (for instance a n–length path might have a weight of $1/n!$ – one over n factorial –). So the path between G and F will be weighted 0.5 while the path between G and I will be weighted roughly 0.166 giving a justification of one of the above sentences. The next step is to answer a (seemingly) meaningless question: "How many paths are there between A and A ?". The answer is: "There are four paths of length two, seven paths of length four and so on". We claim that the sum of the weights of such paths could be a measure of the cognitive interest of the array. This allows us to give a meaning to sentences like "Array A is the most important".

The cognitive meaning of a two length path (namely A–I–A) could be represented with the following sequence of operations: retrieval of the image A – focusing of a particular (namely I) – return to the image A. The topological structure of the representation is clearly subjective, reflecting one's points of view. For instance I know my home much better than my neighbour's one, correspondingly the array representation of my home is far richer.

To solve the third visual abduction problem introduced above (monkey–banana problem) it is easy to use a classical means–ends analysis strategy. The room is represented by an array where all the objects are indicated by symbols on the array. If we describe a primitive distance function, say $dist(x, y)$, we are immediately transported to a new formulation of the goal, that is, dist(monkey,banana) = 0. Let us suppose the "object" monkey is supplied with a set of primitive functions, such as *walk–right*, *walk–left*, *push–left*, *push–right*, *climb*, *grasp*. It is possible to describe visually the effect of every function in the array. A feasible procedure is 1) choose an action 2) if this action leads to a decrease in the distance then 2.1) do it 3) else go to 1). Such an algorithm clearly does not always lead to the goal. It has been improved with a device that is capable of detecting station-

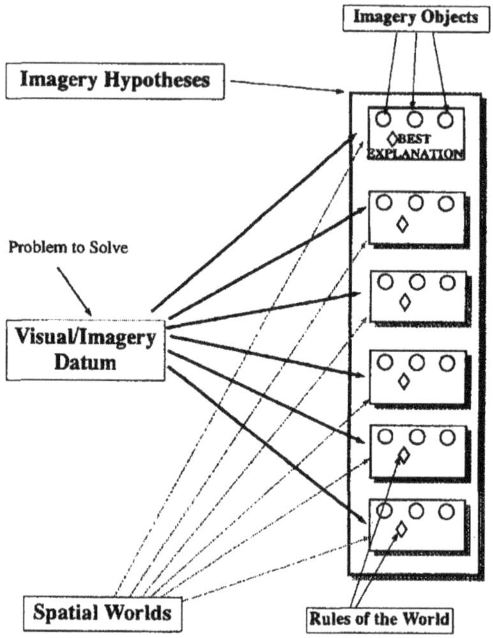

Figure 3: Visual abduction as image-based explanation

ary states which are not the solution of the problem. We can also avoid this impasse by adding to the distance function a penalty function which evaluates the "distance" of the actual state from the set of stationary states.

If we consider each new step leading to the goal (i.e. each new configuration of the array generated by the new positions of the objects) as an *imagery state*, we can say that the "monkey" (or "people" faced with the monkey–banana problem) forms different *imagery hypotheses* devoted to achieving the task. Thus, each step represents a particular imagery world. The configuration of the array in which the goal is achieved performs the image-based best explanation, i.e. the visual abduction: this generated abductive imagery hypothesis is the best explanation of the problem–data, and hence able to perform the planning task (Fig. 3).

Finally, if our spatial world represents a room, it can be supplied with a collection of rules of varying detail (e.g. Newtonian mechanics), yet it can also be supplied with completely different rules. Of course, using less rules results in the objects of the world being less constrained. For example, weakening the rules of Newtonian mechanics can lead to a new kind of spatial world, which is very abstract and considered as being a "virtual" one.

4 Conclusion

This paper has examined a kind of hybrid imagery–linguistic representation used to model image-based hypothesis generation (*visual abduction*): faced to an initial image, in which we have recognized a problem to solve, we have to generate an imagery hypothesis able to explain problem-data.

Although describing a computational model able to "imitate" the real ways the human brain works when it makes visual abductions would be best, our primary concern has been expressiveness, efficiency and inferential adequacy, rather than its explanatory and predictive power as regards psychological research. Notwithstanding this, our system can contribute to discussions in the psychological area of the imagery debate.

References

[1] M.J. Farah (1988), The neuropsychology of mental imagery: converging evidence from brain-damaged and normal subjects. In J. Stiles-Davis, M. Kritchevsky & U. Bellugi (Eds.), *Spatial Cognition - Brain Bases and Development*, Hillsdale, NJ, Erlbaum, 33-56.

[2] R.A. Finke (1989), *Principles of Mental Imagery*, Cambridge, MA, The MIT Press.

[3] R.A. Finke & K. Slayton (1988), Explorations of creative visual synthesis in mental imagery, *Memory & Cognition, 16,* 252-257.

[4] I.J. Glasgow (1993), The Imagery Debate Revisited: a Computational Perspective, *Computational Intelligence, 9(4),* 309-333.

[5] I.J. Glasgow & D. Papadias (1992), Computational imagery, *Cognitive Science, 16(3),* 355-394.

[6] G. Hinton (1979), Some demonstrations of the effects of structural descriptions in mental imagery, *Cognitive Science, 3,* 231-250.

[7] G. Holton (1972), On trying to understand scientific genius, *American Scholar, 41,* 95-110.

[8] J.R. Josephson, B. Chandrasekaran, J.W. jr. Smith & M.C. Tanner (1986), Abduction by classification and assembly. In *PSA 1986*, vol. 1, Philosophy of Science Association, 458-470.

[9] S.M. Kosslyn (1980), *Image & Mind*, Cambridge, MA, Harvard University Press.

[10] S.M. Kosslyn (1983), *Ghosts in the Mind's Machine: Creating and Using Images in the Brain*, New York, W.W. Norton.

[11] S. M. Kosslyn & O. Koenig (1992), *Wet Mind. The New Cognitive Neuroscience*, New York, Free Press.

[12] L. Magnani (1992), Abductive reasoning: philosophical and educational perspectives in medicine. In D.A. Evans & V.L. Patel (Eds.), *Advanced Models of Cognition in Medical Training and Practice*, Berlin, Springer, 21-41.

[13] L. Magnani, S. Civita & G. Previde Massara (1994), Visual Cognition and Cognitive Modeling. In V. Cantoni (Ed.), *Human and Machine Vision: Analogies and Divergences*, New York, Plenum Press, 229-243.

[14] D. Marr & H.K. Nishihara (1978), Representation and recognition of the spatial organization of three-dimensional shapes. In *Proceedings of the Royal Society of London*, 269-294.

[15] A.I. Miller (1989), Imagery and intuition in creative scientific thinking: Albert Einstein's invention of the special theory of relativity. In D.B. Wallace & H.E. Gruber (Eds.) *Creative People at Work. Twelve Cognitive Case Studies*, Oxford, Oxford University Press, 171-187.

[16] N. J. Nersessian (1993), Opening the black box: Cognitive science and history of science. CSL Report 53, January 1993 (Cognitive Science Laboratory, Princeton University, Princeton, NJ).

[17] C.S. Peirce (1931-1958), *Collected Papers*, 8 vols, C. Harstone, P. Weiss & A. Burks (Eds.), Cambridge, MA, Harvard University Press.

[18] I. Peng & I.A. Reggia (1987), A probabilistic causal model for diagnostic problem solving I: integrating symbolic causal inference with numeric probabilistic inference, *IEEE Transactions on Systems, Man, and Cybernetics, 17*, 146-162.

[19] I. Peng & I.A. Reggia (1987), A probabilistic causal model for diagnostic problem solving II: diagnostic strategy. *IEEE Transactions on Systems, Man, and Cybernetics, 17*, 395-406.

[20] H.E. Pople (1973), On the mechanization of abductive logic. In *Proceedings of the International Joint Conference on Artificial Intelligence 8*, 147-152.

[21] Z.W. Pylyshyn (1981), The imagery debate: Analogue media versus tacit knowledge, *Psychological Review, 88*, 16-45.

[22] M. Ramoni, M. Stefanelli, L. Magnani & G. Barosi (1992), An epistemological framework for medical knowledge-based systems, *IEEE Transactions on Systems, Man, and Cybernetics, 22(6)*, 1361-1375.

[23] J.A. Reggia, S.N. Dana & Y.W. Pearl (1983), Expert systems based on set covering model, *International Journal on Man-Machine Studies, 19*, 443-460.

[24] P. Thagard (1988), *Computational Philosophy of Science*, Cambridge, MA, The MIT Press.

[25] P. Thagard (1992), *Conceptual Revolutions*, Princeton, NJ, Princeton University Press.

[26] P. Thagard, D. Gochfeld & S. Hardy (1992), Visual analogical mapping. In *Proceedings of the Fourteenth Annual Conference of the Cognitive Science Society*, Hillsdale, Erlbaum, 522-527.

[27] R.D. Tweney (1989), Fields of enterprise: On Michael Faraday's thought. In D.B. Wallace & H.E. Gruber (Eds.) *Creative People at Work. Twelve Cognitive Case Studies*, Oxford, Oxford University Press, 91-106.

[28] M. Tye (1991), *The Imagery Debate*, Cambridge, MA, The MIT Press.

HYPERACUITY: VERTICAL ASYMMETRY FOR SIZE DISCRIMINATION OF TWO-DIMENSIONAL IMAGES

Svetlana Chukova
Pavlov Institute of Physiology,
Emb.Makarov 6, 199034 St.Petersburg, Russia

The human eye is capable of resolving changes in the relative position of two visual features that are an order of magnitude smaller than the size of a foveal cone. To distinguish these highly precise spatial judgments from measures of visual acuity that assess resolution capability, Westheimer [1] coined the term hyperacuity. "Almost any type of target configuration can be used to measure hyperacuity" [2]: the exquisite sensitivities obtained in hyperacuity tasks are robust to changes in the spatial configuration of stimuli.

Can human observers judge size differences in two-dimensional images with the same exquisite precision? In this work, size discrimination thresholds for high-contrast, outline images were measured in the fovea for different arrangements of the stimuli along the central vertical meridian.

METHOD

The stimuli were black outline squares and rectangles on a bright background (60 cd/m^2). The reference stimuli were 27 squares with sides ranging in length from 8 to 60 arc min, in increments of 2 arc min. Five test rectangles were obtained by decreasing the height of the square by 1.25, 2.5, 3.75, 5.00 and 6.25% (thus, the test stimulus was always smaller than the reference). The outline width for both test and reference was 1/4 of the reference height.

The inspection field was limited by a circular aperture 7 deg in diameter. The reference and test stimuli were presented simultaneously along the central vertical meridian of the visual field (VF). The distance between the internal edges of the stimuli - 30 min arc - was constant for all the sizes. Viewing was central, with no explicit fixation target.

The order of the reference sizes was randomized. For each reference size, the five test-reference pairs were presented 100 times, 50 with the test above and 50 with the test below the reference. A session thus consisted of 500 presentations, whose order was randomized. The exposure duration was 750 ms.

The observer was asked to judge whether the smaller stimulus was presented above or below. Feedback was not provided. The stimuli were viewed monocularly (the left eye) by three experimentally naive observers. The observers had normal or corrected-to-normal Snellen letter acuity.

The numbers of correct responses were plotted separately for test above and test below as a function of size difference, giving two 5-point psychometric functions for each reference size. A threshold size difference was calculated by linear interpolation from each psychometric function corresponding to 75 per cent correct responses, based on 250 responses. The difference between the threshold values was evaluated with Wilcoxon's rank test for dependent values separately for observers, with each of the 27 reference sizes being regarded as independent.

RESULTS AND DISCUSSION

For almost the entire range of sizes, the observers resolved size differences smaller than the size of a foveal cone when the smaller stimulus is presented in the lower visual hemi-field (VHF) and the reference in the upper VHF (Figure 1, open symbols). The lowest threshold value is 3 arc sec (observer S.G.; reference size, 8 arc min). In the reverse arrangement (Figure 1, filled symbols) the thresholds are significantly higher ($p<0.001$ for all three observers, Wilcoxon's rank test).

The variation in threshold with reference size is similar for all the observers: the threshold increases from 3-18 to 90-150 arc sec in the range from 8-48 arc min and then decreases to 20-60 arc sec in the 50-60 arc min range.

Thus, size discrimination does not obey Weber's law when the image size is larger than 48-50 arc min. What are the particular conditions resulting in these low thresholds? The distance between the stimuli to be discriminated was constant for all the sizes used, but the eccentricity of the external sides of the two stimuli changed from 23 to 75 arc min. It appears that even at smaller peripheral distances (somewhat from 65 arc min) the stimulus size matters more than line length does for bisection thresholds [3]. Perhaps the small size increments (2 arc min) make it possible to outline the boundary between the foveal and peripheral discrimination with more precision. Scaling the stimulus side length (50-60 arc min) according to the hyperacuity eccentricity function of Levi et al.[4] ($Length=x(1+E/0.77)$, where E is the eccentricity given in degrees and x is the length of the foveal stimulus) predicts threshold values that are higher by factor of 2.4-2.6 than the experimentally obtained values. In this case, Weber's law holds in the range 50-60 arc min as well.

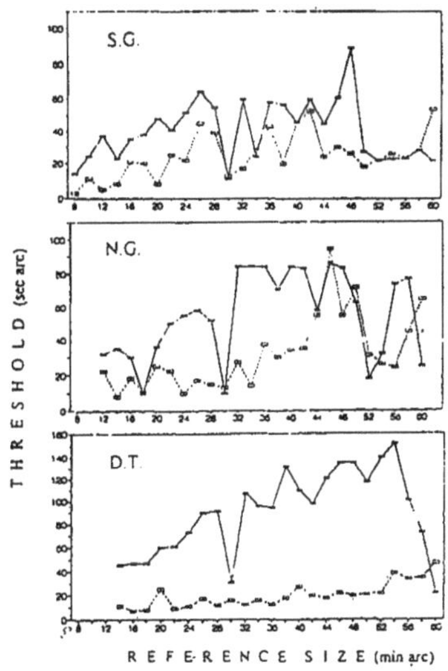

Figure 1. Thresholds for size discrimination as a function of size for the test above the reference and the test below. Results for three observers are shown in separate panels. Each point represents 250 presentations. *Filled symbols*: the test was presented in the upper VHF, the reference - in the lower VHF; *open symbols*: the reversed arrangement. The difference between the threshold values in the range 8-60 arc min evaluated with Wilcoxon's rank test is significant ($p < 0.001$) for all the observers.

But it still remains unclear as to why the thresholds for size discrimination in the range 50-60 arc min are low. When asked about the appearance of the array, all the observers reported that a pair of small images (8-48 arc min) are seen as an unsegmented stimulus. The larger images (50-60 arc min) are perceived separately. These results can be interpreted as evidence that the array is processed either by a single module or by two modules, depending on the image size.

The marked fall in the threshold is observed at the 30 arc min size, where the stimulus size equals the separation between them. Presumably the discrimination is facilitated by the regularity of the array.

The total vertical extent of a pair of stimuli did not exceed 2.5 deg. Given that the cone arrangement is most regular in the central fovea [5], it is unlikely the threshold asymmetry arises from morphological particularities of the retina. A plausible explanation is the perspective illusion: objects in the upper VHF seems to be larger than in the lower VHF because they are implicitly assumed to be at a greater distance [6]. When the smaller stimulus is presented above the reference, the 1.25-2.50% size difference between the stimuli is not enough for successful task performance because of the overestimation of the smaller stimulus in the upper VHF, hence the thresholds are high. When the smaller stimulus is presented in the lower VHF, the apparent size difference is more than quite enough for discrimination owing to the overestimation of the larger stimulus in the upper VHF.

These data suggest that the vertical asymmetry in size perception is an important factor for foveal size discrimination performance as well as for peripheral.

REFERENCES

1. G.Westheimer (1975). Visual acuity and hyperacuity. *Invest. Ophthalmol. Vis. Sci.*, **14**, 570-571.
2. S.P.McKee (1991). The physical constraints on visual hyperacuity. *Vision and Visual Dysfunction. Vol.5. Limits of Vision.* J.J.Kulikowski, V.Walsh and I.J.Murray (Eds.), Boca Raton, CRC Press Inc., 221-233.
3. D.P.Andrews and D.T.Miller (1978). Acuity for spatial separation as a function of stimulus size. *Vision Res.*, **18**, 615-619.
4. D.M.Levi, S.A.Klein and Y.L.Yap (1988). "Weber's law" for position: unconfounding the role of separation and eccentricity. *Vision Res.*, **28**, 597-603.
5. J.Hirsch and W.H.Miller (1987). Does cone positional disorder limit resolution? *J.Opt.Soc.Am.A.*, **4**, 1481-1492.
6. S.Chukova (1994). Size discrimination in different visual fields: magnitude of "equal-images" illusion in the fovea. *Perception*, **23**, Suppl., 103.

THE TEMPORAL AND SPATIAL DYNAMICS OF THE PERCEPTION OF QUANTIFIED IMAGES.

Neeme Kahusk
Tallinn Pedagogical University
Narva Road 25
EE0100 Tallinn, Estonia

Introduction

The results of David Navon [1] started the paradigm of the relative roles of global and local structures in perceiving visual form. Later on several experiments were made on the same subject (see Kimchi [2] for review). The stimuli used in these experiments are mostly ecologically invalid, since letters built from smaller ones can hardly be found in real environment.

An alternative method of studying the relations of global and local levels of form is using spatially quantised stimuli, like in [3 , 4 , 5].

An experiment was made with quantised stimuli to study the problem of global dominance, as postulated by Stoffer in [6]: whether the global dominance can be explained as a phenomenon of perception, that is, global features are extracted earlier or more easily than local ones (see Navon [1,7 ,8]), or as an attentional phenomenon, visual attention is allocated first to global level (Miller [9]).

Experiment.

The aim of the experiment was to test the results of Bachmann [4] in the conditions of spatial precuing. A five-field binocular tachistoscope with semitransparent mirror system was used for stimulus presentation. It was driven by an Apple IIe computer. The viewing distance was equal to 40 cm. Eight volunteers (3 male and 5 female, aged from 22 to 36, including author) served as subjects.

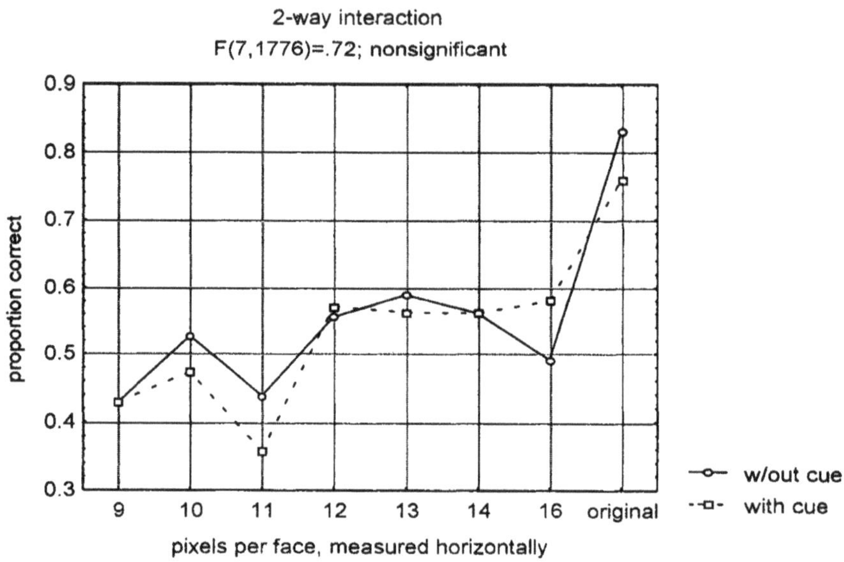

Figure 1. Proportion correct as a function of the number of horizontal pixels per face with cued and non-cued conditions.

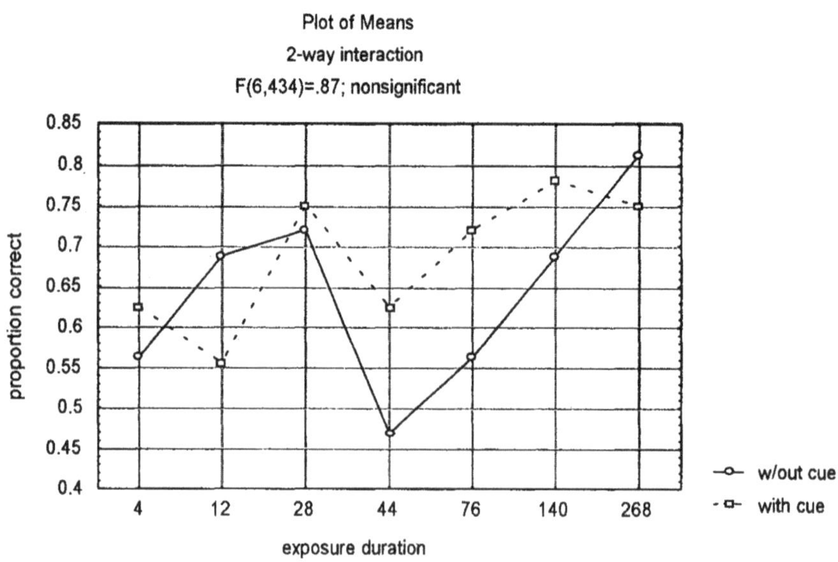

Figure 2. Proportion correct of Landolt ring discrimination as a function of exposure duration with cued and non-cued conditions.

Eight male faces were used as stimuli as described in [4]. The stimuli were quantised and printed on transparency film with Corel PhotoPaint™ program and HP LaserJet III™ at 300dpi, exposed on the luminous background and subtending visual angle of 5.3 × 7.87 deg. Landolt rings were also presented in order to control visual acuity, these trials were randomly intermixed with main trials; the probability of appearing a Landolt ring was 0.25. The subjects were instructed to follow two tasks: recognise faces as the main task and localise the openings of the Landolt rings as the supplementary task.

All stimuli were presented on stimulus field — subtending a square of about 10.7 x 10.7 degrees — at four possible peripheral stimulus positions around central fixation point: in the four corners of the stimulus field. The luminance of the stimulus field was 19.6 cd/m^2.

Each trial consisted of the successive exposure of (1) fixation point (a luminous dot on a dark background) and exposures of (2) cue-and-stimulus or stimulus alone depending on condition. When stimulus was preceded with cue (a small — 0.3 deg. — black dot), its exposure duration equalled 20 ms and SOA equalled 120 ms. Eight exposure durations (1, 4, 12, 28, 44, 76, 140 and 268 ms) were used for stimuli in counterbalanced order.

It was hypothesised that spatial precuing should increase the efficiency of recognition of spatially quantised stimuli due to preliminary accumulation of attentional recourses on the locus of expected stimulus.

Results and Discussion.

The results are expressed as proportion of correct responses as a function of quantisation with the presence of cue as the parameter.

There was statistically significant main effects of exposure duration ($F_{7,1920}$ = 10.15; p<.0000), cue ($F_{1,1920}$ = 9.49; p<.002) and quantisation level ($F_{7,1920}$ = 15.03; p<.0000); also 2-way interaction of cue and exposure duration was significant ($F_{7,1920}$ = 5.87; p<.0000). Since these effects may have been the result of the masking effect due to the integration of the cue-stimulus field and a weak stimulus field at 1 ms exposure condition, then the 1 ms condition was excluded from the further analysis. As a result only the main effect of quantisation level ($F_{7,1680}$ = 13.57; p<.0000) was finally found (see Figure 1.).

The absence of the effect of cueing contradicts the results of previous works with precuing (See [10] for review). It may be hypothesised that the absence of the facilitative effect of the cue is the result of the differences between of the relative sizes of the cue and the stimulus. However the effect of precuing is not significant also with relative small (0.7 deg.) Landolt rings (Main effect of cueing $F_{1,\ 434}$ = 0.96; p<.33, of cue and exposure duration the interaction $F_{6,434}$ = 0.87; p<.52); see Figure 2. There was significant main effect of exposure duration, however ($F_{6,434}$ = 2.15; p<.047).

Another possible explanation is that the positions of the stimuli on the stimulus field were relatively too close to the central point for the effect of cueing to manifest. The possible reasons of these unexpected results are subject for future studies.

References

1 D.Navon (1977) Forest before trees: The precedence of global features in visual perception. *Cognitive Psychology* 9, 353 – 383

2 R.Kimchi (1992) Primacy of holistic processing and global/local paradigm: A critical review. *Psychological Bulletin* 112, 24 – 38

3 T.Bachmann (1987) Different trends in perceptual pattern microgenesis as a function of the spatial range of local brightness averaging. *Psychol Res.* 49, 107 – 111

4 T.Bachmann (1991) Identification of spatially quantised tachistoscopic images of faces: how many pixels does it take to carry identity? *European Journal of Cognitive Psychology* 3, 87 – 103

5 N.P.Costen, D.M.Parker, I.Crow (1994) Spatial content and spatial quantisation effects in face recognition. *Perception* 23, 129 – 146

6 T.H.Stoffer (1993) The duration course of attentional zooming: A comparison of voluntary and involuntary allocation of attention to the levels of compound stimuli. *Psychol. Res.* 56, 14 – 25

7 D.Navon (1981) The Forest revisited: more on global precedence. *Psychol Res.* 43, 1 – 32

8 D.Navon (1983) How many trees does it take to make a forest? *Perception* 12, 239 – 254

9 J.Miller (1981) Global precedence in attention and decision. *Journal of Experimental Psychology: Human Perception and Performance* 7, 1161 – 1174

10 A.H.C. van der Heijden (1992) *Selective Attention In Vision*. London-New York. Routledge

CHARACTERISTICS OF THE CAT LGB RESPONSE OSCILLATIONS DEPENDENT ON THE VISUAL STIMULUS PROPERTIES.

N.A. Scherbakova', N.F. Podvigin', E. Poppel", I.N. Bogacheva',
A.N. Chizh', N.B. Kiselyeva'

'Laboratory of Neurophysiology of Sensory Motor Functions,
I.P. Pavlov Institute of Physiology, Russian Academy of Sciencies,
199034, St.Petersburg, Russia,
"Research Center Julich (KFA),52425 Julich, Germany.

The periodic oscillations of potential in the visual systems has been described since the beginning of this century, when oscillations of about 40 Hz (the so-called gamma type) were registered in the retina of the octopus [2]. Later they were recorded and studied by many investigators not only in the retina, but in the higher levels of the visual systems of many vertebrate and invertebrate species, as well as in the other sensory and motor systems. For example, such oscillatory potentials were described in the olfactory system [1]. It is possible that the numerical coincidence of the oscillation frequencies, revealed in different physiological systems, points to the common character of their neuron mechanisms and probably common functional significance as well. Different notions on the physiological significance have been advanced. Two of them are discussed more widely: the first idea, that gamma-oscillations provide the synchronization of neuronal activity in parallel circuits of information processing in neuronal structures of the visual system [4,6,8,11]; and the second one concerned with current time quantization, which is provided by oscillatory processes, the subjective time perception system being based on this mechanism [8,6]. It is possible that the quantum time mechanism and the synchronization mechanism are tightly related because the multi-channel system of parallel information processing needs the synchronization of the quantum time in parallel circuits.

However, some questions are still unsolved. First, there is the question of the functional significance of gamma-oscillations in brain activity implementation and, particulary, in the visual system processing. For the solving of this problem it is necessary to determine the quantitative characteristics of the oscillations on different levels of the visual system from retina to cortex. Furthermore, the question of the levels, where the oscillations actually occur, is under discussion. According to some investigators of cortex neuron activity [3], the generation of gamma- oscillatory responses is a cortical phenomenon.

In our previous works [5,6,7] we described the oscillatory processes in the responses of neurons at the subcortical levels of the visual system, particulary in the lateral geniculate body (LGB), to an applied visual stimulus. The purpose of the present study is to determine whether the spatio-brigthness stimulus properties, essential for LGB, influence the oscillation frequency characteristics. In this research we analysed the responses of cat LGB neurons, which were stimulated by presentations of images of black-and-white or shaded geometrical figures. The stimulus was moved in steps across the receptive field (RF) of neurons in order to construct a response contour or, in other words, the pattern of activity that arises in response to the object presentation [9]. The test image was presented 30 times in every position of matrix, conditionally superimposed on the RF (Fig. 1), for 500 msec ON-condition interrupted by dark periods of 500 msec OFF-condition. Based on response contour we distinguish between the test points, corresponding to the internal part of the figure (F-area), the edge of figure (E-area), and the background (B-area).

We estimated frequency spectra using the Fourier transformation of the autocorrelation function of each recorded PSTH (2 msec bin width) for ON- and OFF- conditions separately. The oscillations were found in 32% of 86 both ON and OFF tonic neurons. The spectra, corresponding to F-, E- and B-areas, were averaged. Thus, we have obtained six mean spectra as a result of analysis for every testing. The example of data for ON neuron is shown on the Fig. 1. The spectra indicated that the oscillation frequency depends on stimulus position related to the RF.

Three main domains of dominant frequencies were found: 10-20 Hz (alpha range), 30-80 Hz (gamma range) and 85-100 Hz. These domains are similar to the three types of oscillations in cat retinal horizontal cell activity [10]. A significant influence of the stimulation on the alpha-frequency oscillations was not revealed, except for a slight reduction in spectral power in this band for the F-area. The 80-120 Hz oscillations were observed mainly for areas E and B.

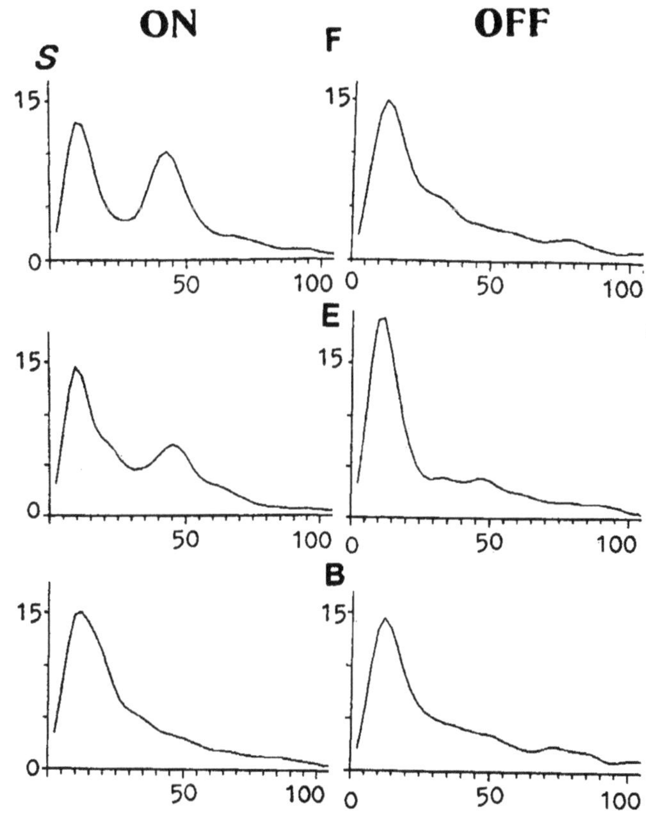

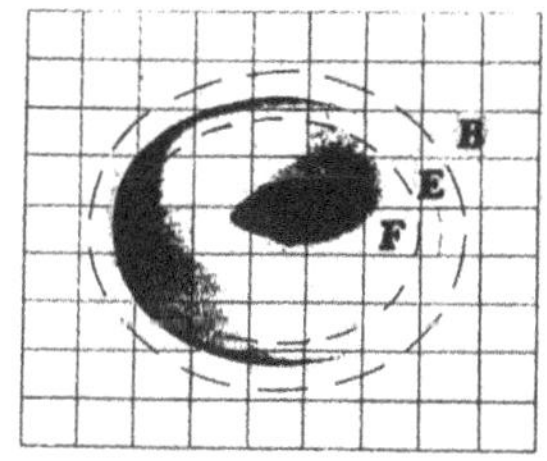

Fig. 1. Left- Examples of averaged spectra correspon-ding to ON and OFF stimulus conditions for ON-neuron stimulation.
s: power, f: frequency, Hz.

Above- Stimulus: the shading image of a toroid mapped in scale with the matrix of testing.
Dotted lines: the contours of F-, E-, B-areas.

The oscillations of about 40 Hz occur in the neuron responses to the presentation of a stimulus adequate for this neuron (ON stimulating conditions for ON type neurons and OFF conditions for OFF neurons), and only in case F, when the stimulus was placed inside the RF. The oscillation frequency distribution within this range depends on stimulus spatio-brightness characteristics. When the binary images with constant brightness were presented, the dominant frequencies of oscillations did not vary at all or varied within a narrow frequency range (σ = 5Hz). The oscillation frequency in response to a shaded stimulus varied over a wide range (σ = 15 Hz). Moreover, the frequency variation was correlated with the shade distribution [5,7].

A further analysis of these facts is required, but, using the information available, we suppose that stimulus-dependent changes in oscillation properties within the three ranges mentioned above, are perhaps the indication of the different functional roles of these three types of oscillations in information processing. The fact that gamma-oscillations were revealed only in responses to adequate stimuli points to the significance of this type oscillations in stimulus information processing. Considering the possible mechanisms of synchronization of neuron activity and time quantization during information processing at the subcortical levels of the visual system, we assume that there are gamma-oscillations which can provide these mechanisms.

1. E.D.Adrian (1953) Sensory messages and sensation: the response of the olfactory organ to different smells. ***Acta physiol. Scand.***, 29,5-14

2. F.W.Frohlich (1914) Beitrage zur allgemeinen Physiologie der Sinnesorgane. ***Z. Psychol. and Physiol. Sinnesorg. II. Abt. Sinnesphysiol.***, 48,28-164

3. C.Gray, P.Konig, A.K.Engel, W.Singer (1989) Oscillatory responses in cat visual cortex inter-columnar synchronization which reflects global stimulus properties. ***Nature***, 338, 334-337

4. N.F.Podvigin (1979) ***Dynamic Properties of Neuronal Structures of the Visual System.*** Leningrad, Nauka, 157

5. N.F.Podvigin, H.Jokeit, E.Poppel, A.N.Chizh, N.B.Kiselyeva (1992) Stimulus-dependent oscillatory activity in lateral geniculate body of the cat. ***Naturwissenschaften***, 79, 428-431

6. N.F.Podvigin, E.Poppel (1994) Characterictics and functional meaning of oscillatory processes in retina and lateral geniculate body. ***Sensory systems***, 8,2,61-70

7. N.F.Podvigin, E.Poppel, H.Jokeit, A.N.Chizh, N.B.Kiselyeva, I.N.Bogacheva, N.A.Scherbakova (1994) Dependence of the parameters of the cat LGN responses oscillation on the visual stimulus. ***Sensory Systems***, 8,1,42-49

8. E.Poppel (1978) ***Handbook of Sensory Physiology.*** Berlin, Springer-Verlag, 8,714-729.

9. R.Ratliff, H.K.Hartline (1959) The responses of *Limulus* optic nerve fibres to patterns of illumination of the receptor mosaic. ***J. Gen. Physiol.***, 42,1241-1255

10. N.A.Scherbakova, N.F.Podvigin, E.Poppel, I.N.Bogacheva, A.N.Chizh, N.B.Kiselyeva (1994) Oscillatory responses in the cat LGB stimulated with binary and shaded images. ***Perception***, 23,31

11. W.Singer, C.M.Gray, A.Engel, P.Konig (1988) Spatio-temporal distribution of stimulus-specific oscillations in the cat visual cortex. II. Global interactions. ***Soc. Neurosci. Abstr.***, 14,362,13

MATHEMATICAL MODEL OF THE DEPTH EFFECT IN THE TRANSLATORY ALTERNATING MOVEMENT
Part A: 3-D perception of length amplification

E. Xausa+, L. Beghi+, M. Zanforlin *

+ Dipartimento di Matematica Pura e Applicata, Università di Padova, Via Belzoni 7, Padova 35131, Italy
** Dipartimento di Psicologia Generale, Università di Padova, Piazza Capitaniato 3, Padova 35139, Italy*

The hypothesis is based on a process that minimizes the difference in velocities of points in the moving pattern, i.e. if two points move on the frontal plane with different velocities, their velocities can be equalized in absolute value by adding to them an orthogonal component, in such a way that the lengths of the two resulting vectors are equal and the movement of the points in the frontal plane can be considered the projection of a movement in depth [1,2].

When not all the velocities of a complex pattern can be equalized, we assume that a 3-D pattern of minimum differences can be found.

Let us consider two vertical bars with the same length, located on the frontal plane π of the observer; let us further assume that one of them is stationary, while the other moves with constant speed v in the plane π from left to right (expanding phase) and from right to left (contracting phase), during time intervals of constant amplitude T [3].

We can refer to a one-dimensional representation of this motion by considering a section of the two vertical bars with a plane δ orthogonal to them; we further introduce an orthogonal reference system Oxyz, where the origin O is the intersection point between the δ plane and the stationary bar, the x axis is parallel to the direction of movement, originates in O and is oriented from left to right, the z axis is parallel to the bars and is oriented upwards, and finally the y axis is orthogonal to the Oxz plane and is oriented "in depth" (see Fig. 1).

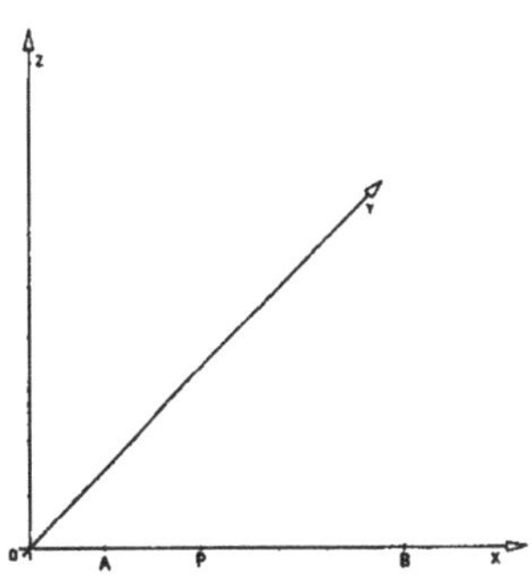

Fig. 1

On the x axis, let P be intersection of the moving bar with the δ plane at time t, and let A, B be the extreme positions, alternately reached by P during its displacements. If a and b are the x-coordinates of A and B, the x-coordinate of P will be

$$x_P = a + v\,t \tag{1}$$

during the expanding phase, and

$$x_P = b - v\,t \tag{2}$$

during the contracting phase. The semi-period of obscillation will be $T = (b - a) / v$.

We define now a velocity vector field on the segment OP (referring for simplicity to the first expanding phase : $0 \le t \le T$), according to the following criteria.

Let us consider first the middle point M of this segment; at time $t = 0$, M has an x-coordinate $x_M(0) = a/2$, while at time $t = T$ its coordinate will be $x_M(T) = b/2$.

As a consequence, the mean translation speed of M during the time interval [0, T] will be $(b/2 - a/2)/T = v/2$. We can also define the instantaneous velocity of point M at time t; it is a vector parallel to the x-axis, oriented toward the increasing values of x, and its length $v_{Mx}(t)$ is defined by the limit:

$$v_{Mx}(t) = \lim_{\Delta t \to 0} [\, x_M(t + \Delta t) - x_M(t)\,] / \Delta t = v/2,$$

where

$$x_M(t + \Delta t) = [a + v\,(t + \Delta t)] / 2 \quad \text{and} \quad x_M(t) = [a + v\,t\,] / 2 .$$

More generally, if Q is a point of segment OA, with abscissa $x_Q(0) = \lambda a$, where $0 \le \lambda \le 1$, during time T it will undertake a displacement along the x-axis, finally reaching the position with abscissa $x_Q(T) = \lambda b$; the mean translation speed of Q will be $\lambda(b-a)/T = \lambda v$.

To point Q we can associate also an instantaneous velocity vector of amplitude $v_{Qx}(t)$, parallel to the x-axis, as done for point M.

If $x_Q(0) = \lambda a$, then we assume that $x_Q(t) = \lambda x_P(t)$ as t varies, i.e. $v_{Qx}(t) = \lambda v$, coinciding with the mean translation speed of Q; more generally $v_{Qx}(t) = \lambda v_P(t)$. The relative velocity of Q with respect to the point M, v^*_{Qx}, is:

$$v^*_{Qx} = v_{Qx} - v_{Mx} = v[x_Q / x_P - 1/2\,]. \tag{3}$$

The maximal absolute value of v^*_{Qx} is $v^*_{max} = v/2$. The component v^*_{Qx}, as a function of x_Q, linearly varies from $-v/2$ to $v/2$; at different times this velocity distribution will be different; the plot of v^*_{Qx} is given in Fig. 2.

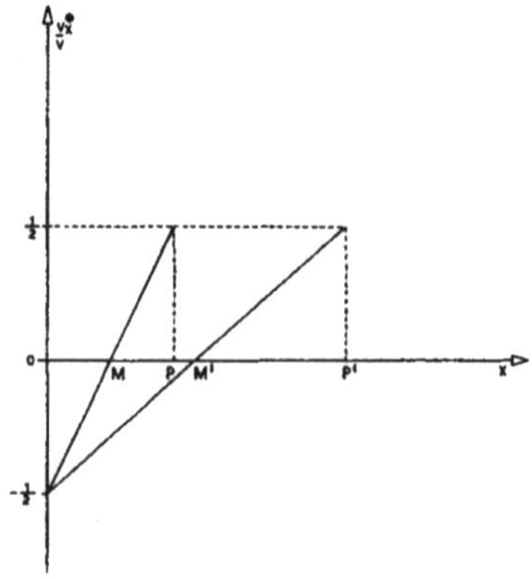

Fig. 2

In order to explain the 3-D perception of length "amplification " (i.e. the amplification of the distance between the stationary and the moving bar or, equivalently, of the length of segment OP), we introduce now the assumption of "minimization of the differences between the absolute values of the velocity vectors" of the points of the OP segment with respect to the central point M [1,2].

This leads to a modification of the velocity field on the OP segment, performed by associating to each point Q an additional velocity component v_{Qy} along the y-direction (i.e. along the "depth" direction), whose amplitude $|v_{Qy}(t)|$ will be determined according to the relation:

$$v^*_{Qx}(t)^2 + v_{Qy}(t)^2 = v_{max}^2 = (v/2)^2 \tag{4}$$

from which it follows:

$$|v_{Qy}(t)| = v\sqrt{[x_Q/x_P(t)]\,[1 - x_Q/x_P(t)]}\,. \tag{5}$$

For Q=M, it will be

$$|v_{My}(t)| = v/2.$$

We introduce now the relative velocity component of point Q along the y axis with respect to point M, $v^*_{Qy} = v_{Qy} - v_{My}$, whose absolute value is given by the relation:

$$|v^*_{Qy}| = |v\sqrt{[x_Q/x_P(t)][1 - x_Q/x_P(t)]} - v/2\,| \tag{6}$$

To v^*_{Qy} we could associate a sign according to the following criterium: v^*_{Qy} will be negative for $[x_Q/x_P(t)] < 1/2$ and it will be positive for $[x_Q/x_P(t)] > 1/2$. Indeed we perceive a counterclockwise rotation around a middle point M; when the point Q precedes M (i.e. its x-component is $x_Q < [x_P(t)/2] = x_M$), its velocity v^*_Q has negative components. When the point Q follows M (i.e. its x-component is $x_Q >$ $[x_P(t)/2]$) its velocity v^*_Q has positive components. The plot of v^*_{Qy} is given in Fig. 3.

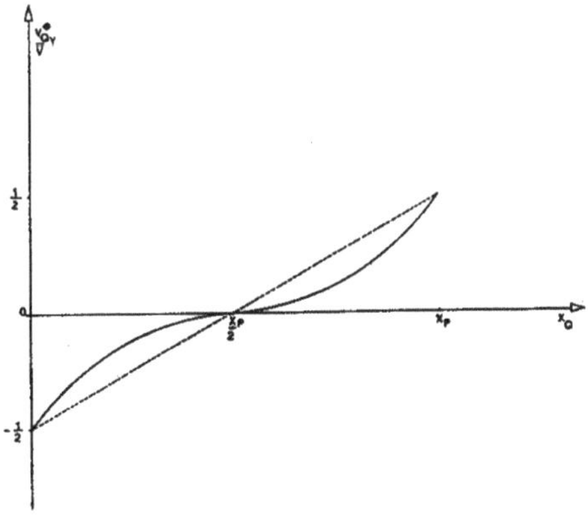

Fig. 3

In order to estimate the "mean expansion coefficient" $\overline{\alpha}_P$ of the perceived distance between the two vertical bars when one of them is moving with respect to the other on the frontal plane of the observer and also the mean inclination of segment OP* with respect to the Oxz plane during the expanding phase, we could solve the following system of differential equations:

$$dx_Q = v\,|x_Q/x_P(t) - 1/2|\,dt, \qquad dy_Q = v\sqrt{[x_Q/x_P(t)][1 - x_Q/x_P(t)]} \tag{7}$$

with initial conditions $x_Q(0) = 0$, $y_Q(0) = 0$ and $0 \le t \le T$.

A numerical solution of this system, for a= b/2, is given in Fig. 4.
The plot of y(x) represents an increasing straight line whose inclination angle with respect to the x axis is about of 45 degrees; the "mean expansion coefficient" of segment OP* with respect to segment OP is about $\sqrt{2} \approx 1.42$.
Substituting the velocity component along the x-direction with the velocity of the middle point M of the OP segment (i.e. by putting $d\,x_Q/dt = v/2$) and assuming $x_P = b$ we would obtain the following analytical approximation of the solution:

$$y_Q(t) = b\{\pi/4 + \lambda/(1+\lambda^2)^2 - \lambda/[2(1+\lambda^2)] - (1/2)\tan^{-1}(\lambda)\} \tag{8}$$

where $\lambda = \sqrt{(1-x/b)/(x/b)}$.

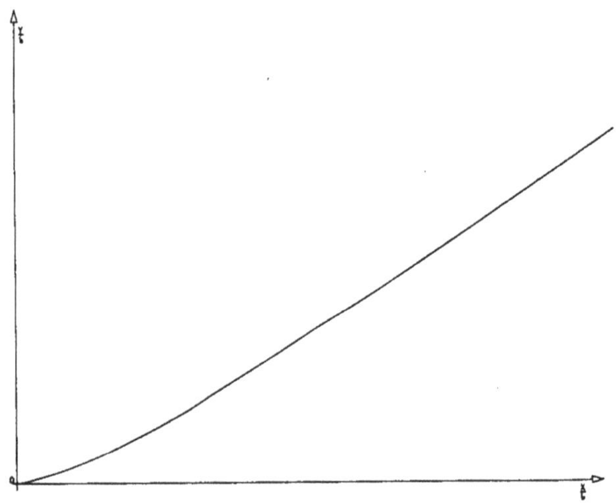

Fig. 4

<u>References</u>

[1] L. Beghi, E. Xausa, M. Zanforlin (1991) Analytical determination of the depth effect in stereokinetic phenomena without a rigidity assumption, *Biol. Cybern.*, 65, 425-432

[2] L.Beghi, E. Xausa, C. De Biasio, M. Zanforlin (1991) Quantitative determination of the 3-D appearances of a rotating ellipse without a rigidity assumption, *Biol. Cybern.*, 65, 433-440

[3] M. Zanforlin, L. Beghi, E. Xausa, L. Tomat (1994) The depth effect in the stereokinetic phenomenon of "swinging gate", *in this volume*.

MATHEMATICAL MODEL OF THE DEPTH EFFECT IN THE TRANSLATORY ALTERNATING MOVEMENT.
Part B: "swinging gate" phenomenon

L. Beghi+ , E. Xausa+, M. Zanforlin *

+ *Dipartimento di Matematica Pura e Applicata , Università di Padova, Via Belzoni 7, Padova 35131 , Italy*

* *Dipartimento di Psicologia Generale, Università di Padova, Piazza Capitaniato 3 , Padova 35139 , Italy*

In what follows, we refer to [1,2]. For the sake of clarity, we summarize the phenomenon : two vertical bars of equal length lie in the frontal plane of an observer; one of them is stationary while the other oscillates from left to right (expanding phase) with constant velocity v, and from right to left (contracting phase)with constant velocity -v, alternately moving apart from and coming closer to the first. The two bars appear to the observer as lines delimiting a "gate", tilted in depth and swinging around its pivot, which is constituted by the stationary bar. In order to build up a mathematical model of the phenomenon, we consider first the intersection of the frontal plane π with an horizontal plane δ, orthogonal to the former and cutting the two bars at points O (stationary point) and P (moving point). A cartesian reference system Oxyz is then introduced, with the Ox axis lying at the intersection of the two orthogonal planes and oriented from O toward P, the Oz axis containing the vertical segment constituting the stationary bar and oriented "upwards", the Oy axis orthogonal to the frontal plane and oriented "inwards". For the sake of simplicity we will take into consideration exclusively the expanding phase: point P, whose coordinate is $x_P(t)$, moves from the initial position A ($x_A = a$) to the final position B ($x_B=b$) with constant velocity v , during the time interval T = (b-a)/v. According to our hypotheses the visual system associates to each point Q of segment OP a velocity component, oriented along the Ox direction,

$$v_{Qx}(t) = v \cdot x_Q(t)/x_P(t) \ , \qquad (1)$$

where $x_Q(t)$ is the coordinate of Q; denoting by M the middle point of the OP segment ($x_M(t) = x_P(t)/2$) , its velocity will be $v_{Mx}(t) = v/2$. The relative velocity of point Q with respect to M along the x direction will be :

$$v^*_{Qx}(t) = v \cdot x_Q(t)/x_P(t) - v_{Mx}(t) = v\,[x_Q(t)/x_P(t) - 1/2\,] \,. \qquad (2)$$

This velocity component could be equally well associated to a virtual point Q* of the Oxy plane, having point Q as orthogonal projection on the Ox axis. Making now use of the principle of "minimization of the differences between the absolute values of the relative velocity vectors (with respect to the perceptual centre of the moving image)" , as performed by the visual system, to each point Q we associate an additional velocity component along the Oy direction, $v_{Qy}(t)$, defined in such as to satisfy the invariance condition :

$$v^*_{Qx}(t)^2 + v_{Qy}(t)^2 = \text{constant} = (v/2)^2 \qquad (3)$$

from which it follows :

$$|v_{Qy}(t)| = v\ \sqrt{[x_Q(t)/x_P(t)\,]\,[\,1 - x_Q(t)/x_P(t)]} \tag{4}$$

$$|v\,^*_{Qy}(t)\,| = |v_{Qy}(t) - v_{My}(t)\,| = v\,|\sqrt{[\,x_Q(t)/x_P(t)]\,[1 - x_Q(t)/x_P(t)} - 1/2\,| \tag{5}$$

To $v\,^*_{Qy}(t)$ we attribute the positive sign for $x_Q(t)/x_P(t)) < 1/2$, the negative sign for $x_Q(t)/x_P(t)) > 1/2$.

As done in the case of $v^*_{Qx}(t)$, this velocity component could be associated to the same virtual point Q* of the Oxy plane, having Q as orthogonal projection on the Ox axis. The behaviour of $v^*_{Qy}(t)$ could be described by the plot of Fig.1 (where the dashed line constitutes the linear approximation of $v^*_{Qx}(t)$ around the "central" point with coordinate $x_M(t) = x_P(t)/2$).

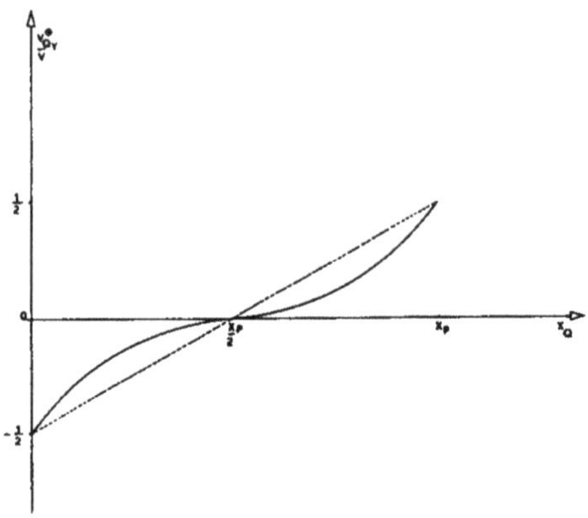

Fig.1

In order to explain the "swinging gate" phenomenon we make now the following consideration:

the "additional" relative velocity component $|v^*_{Qy}(t)|$, associated by us to the virtual points Q* of the Oxy plane, accounts for the "equalization" of the absolute values of the velocities of these points at a given time t, but the distribution of this velocity field varies with time. This means that, for a given coordinate $x_Q(t)$, at time t ($0 \leq t < T$), we can define an acceleration vector

$$a^*_Q(t) = \partial v^*_Q(t)/\partial t + \partial v^*_Q(t)/\partial x_Q\ v^*_{Qx} \tag{6}$$

whose components are

$$a^*_{Qx} = -v^2/[2x_P\,], \tag{7}$$

$$a^*_{Qy} = v^2/[2x_P]\cdot[x_Q/x_P - 1/2\,]\,/\sqrt{[\,x_Q/x_P]\,[1 - x_Q/x_P\,]} \tag{8}$$

The absolute value of vector $\mathbf{a}^*_Q(t)$ is :

$$|\mathbf{a}^*_Q(t)|=v^2/[4x_P]\cdot 1/\sqrt{[x_Q/x_P][1-x_Q/x_P]} \tag{9}$$

The acceleration vector $\mathbf{a}^*_Q(t)$ can be decomposed into the sum of two orthogonal vectors of the δ plane, $\mathbf{a}^{(c)}(t)$, $\mathbf{a}^{(s)}(t)$: $\mathbf{a}^{(c)}(t)$ is oriented as segment Q*M* and is due to the rotation of point Q* around point M* in the counterclockwise direction (centripetal acceleration), while $\mathbf{a}^{(s)}(t)$ (which is orthogonal to $\mathbf{a}^{(c)}(t)$ and has positive x and y components) is due to the "elastic stretching " of vector M*Q* along the radial direction (Coriolis acceleration) (see Fig.2 in which M*x'y'z' is a orthogonal reference system with the axes x', y', z' parallel to x, y, z respectively).

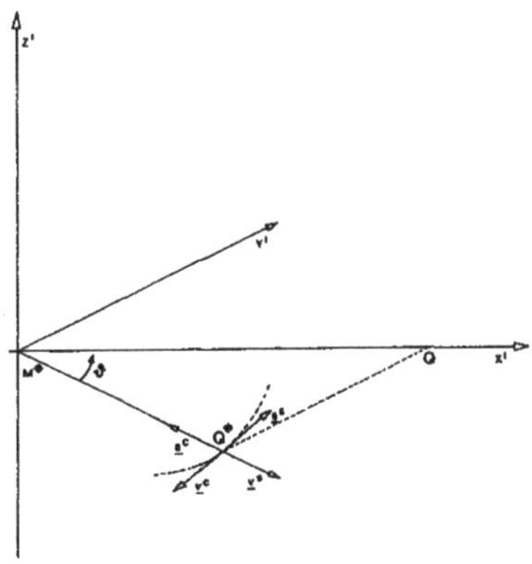

Fig.2

Under these assumptions, $|\mathbf{a}^*_Q(t)|$ can be decomposed in the following way :

$$|\mathbf{a}^*_Q(t)| = \sqrt{[\Omega^2 R^*_Q]^2 + [2\,\Omega\, v_Q^{(s)}]^2} \tag{10}$$

where

$$\Omega = v/[\sqrt{2}\, x_p], \tag{11}$$

$$R^*_Q = \sqrt{2}\,[x_p/2 - x_Q] \tag{12}$$

and $v_Q^{(s)}$ is the velocity component of point Q* due to the "elastic stretching".

As far as $v_Q^{(s)}$ is concerned, it seems reasonable to assume that it is proportional to $[x_Q/x_P]\cdot v$, what amounts to putting :

$$v_Q^{(s)}=\Omega\,\gamma(x_Q,t) \tag{13}$$

where

$$\gamma(x_Q,t) = (x_p/2)\sqrt{[(v/2)/v_{Qy}]^2 - [v^*_{Qx}/(v/2)]^2/2} \tag{14}$$

We could approximate v^*_{Qy} by a linear function of x_p, varying from $-v/2$ to $v/2$ (see dashed line in Fig. 1) .The distribution of velocity vectors $\mathbf{v}^*_Q$ (with components v^*_{Qx}, v^*_{Qy}) could be then explained as due to the rotation of a rigid bar O*P*, on the Oxy plane, around its middle point M*: the "stretching" of the rotating bar would be neglected.

According to formula (10), point Q* belongs to the straight line connecting M* with P*; the segment MP is amplified according to "amplification coefficient"$\sqrt{2}$, corresponding to an inclination angle of 45° of the M*P* segment with respect to the x-axis.

In reality, the inclination angle θ_P varies with time. Recalling that

$$\alpha_P = \sqrt{1 + \tan(\theta_P)^2} = 1/\cos(\theta_P), \qquad \Omega = d\theta_P/dt$$

and assuming $\Omega = v/[\alpha_P(t).x_P(t)]$ we obtain the following differential equation :

$$d\theta_P = [\cos(\theta_P)]\,[v/(a+vt)]\,dt \quad , \qquad 0 \le t \le T\,,\ \theta_P(0) = \theta_o\,, \tag{15}$$

for the expanding phase (θ_o is initial value to be determined experimentally) .

An approximate solution of (15) is given in Fig. 3 , under the assumption that $\theta_o = -\pi/6$ (i.e. 30°) and a = 0.5 b.

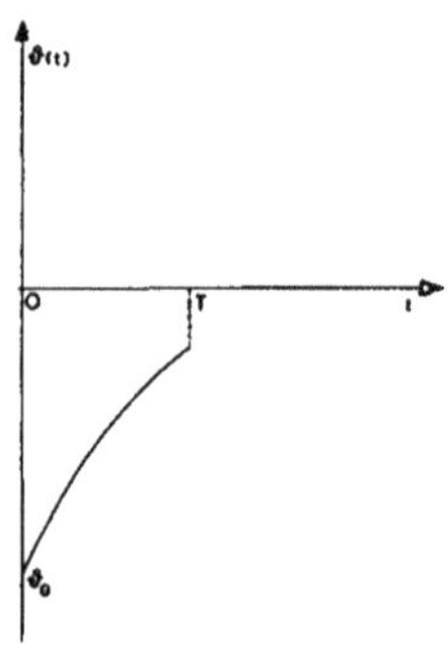

Fig.3

It should be noticed that the choice of a negative sign for $\theta(t)$ is due to the fact that, during the expanding phase, the observer perceives the moving gate as rotating around the z axis "from the observer" toward the frontal plane .

<u>References</u>

[1] M. Zanforlin, L. Beghi, E. Xausa, L. Tomat (1994) Experimental determination of the depth effect in the stereokinetic phenomenon of the swinging gate, *in this volume.*

[2] E. Xausa, L. Beghi , M. Zanforlin (1994) Mathematical model of the depth effect in the translatory alternating movement. Part A : perception of length amplification, *in this volume.*

THE DEPTH EFFECT IN THE STEREOKINETIC PHENOMENON OF "SWINGING GATE "

M. Zanforlin *, L. Beghi +, E. Xausa+, L. Tomat *

+ *Dipartimento di Matematica Pura e Applicata, Università di Padova, Via Belzoni 7, Padova 35131, Italy*

* *Dipartimento di Psicologia Generale, Università di Padova, Piazza Capitaniato 3, Padova 35139, Italy.*

When a person moves his/her head sideways, the vertical objects of the landscape project onto the retina a series of images that move at different speeds depending on the distance of the objects.

A schematic representation of this situation can be obtained in the laboratory with the stereokinetic phenomenon of the swinging gate.

When a vertical bar moves on the frontal plane, alternatively approaching and receding from a stationary and parallel bar, a depth effect is observed: the two bars appear to be rigidly connected, forming the extremities of a swinging gate that oscillates in depth. The distance separating the two moving bars, i.e. the length of the gate, appears well defined and much longer than the maximum distance between the two bars when they are stationary on the frontal plane [14].

This latter aspect of the phenomenon appears of particular theoretical relevance because, as it is well known, the perception of a 3-D structure from 2-D moving patterns poses the so-called inverse problem of projective geometry: i.e. a problem that has no unique solution [7]. To constrain the number of possible solutions Ullman [8,9] proposed a "rigidity assumption": i.e. if we can assume that the moving pattern is the 2-D projection of a "rigid object", then, in many cases (depending on the stimulus conditions) a unique 3-D solution is possible. Various other theories of 3-D from 2-D moving patterns are also based on a rigidity assumption [4,3,5].

But the rigidity assumption cannot explain the perceived length of the swinging gate because, as can be seen in Fig.1b, the two bars can be the 2-D projection of a tilted gate of any length.

a

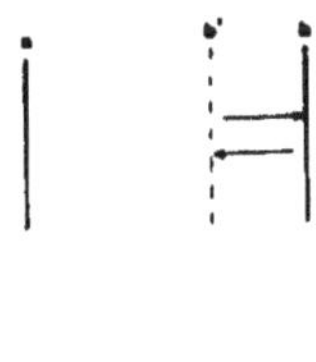

b

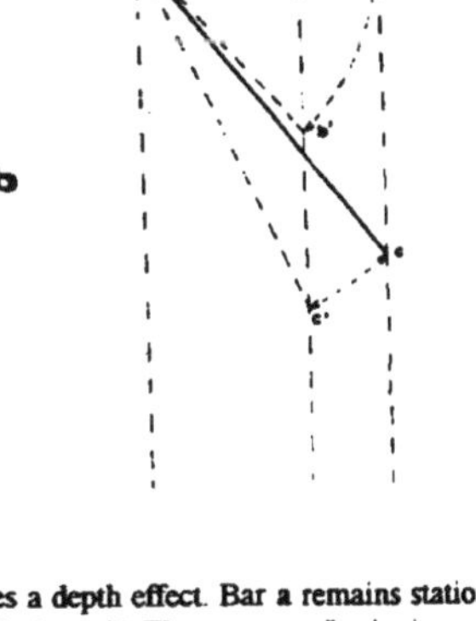

Fig. 1 a) Stimulus configuration that produces a depth effect. Bar a remains stationary while bar b moves alternately from b to b' on a frontal plane. b) The apparent "swinging gate" seen from above. Although the stimulus configuration in a) may be the planar projection of a "gate" of any length (see for example a-b. a-c). the swinging gate is perceived to have a well definited length depending on the maximum distance on the frontal plane of the two bars a and b in Fig. 1a.

The rigidity assumption is clearly derived from our daily experience with the real 3-D objects. But Gestalt theory postulated long ago that regardless of our daily experience, perceptual processes follow a minimum principle[12].

In an attempt to apply this general principle to moving stimuli a formal hypothesis was recently advanced by Beghi, Xausa and Zanforlin [1,2,11,12,13] to explain the depth effect in stereokinetic phenomena produced by rotatory movements. None of these phenomena can be explained by the rigidity assumption. The hypothesis based on a procedure of velocity difference minimization, allows a unique 3-D theoretical solution that has proved to be very close to the experimental results [1,2,11,13].

To our knowledge, no other hypothesis has yet been advanced which allows a single solution for the depth effect in these phenomena.

To see whether the hypothesis of velocity difference minimization could also explain the depth effect of the swinging gate, a more accurate experimental analysis of the phenomenon was carried out. The experimental results of the depth effect were then compared with the theoretical predictions worked out by Beghi, Xausa and Zanforlin [3,10] on the basis of the mathematical model of velocity difference minimization.

Experiment

As the materials and methods are the same as those described in detail in a previous experiment [14] they will be summarized here.

The configurations consisted of two vertical bars. One was kept stationary while the other moved horizontally at uniform speed with an oscillation half the maximum distance reached by the moving bar from the stationary one.

The moving bars were produced on the screen of an IBM type computer on which the various parameters could be modified independently. The variables considered were: 1) maximum distance of the two bars (3 different distances), 2) movements of expansion and contraction considered separately (i.e. the configuration disappeared after the end of the expansion movement and vice-versa for the contraction).

The subjects were seated in front of the screen with their head coaxial with it and 2 meters away. They were asked to look at the configuration with one eye only and to evaluate:

1) the apparent movement in space described by the bars;

2) the apparent maximum distance of the two extreme bars or the length of gate, by comparing it with a series of line segment of standard length drawn horizontally on a piece of cardboard at the side of the screen;

3) the minimum and the maximum angles of inclination of the moving gate with the frontal plane. This was done by positioning a clock hand on a cardboard disk that the subjects had in front of them.

The subjects were 12 university students, naive as to the aims of the experiment and with normal or corrected to normal vision.

Results

All subjects described the two bars as an almost rigid surface or gate oscillating in depth with an apparent rotation movement. The apparent rotation in depth appeared to all subjects even when there was only the expansion or only the contraction movement.

The quantitative results for the apparent length of the gate are reported in Fig.2 where they are compared with the predictions obtained from the mathematical model. No statistical difference was observed between the apparent length of the gate with configurations that presented a continuous oscillation and those that were only expanding or only contracting.

In both types of configurations the maximum and the minimum angles were statistically different and they had an average estimation of 46 and 23 degrees respectively with a S.D. of 17 and 15 degrees respectively. Yet again no difference was observed among the two types of configuration.

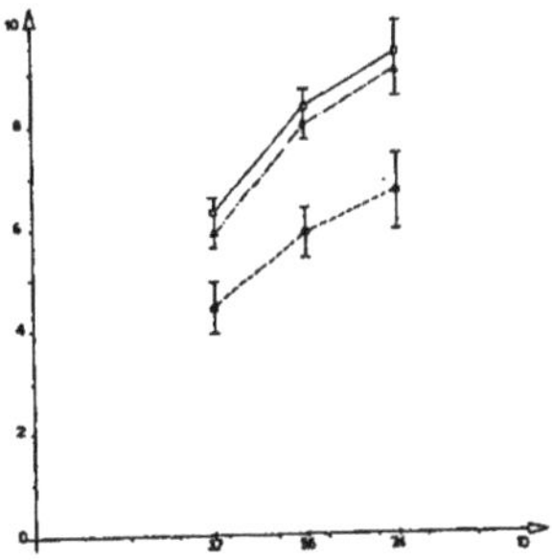

Fig.2

Fig.2 Average values of the estimated length of the swinging gate compared with the predicted values.(Δ) Observed length of the swinging gate; (□) Predicted values; (•) Estimated distance of the two stationary bars.The x-axis represents in cm the physical distance between the stationary bars; the y-axis represents in cm the estimated and theoretical values.

Conclusions

The experimental results previously reported [11] showed that the perceived length of the swinging gate depends only on the maximum distance on the frontal plane of the moving bar from the stationary one. The ratio of the apparent length of the swinging gate to this maximum distance was 1.35 and is quite close to the ratio predicted by the mathematical model and to that obtained in the present experiment.

Two other predictions of the mathematical model were confirmed: the apparent rotation of the gate during the expanding movement, when the visual system had no information as to how far the bar would move; and the apparent angles made by the swinging gate with the frontal plane, that are quite close to the predicted ones.

It thus appears that the hypothesis of velocity difference minimization is quite well supported by the experimental results.

References

1) L.Beghi, E. Xausa, M. Zanforlin (1991) Analytical determination of the depth effect in stereokinetic phenomena without a rigidity assumption. *Biol. Cybern.*, 65, 425-432.

2) L.Beghi, E. Xausa, C. De Biasio, M. Zanforlin (1991) Quantitative determination of the three-dimensional appearance of a rotating ellipse without a rigidity assumption. *Biol. Cybern.*, 65, 433-440

3) E. Xausa, L. Beghi, M. Zanforlin (1994) Mathematical model of the depth effect in the translatory alternating movement. Part A: 3-D perception of length amplification, *in this volume.*

4) E.C. Hildreth, N.M. Grzywacz, E.H. Adelson and V.K. Inada (1990) The perceptual buildup of three-dimensional structure from motion. *Percep. and Psychophys.*, 48, 19-36.

5) H.C. Longuet-Higgins (1986) Visual motion ambiguity. *Vis. Res.*, 26, 181-183.

6) K. Nakayama (1985) Biological image motion processing: a review. *Vis. Res.*, 25,(5), 625-660.

7) T. Poggio and C. Koch (1985) Ill-posed problems in early vision: from computational theory to analogue networks. *Proc.of the Roy.Soc. of L., B*, 226, 303-323.

8) S. Ullman (1979) The interpretation of visual motion, *MIT Press*, Cambridge Mass.

9) S. Ullman (1984) Maximizing rigidity. The incremental recovery of 3-D structure from rigid and non-rigid motion. *Perception*, 13, 255-274.

10) L. Beghi, E. Xausa, M. Zanforlin (1994) Mathematical model of the depth effect in the translatory alternating movement. Part B: "swinging gate"phenomenon, *in this volume*.

11) M. Zanforlin (1988) The height of a stereokinetic cone: a quantitative determination of a 3-D effect from a 2-D moving pattern without a "rigidity assumption". *Psychol. Res.*, 50, 162-172.

12) M. Zanforlin (1986 b) Stereokinetic phenomena as good Gestalts. The minimum principle applied to circles and ellipses in rotation: a quantitative analysis and theoretical discussion.*Gestalt Theory*, 10, 187-214.

13) M. Zanforlin and G. Vallortigara (1988) Depth effect from a rotating line of constant length. *Percep.and Psychopys.*,44, 493-499.

14) M. Zanforlin and L. Tomat (1993) Depth effect from line segments in relative movements on the frontal plane. *Atti e Memorie Acc. Pat. di SS. AA.* vol. CIV, parte II, pp. 33-44.

(FROM ANIMAL VISION TO) MACHINE VISION

View–Based Navigation and Cognitive Maps in Man and Machine[1]

Hanspeter A. Mallot
Max-Planck-Institut für biologische Kybernetik
Spemannstr. 38, 72076 Tübingen, Germany
ham@mpik-tueb.mpg.de

Abstract

The visual apprehension of three–dimensional space both in animals and machines is based on flat (two–dimensional) images or views. It is often assumed that visually guided behaviour in the 3D world relies on 3D models that are contructed from the images. In this paper, it is argued that such models are not necessary for a number of spatial tasks including obstacle avoidance and maze exploration. An analysis of the minimal amount of information required to perform these tasks shows that view–based processing suffices. The results are related to current models of rotation invariant object recognition that also suggest view–based processing.

1 View–based vs. model–based perception of spatial relations

Information on spatial relations in the environment is crucial for the generation and control of most kinds of behaviours both in living beings and in robots. One source of such information is the pair of retinal images which usually contains a number of cues to depth, including shading, texture gradients, stereoscopic disparities and motion parallax, among others. Following the ideas of David Marr (1982), extensive research efforts have been directed towards extracting the 3D information included in these cues in an *explicit* form, i.e. to construct various representations of depth from them. Examples of such representations include Marr's $2\frac{1}{2}$D–sketch for surfaces, generalized cylinders for objects (Marr and Nishihara, 1978), or cognitive maps for representing the environment (see Gallistel 1990, O'Keefe 1991).

One way to think about representations is as a piece of information that has been made explicit in the brain. If the 3D structure of an object is explicitly known, it should in principle be possible to predict novel views of this object. Recent psychophysical work shows that this is not the case: even if ample 3D information is provided (e.g., by stereo or the kinetic depth effect), recognition is much harder for novel views than for familiar views included in the training set (Bülthoff and Edelman 1992). One possible interpretation of this result is that 2D views rather than some explicit representation of 3D shape are stored in the brain.

Instead of making all available image information explicit and storing it in a representation, one could think of picking just the right pieces of information required for a given behavioural task. In this view, perception is part of a perception–action cycle that uses image information with the least possible amount of intermediate computation. The ultimate goal of neural information processing is behaviour, not representation. In the behaviour, however, information extracted by the perception system is incorporated again in an implicit form; it is therefore not quite clear why it should be made explicit in between. A particularly interesting way to state the underlying question is this: how complicated can spatial behaviour get without using explicit representations of space? In this paper, this question is explored in two problems from the field of navigation in space, obstacle avoidance and maze exploration.

2 Obstacle avoidance

The avoidance of obstacles is a simple task that has to be handeled prior to the more complicated information processing required to recognize or classify the obstacle as a particular object. For a terrestrial animal or robot, the simplest definition of an obstacle that does not involve object recognition is this: any deviation of the environment from the ground plane which occurs in the vicinity of the intended path is an obstacle. This definition refers to the 3D structure of the environment since

[1] I am grateful to Heinrich H. Bülthoff, Sabine Dartsch and Bernhard Schölkopf for valuable contributions.

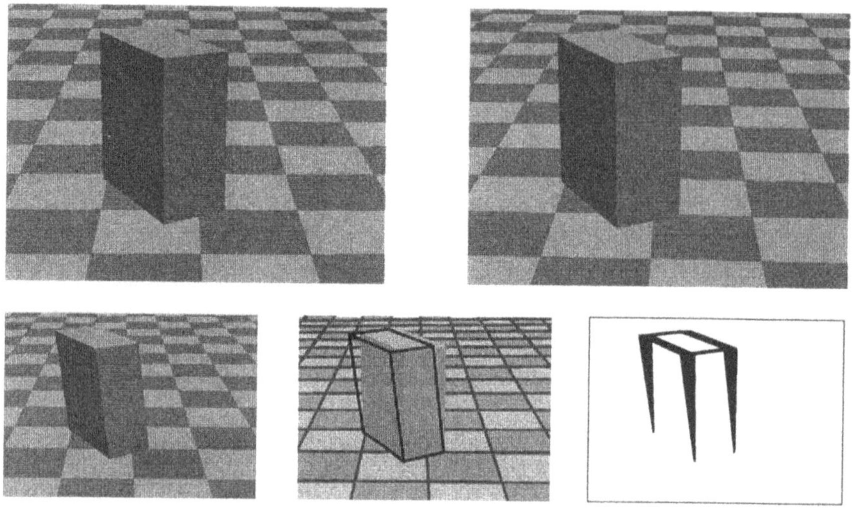

Figure 1: Obstacle avoidance by inverse perspective mapping. The top row shows a simulated stereo pair picturing a groundplane with an obstacle. Bottom left: right view warped by inverse perspective mapping so as to predict the left camera's view. Bottom middle: actual and predicted views of the left camera superimposed (the actual view is shown in grey-tones, the predicted view by edges only). Bottom right: Thresholded image difference of the actual and the predicted view. The obstacle stands out clearly since the prediction is correct only for the ground plane.

"deviations" are surface variations in the vertical direction. The groundplane can be considered as the plane spanned by the (terrestrial) animal's degrees of freedom for movement.

One way to implement an obstacle avoidance system based on this definition is to construct a 3D model of the environment, e.g., from stereopsis. This is a rather costly procedure and results in a considerable amount of excess information which is not needed for obstacle avoidance.

A simpler way to answer the question whether or not a certain path is blocked by an obstacle is sketched in Fig. 1. Starting with a stereo pair of views, the hypothesis of a free path can be tested in the following way: if no obstacles are present, the view of the left eye can be predicted from that of the right eye. The idea is to formally project the right image back onto the hypothesized ground plane and view this projection from the other eye. If only the ground plane is there, the actual view and the prediction from the other eye will coincide. If differences are found, the hypothesis of a free path must be rejected, i.e. an obstacle in the sense of the above definition has been detected.

The prediction is calculated by a perspective collineation dubbed "inverse perspective mapping" and described in detail by Mallot et al. (1988, 1991). While this work gives an analytical account based on the camera parameters and projective geometry, a system learning the inverse perspective map from examples can easily be conceived of. The scheme has a number of attractive properties including its robustness with respect to ground plane texture or corrugations, the increase of the the image difference with obstacle elevation in a body-scaled manner, and the low computational cost. For implementations on autonomous vehicles, see Storjohann et al. (1990), Bohrer et al. (1992), and Mayhew et al. (1992).

With respect to the topic of this paper, it should be stressed that inverse perspective mapping is a view-based rather than a model-based scheme. Binocular depth information is used in the form of mere image differences, without solving the stereo correspondence problem. Therefore, no excess information irrelevant to the task of obstacle avoidance is collected. Interestingly, evidence for the existance of an image difference mechanism in human stereovision was presented recently by Arndt et al. (in press).

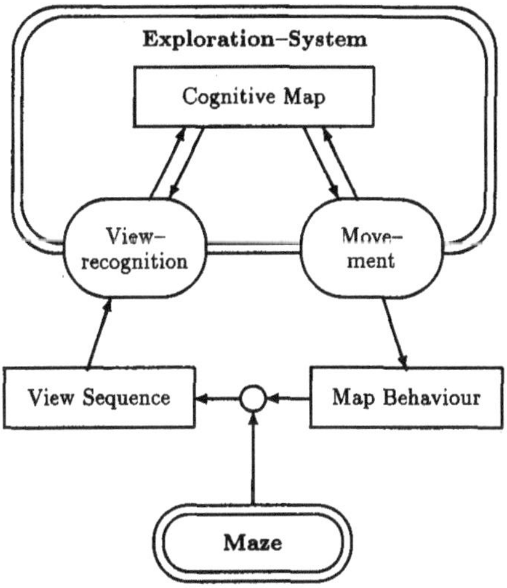

Figure 2: Cognitive maps and the perception–action cycle. The sensory input available to the system is a sequence of views reflecting the structure of the maze and the observers movement decisions. The cognitive map stores information on the interrelation of the view sequence and the movements taken (upward arrows). It also generates "expectations" in the sense that (i) similar views can be distinguished by their position in the maze and that (ii) movement decisions are guided by the cognitive map (e.g., for exploration or goal–finding). For further explanations see text.

3 Cognitive maps[2]

3.1 The perception–action cycle for map behaviour

A *cognitive map* is a neural mechanism which enables its user to solve navigation and orientation tasks as if using a real map of the environment. Examples of map behaviour include: repetition of previously travelled routes, finding detours around obstacles, approaching a goal from a novel starting point, finding a path between two arbitrary points of the environment (topological use of the map), finding the shortest path (topological and metrical use of the map), etc.

A cognitive map contains two types of information, concerning (a) the recognition of places and (b) the connections between them. Previous approaches have often focused on the first problem and tried to learn paths as an association between recognized views or places and motor commands (for review, see O'Keefe, 1991). In this paper, we give an explicit account of the connectivity structure of the explored environment in terms of the view–graph.

The notion of a cognitive map pursued in this paper is illustrated in Fig. 2 as part of an action–perception cycle. Since we restrict ourselves to cognitive maps of mazes, sensory input can be described as a sequence of views reflecting the connectivity structure of the maze together with the movement decisions taken by the system exploring the maze. Another advantage of mazes is the fact that there are only a small number of possible (egocentric) movements, such as "go left", "go right", "go back", "turn left", etc. The cognitive map does not represent the complete information of the maze but only those aspects that are relevant to (i) the interpretation of view–sequences or (ii) the generation of map behavior. The left double–arrow in Fig. 2 indicates the interaction of view recognition and the cognitive map: if a view is recognized, the current position in the map can be updated. Vice versa, knowledge on the current position in the map helps distinguish similar views occuring at different locations. The right double–arrow indicates the fact that movement decisions must be associated with the resulting view changes in order to make the cognitive map predictive; also, map information will be used to make movement decisions with respect to a given plan (such as exploration, approach to a target, etc.).

In Sect. 3.2, we present the view–graph as a data–structure that is built on concepts such as views and movements and is able to represent all relevant information on mazes. The argument will be based mostly on mathematical graph theory. Sect. 3.3 deals with an artificial neural network reconstructing a cognitive map (i.e. a view–graph) from a sequence of views and motion decisions.

[2] A more detailed description of this work including further simulations is presented by Schölkopf and Mallot, 1994.

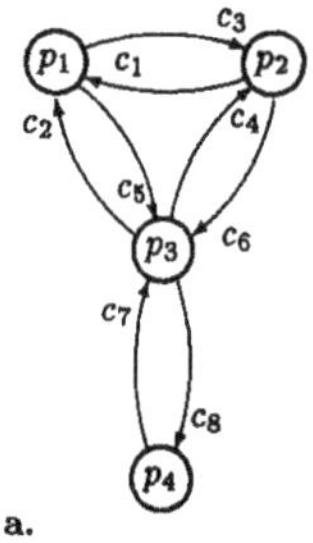

a.

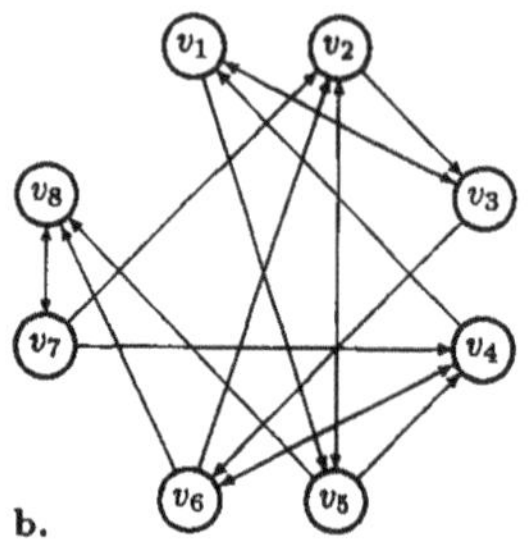

b.

to view \ from view	v_1	v_2	v_3	v_4	v_5	v_6	v_7	v_8
v_1	t_o	–	g_b	g_l	–	t_r	–	–
v_2	–	t_o	–	t_l	g_b	g_r	g_l	t_r
v_3	g_b	g_r	t_o	–	t_l	–	–	–
v_4	–	t_r	–	t_o	g_l	g_b	g_r	t_l
v_5	g_l	g_b	t_r	–	t_o	–	–	–
v_6	t_l	–	g_r	g_b	–	t_o	–	–
v_7	–	–	–	–	–	–	t_o	g_b
v_8	–	t_l	–	t_r	g_r	g_l	g_b	t_o

c.

Figure 3: **a.** Simple maze shown as a directed graph with places p_i and corridors c_j. **b.** Associated view–graph where each node v_i corresponds to one view, i.e. one directed connection in the place graph. Only the edges with go–labels are shown. In graph theory, b. is called the *interchange graph* of a. Simpler plots of b. are possible but not required for our argument. **c.** Adjacency matrix of the view–graph with labels indicating the movement leading from one view to another. Go–labels (involving a locomotion from one place to another): g_l (go left), g_r (go right), g_b (go backward). Turn–labels (e.g., probing of corridors): t_l (turn left), t_r (turn right), t_o (stay).

3.2 The view–graph as a sufficient representation of mazes

3.2.1 Places, views, and movements

Consider a simple maze composed of *places* $p_1, ..., p_n$ and *corridors* $c_1, ..., c_m$ (Fig. 3a). One way to think of this maze is a graph where the places are the nodes and the corridors are the edges. We consider all corridors to be directional but allow for the existence of two corridors with opposite directions between any two nodes. Throughout the paper, we will assume that there are no isolated places in the maze, i.e., each place can be reached from any other place by at least one sequence of corridors.

When exploring a maze, the observer generates a sequence of movement decisions defining a path through the maze. In doing so, he encounters a sequence of *views* from which he wants to recover the place–graph. In order to study the relation of views, places and movements, we make the following simplifying assumptions.

Views: There is a one–to–one correspondence between directed corridors and views. All views are distinguishable and there are no "inner views" in a place that do not correspond to a corridor.

Movements: At each time step, one movement from a finite (usually small) set is selected. Thus, the observer knows if he simply probed a corridor (by turning towards it without moving) or if he actually walked it. Clearly, the same view would be encountered in both cases.

With these assumptions, we can construct the *view–graph* that an observer will experience when exploring a maze (Fig. 3b,c). Its elements are:

- The **nodes** of the view–graphs are the views v_p, which, from the above assumption, are simply identical to the corridors in the place–graph. We denote the start and target place of a view v by $P_{out}(v)$ and $P_{in}(v)$, respectively. Of course, the functions P_{in} and P_{out} are not known when exploring the maze.
- The **edges** of the view–graph indicate temporal coherence: two views are connected, if they can be experienced in immediate temporal sequence. The edges are labelled with the movements resulting in the corresponding view sequence. It is convenient to distinguish two types of labels:
  - "Go–labels" specify movements leading from one place to another. In Fig. 3c, three types of go–labels are shown, indicating leftward, rightward and backward movements.
  - "Turn–labels" specify movements within one place, i.e. probing movements where a view is looked at but not walked to.

The resulting adjacency matrix with movement labels is depicted in Fig. 3c. Note that all edges starting from the same node will have different movement labels.

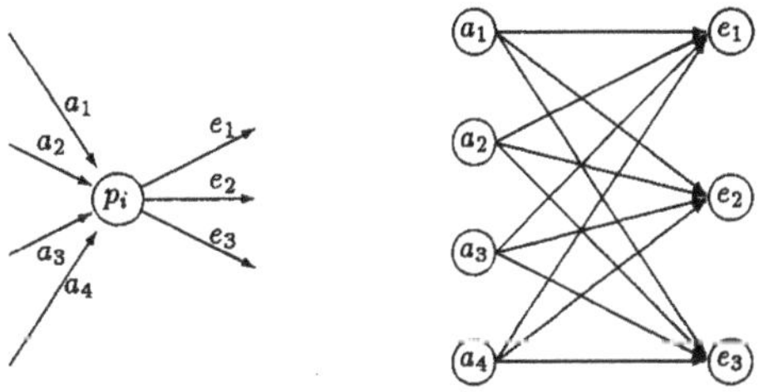

Figure 4: Subgraphs of maze– and view–graph corresponding to one place. The subgraph of views is a complete bipartite graph since every entry view is connected to every exit view (i.e., the successors of all entry views are the same). The complete bipartite structure of the subgraph has consequences for the redundancy and the planarity of the view–graph.

For the graph–theoretic argument in the rest of this section, we identify all go–labels and neglect the turn–labels. Thus, two views v_p and v_q are connected iff $P_{in}(q) = P_{out}(p)$. We denote by $A = (a_{pq})$ the simplified *adjacency matrix* of the view–graph:

$$a_{pq} = \begin{cases} 1 & P_{in}(q) = P_{out}(p) \\ 0 & \text{otherwise} \end{cases} \tag{1}$$

It can be obtained from the adjacency matrix of Fig. 3c by replacing all go–labels by ones and the turn–labels and bars by zeros. Taken together, the (now unlabelled) view–graph defined here is the *interchange graph* (e.g., Wagner 1970) of the place–graph.

3.2.2 Which information does the view–graph represent?

In this Section, we assume that the view–graph has been learnt, e.g. by the neural network presented in Sect. 3.3. The question then is: does the view–graph contain enough information to guide map behaviour? To answer this question, we will now mathematically reconstruct the underlying maze–graph from the view–graph. It turns out that this can be done already with the simplified (unlabelled) view–graph introduced in the previous section.

It should be clear that the place–graph can be recovered from the view–graph only up to permutations of the place– and view–numbers. Strictly speaking, we will construct a third graph whose nodes are sets of views leading to one place. This graph will be isomorphic to the original place–graph. We start by defining the *successor* $\mathcal{S}(v_p)$ of a view v_p as the set of views that can follow v_p:

$$\mathcal{S}(v_p) = \{v_r | a_{rp} = 1\} \tag{2}$$

The reconstruction method rests on the idea that two views lead to the same place, if they have the same successor:

$$P_{in}(v_p) = P_{in}(v_q) \Longleftrightarrow \mathcal{S}(v_p) = \mathcal{S}(v_q). \tag{3}$$

To prove this equivalence, we substitute from Eqs. 2, 1 and obtain:

$$\begin{aligned} \mathcal{S}(v_p) = \mathcal{S}(v_q) & \iff \{v_r | a_{pr} = 1\} = \{v_s | a_{qs} = 1\} \\ & \iff \{v_r | P_{out}(r) = P_{in}(p)\} = \{v_s | P_{out}(s) = P_{in}(q)\} \end{aligned}$$

The last two sets cannot be empty, because the maze does not contain isolated places. Since the indices r and s refer to the same arbitrary numbering of the set of views, it is easy to see that the final equality implies $P_{in}(p) = P_{in}(q)$, which proves our proposition. Note that in addition to Eq. 3, we have proved $\mathcal{S}(v_p) \cap \mathcal{S}(v_q) \neq \emptyset \Longleftrightarrow \mathcal{S}(v_p) = \mathcal{S}(v_q)$: Successors are either disjoint or identical.

While we cannot recover the functions P_{in} and P_{out} explicitly, Eq. 3 shows that the equality of their successors partitions the set of all views into n subsets correponding to the places $\{p_i | i \leq 1 \leq n\}$. These are the nodes of the recovered graph. The corridors can be found by inspection of the view–graph.

The above results were obtained for the unlabelled version of the view–graph. Since this can always be derived from the labelled view–graph, the results hold for the richer graph *a forteriori*. It should be clear that the reconstruction described here need not be carried out in the brain; it simply illustrates the sufficiency of the representation.

3.2.3 Redundancy of the view–graph and the completion matrix

The fact that all views leading into the same place have the same successor can be restated by saying that each place in the maze corresponds to a complete bipartite subgraph in the view–graph (see Fig. 4). A complete bipartite graph consists of two subsets of nodes such that each node from one subset is connected to each node of the second subset while there are no connections within subsets. Here, the two subsets are the views leading into one place and the views leading away from this place. This second set is the successor of each member of the first set.

This structure of the view–graph entails an interesting type of redundancy which might be useful in exploration. Suppose that view a_1 is known to have the views e_1, e_2 in its successor and view a_2 is known to have e_1 in its successor. It can than be predicted that the sequence $a_2 \rightarrow e_2$ should also exist. If it doesn't, the views e_1 as seen in the sequences $a_1 \rightarrow e_1$ and $a_2 \rightarrow e_1$ are probably confused and must be distinguished by further examination.

The bipartite structure of the subgraphs has an interesting counterpart in the adjacency matrix. The successor of view v_p (Eq. 2) corresponds to the pth column of the adjacency matrix A; its size is the number of ones in that column. The reconstruction result (Eq. 3) therefore implies that A can have only up to n (the number of places) different columns. Two columns of A are either identical or orthogonal (no coinciding ones). Therefore, for a suitable permutation of the views, the symmetrical matrix $A^{\top}A$ will have block structure with n blocks each corresponding to one bipartite subgraph of views or to one place in the maze. If A is interpreted as a transition matrix, $A^{\top}A$ describes a step forward followed by a step backward, not necessarily the reverse of the step forward. Starting with some view, this movement will reach all views leading to the same place. We call $A^{\top}A$ the *completion matrix* of the view–graph. The size of the blocks equals the frequency of a particular column (successor) in a matrix, i.e., the in–degree of the corresponding place. The values taken by the coefficients inside each block are constant and correspond to the out–degrees of the corresponding places. Outside the blocks, the matrix takes the value zero (cf. Fig. 6).

3.2.4 Is the view–graph planar?

In the previous section, we have shown that the view–graph contains all information required to reconstruct the place–graph. If we want to represent view–graphs in a neural network, it is important to find a network topology in which they can actually be embedded. For example, two–dimensional grids are best suited for planar graphs, i.e., graphs that can be drawn on a sheet of paper without intersecting edges. Since the problem of planarity is completely solved for bipartite graphs (e.g., Wagner, 1970), we can give a simple sufficient criterion for non–planarity of the view–graph: Complete bipartite graphs are non–planar if both subsets contain at least three nodes. Therefore, the view–graph of a maze containing a place where three or more corridors meet, cannot be planar. This is the case for all interesting mazes.

3.2.5 Paths and movement sequences

So far, we have shown that the (unlabelled) view–graph contains all information contained in the place–graph. However, for path planning some additional information is required. This can be seen from the fact that simple transformations such as mirroring applied to the maze would affect neither its graph structure nor the structure of the view–graph. It would, however, strongly affect the paths taken through the maze since all left/right decisions would have to be reversed.

In the original maze, directional information can be included by allocentric or world centered direction labels assigned to the corridors (e.g., in Fig. 3a, corridor c_1 would be labelled "west", corridor c_4 "north–east", etc.). In the view–graph, we can now use the egocentric labelling system introduced in Sect. 3.2.1. The reconstruction method given above was based solely on the go–labels and ignored the connections having turn–labels. Of course, the turn–labels will be useful for reconstruction of the places. It is an empirical problem to find out the relative importance of these labels in biological maze learning.

For the generation of movement sequences, the turn–labels are irrelevant. If a path is given connecting two views in the view–graph by edges carrying go–labels, a corresponding movement sequence can easily be generated by simply listing the labels along the path.[3]

[3] If two views are connected by a turn–label, they will also be connected by a two–step path using go–labels.

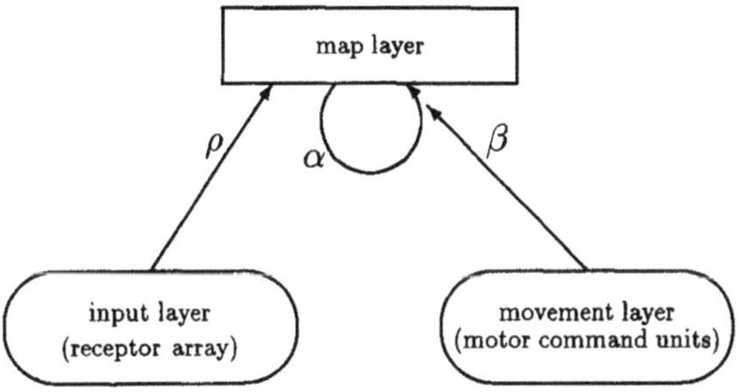

Figure 5: Structure of the neural network. ρ: input weights, α: map layer weights, β: axo-axonic weights from the movement units. Note the similarity to the intra-organismic part of the action-perception cycle depicted in Fig. 2.

3.3 Learning mazes from view sequences

3.3.1 Self-organizing sequence map

We construct a neural network consisting of three sets of units: the input, the movement, and the map layer. In order to support orientation behaviour in a maze, the network must represent information concerning two problems:

P1 Each view must be identified and associated with a particular location in the map. In our network, this is achieved by a set of input weights ρ_{ij}. After learning, each view will be represented by activity in the associated map unit, i.e. the unit whose input weights are most closely tuned to the presented view. Neurons representing particular places have been recorded from rats (see O'Keefe, 1991). For results on directional place cells, which more closely correspond to our views, see McNaughton et al. (1983).

P2 The topology of the maze is represented by weights α_{ik} connecting units within the map layer. Weights between units will be assigned according to the temporal sequence of the views represented by these units. Movement labels are associated to the connections in the map layer by presynaptic facilitating connections from the corresponding movement unit.

While the proposed network is reminiscent of a standard Kohonen map (Kohonen 1982), there are two important differences: First, nearness in the map corresponds to temporal adjacency, not to featural similarity. In fact, similar views can occur at great distances in the maze and must not be confused. Second, distance in the map is measured as the minimal number of synapses that must be passed between two units (the "combinatorial distance" in graph theory). Therefore the topological structure of the resulting "map" is not limited (see Martinetz and Schulten 1994). This is desirable since view-graphs need not be planar.

3.3.2 Description of the model

Network structure. The network consists of an *input layer* (J units) with activity variables f_j^t, a *map layer* (I units) with activity variables e_i^t and thresholds θ_i^t, and a *movement layer* (K units) with activity variables m_k^t; superscripts denote time steps. The input layer is fully connected to the map layer via the input weights ρ_{ij}^t. For each map unit i, the weight vector $\mathbf{r}_i := (\rho_{i1}, ..., \rho_{iJ})$ is called its "receptive field". As a consequence of the Cauchy-Schwarz inequality, the map unit will be activated most strongly by an input vector identical to its receptive field. Within the map layer, synaptic weights α_{in}^t are initialized to zero but may evolve during learning. The movement layer contains one unit for each possible movement (label of the view-graph connections). Each unit will establish facilitating presynaptic connections to all appropriate map layer connections. The map layer weight α_{in}^t may be facilitated by a presynaptic weight $\beta_{in,k}^t$, if m_k is the movement type associated with view connection α_{in}.

Input sequence. A sequence of movement signals is given as external input to the movement units. At the same time, a sequence of input vectors, $\mathbf{f}^t = (f_1^t, ..., f_J^t)^\top, t \in \mathbf{N}$, is simulated from the corresponding movements in some underlying maze-graph and presented to the input layer. In the first series of simulations, the movement decisions were chosen at random with equal probability, resulting in a random walk through the maze. For each view (directed corridor in the maze) a fixed view vector is chosen which is fed to the input layer of the network each time the random walk passes by. The view vectors can be either canonical base vectors (only one component different from zero) or random. In the first case, problem P1 above is assumed to be solved by some ideal preprocessing; in the second case, P1 and P2 are approached simultaneously.

Activation dynamics. The acitivity of the map layer units is described by

$$e_n^t = g\left(-\theta_n^{t-1} + \sum_{j=1}^{J} \rho_{nj}^t f_j^t + \sum_{i=1}^{I} \tilde{\alpha}_{ni}^t e_i^{t-1}\right), \tag{4}$$

where $g : \mathbf{R} \to [0,1]$ denotes the logistic function. Let $w(t)$ denote the index of the most active cell (the winner cell), i.e.: $e_{w(t)}^t = \max_i\{e_i^t\}$. It represents the view currently perceived in the maze. The effects of incoming information are biased by an intrinsic term (the second sum in Eq. 4) such that the current winner is likely to be a unit connected to the last winner via a strong weight. By $\tilde{\alpha}$, we denote the joint effect of map layer weight and the presynaptic facilitation from the movement layer:

$$\tilde{\alpha}_{in}^t = \alpha_{in}^t + \sum_{k=1}^{K}(1-\alpha_{ik}^t)\beta_{in,k}^t m_k. \tag{5}$$

Weight dynamics. The input weights ρ_{ij} are randomly initialized to values between 0 and 1, with a subsequent Euclidean normalization of each receptive field $\mathbf{r}_{w(t)} := (\rho_{i1}, \ldots \rho_{iJ})$. The receptive field of the winner unit approaches the presented input via the competitive learning rule with learning rate λ_1 (see Kohonen 1982)

$$\mathbf{r}_{w(t)}^{t+1} = \frac{\mathbf{r}_{w(t)}^t + \lambda_1 \mathbf{f}^t}{||\mathbf{r}_{w(t)}^t + \lambda_1 \mathbf{f}^t||}. \tag{6}$$

The input weights of the other units remain unchanged. Connections within the map layer are established between the last two winners and represent a transition between the last two views in the maze:

$$\alpha_{w(t),w(t-1)}^t = (1-\lambda_2)\alpha_{w(t),w(t-1)}^{t-1} + \lambda_2 \alpha_{max} \tag{7}$$

Here, λ_2 is a learning rate and α_{max} an upper limit for the weights.

Finally, the weights of the presynaptic connections $\beta_{in,k}$ are set to some constant $\phi \in (0,1]$ if movement m_k coincided with the last increase of α_{in}; otherwise, $\beta_{in,k}$ is zero.

Threshold control. Due to the intrinsic term in the activation dynamics, the network could converge to a state where just two strongly connected map units are the winner units for all view presentations. In order to overcome this problem, the thresholds of winner units increase according to the *threshold dynamics* (λ_3: learning rate, θ_{max}: maximal threshold):

$$\theta_{w(t)}^t = (1-\lambda_3)\theta_{w(t)}^{t-1} + \lambda_3 \theta_{max}. \tag{8}$$

3.3.3 Simulations

In this Section, we briefly summarize a number of simulation results obtained with the neural network model. A more thorough account of the network's computational capabilities will be given in a forthcoming paper by Schölkopf and Mallot (1994).

- **Convergence.** For the toy–maze of Fig. 3a, a network with $J = 8$ input–units and $I = 20$ map–units (no movement units) converges in 60 presentation steps. By this time, specific view–units together with the approriate map–layer connections have evolved.

- **View Recognition.** The intrinsic term of the map layer activation function (Eq. 4) helps recognize the views. If noise is added to the views, the map layer weights reduce the signal–to–noise ratio required for recognition by a factor of about 2 (3 dB). This indicates that the topological structure stored in the map layer weights is used to distinguish similar, but distant views.

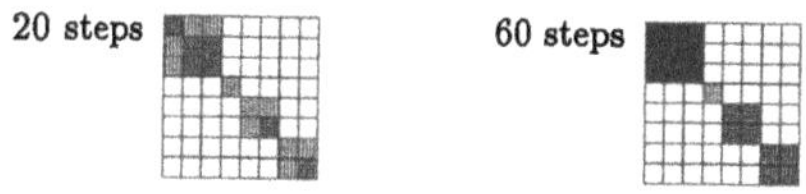

Figure 6: Completion matrices derived from the weight matrices after 20 and 60 learning steps. Grey-levels correspond to the values 0 (white) to 3 (black). The block structure is clearly visible already after 20 steps, the underlying maze can be inferred already from this stage.

- **Maze Reconstruction.** From the weight matrix, we derive an estimate of the adjacency matrix A of the view graph by deleting the all-zero rows and columns, thresholding the remaining entries, and suitable reordering of the rows and columns. The estimated completion matrix after 20 and 60 learning steps is shown in Fig. 6. The block structure is already fully developed after 20 learning steps. In addition, from the inhomogenities within the blocks, optimal strategies for further exploration of the maze can be derived.

4 Discussion

The examples presented in this paper indicate that view-based approaches can go a long way in the processing of spatial information. The information implicitly present in the images can support many behavioral competences without being transformed into explicit spatial representations. Of course, explicitness of internal representations is a graded property. In the case of obstacle detection, the representation is a pair of topographically transformed images which are very close to the original views. In the case of maze exploration, much more information is represented in the view-graph, even though it can be regarded as an essentially two-dimensional storage.

The view-graph can also be used to represent three-dimensional objects by their various views. Edelman and Weinshall (1991) proposed a neural network model for rotation-invariant object recognition which recovers view-graphs of objects (called "footprints" in that paper). The authors argue that views perceived in close temporal sequence are likely to belong to the same 3D object and should be linked in the view-graph. The predictions for view extrapolation made by view-based models have been confirmed in psychophysical experiments by Bülthoff and Edelman (1992). In maze reconstruction, the problem is slightly more complicated: again, views perceived in temporal sequence are connected, but do not in general belong to the same place. Object recognition is thus analogous to the reconstruction of the places of a maze from connections with "turn-labels". In order to recover the connections between places, "go-labels" are required as well (Sect. 3.2.2).

In general, the amount, type, and explicitness of the information represented at a stage depends on the information processing streams interacting at this stage. If, for example, maze information could be gathered from additional sources such as global compass, general guidances and world knowledge (e.g.: "if you want to reach the water, go downward"), acoustic information, or even communication with other exploring systems, a common stage would be needed where all these inputs could be compared. A "good" representation is therefore not so much characterized by its explicitness but by its ability to integrate data from different streams. To what degree explicit representations of space are required to make the various inputs and outputs commensurable, is an open question.

References

[1] P. A. Arndt, H. A. Mallot, and H. H. Bülthoff. Human stereovision without localized image-features. *Biological Cybernetics*, in press.

[2] S. Bohrer, M. Brauckmann, and W. von Seelen. Visual obstacle detection by a geometrically simplified optical flow approach. In *Proc. 10th European Conference on Artificial Intelligence, ECAI-92*, 1992.

[3] H. H. Bülthoff and S. Edelman. Psychophysical support for a two-dimensional view interpolation theory of object recognition. *Proceedings of the National Academy of Sciences, USA*, 89:60 - 64, 1992.

[4] S. Edelman and D. Weinshall. A self-organizing multiple-view representation of 3D objects. *Biological Cybernetics*, pages 209 – 219, 1991.

[5] C. R. Gallistel. *The organization of learning*. The MIT Press, Cambridge, MA, USA, 1990.

[6] T. Kohonen. Self-organized formation of topologically correct feature maps. *Biological Cybernetics*, 43:59 – 69, 1982.

[7] H. A. Mallot, H. H. Bülthoff, J. J. Little, and S. Bohrer. Inverse perspective mapping simplifies optical flow computation and obstacle detection. *Biological Cybernetics*, 64:177 – 185, 1991.

[8] H. A. Mallot, E. Schulze, and K. Storjohann. Neural network strategies for robot navigation. In L. Personnaz and G. Dreyfus, editors, *Neural Networks from Models to Applications*, pages 560 – 569, Paris, 1989. I.D.S.E.T.

[9] D. Marr. *Vision*. W. H. Freeman, San Francisco, 1982.

[10] D. Marr and H. K. Nishihara. Representation and recognition of the spatial organization of three-dimensional shapes. *Proceedings of the Royal Society (London) B*, 200:269 – 294, 1978.

[11] T. M. Martinetz and K. Schulten. Topology representing networks. *Neural Networks*, 7:507 – 522, 1994.

[12] J. E. W. Mayhew, Y. Zheng, and S. Cornell. The adative control of a four-degrees-of-freedom stereo camera head. *Proceedings of the Royal Society (London) B*, 337:315 – 326, 1992.

[13] B. L. McNaughton, C. A. Barnes, and J. O'Keefe. The contributions of position, direction, and velocity to single unit activity in the hippocampus of freely moving rats. *Experimental Brain Research*, 52:41 – 49, 1983.

[14] J. O'Keefe. The hippocampal cognitive map and navigational strategies. In J. Paillard, editor, *Brain and Space*, pages 273 – 295. Oxford University Press, Oxford, 1991.

[15] B. Schölkopf and H. A. Mallot. View-based cognitive mapping and path planning. Technical Report 7, Max-Planck-Institut für biologische Kybernetik, Tübingen, Germany, 1994. (Available as /pub/mpi-memos/TR-7.ps.Z via anonyoums ftp from ftp@mpik-tueb.mpg.de.).

[16] K. Storjohann, Th. Zielke, H. A. Mallot, and W. von Seelen. Visual obstacle detection for automatically guided vehicles. In *Intl. Conf. Robotics and Automation*, pages 761 – 766, Los Alamitos, CA, 1990. IEEE.

[17] K. Wagner. *Graphentheorie*. Bibliographisches Institut, Mannheim, Wien, Zürich, 1970.

ACTIVE VISION AND REPRESENTATION

Massimo Savini

Dipartimento di Informatica e Sistemistica
Universita` di Pavia
via Abbiategrasso, 209
27100 Pavia - ITALY
savio@ipvvis.unipv.it

Abstract

We try here to show what is meant by "active" vision and what are the benefits emerging from this approach, both in the reconstructionist and the purposive paradigms. Beyond that, we will try to place the active approach to vision in the perspective of an intelligent autonomous system, and to suggest the importance of the role of active vision in the debated problem of representation.

Problems with the passive approach to vision

Useful computer vision has to face the difficulties posed by non trivial, real world images. One image provides a significant amount of data, that becomes enormous if we have to deal with an image sequence in a situation which involves motion and we want to approach real time performance. Moreover, anything can happen in an image; not only noise, but the extreme variability of the real world constitute a very difficult problem at the computational theory and algorithmic levels. So, in search for a solution, a heavier and heavier burden is laid on the available computing resources. For these reasons the past predominant approach considered a static observer trying to recognise objects by analysing monocular images or, at most, stereo pairs. It was thought that observer motion would further complicate the already difficult job. This intuition would have been proved wrong.

Another reason for not considering motion stemmed from the origin of computer vision, in the early sixties, rooted in symbolic artificial intelligence. According to the traditional AI perspective, vision was then mostly seen as an input functionality for the higher reasoning processes. In this way the perception process was decomposed in successive layers of computations, organised in an open loop hierarchy. At the lowest level were processes devoted to noise reduction, edge points extraction, disparity calculation, etc. At an intermediate level were processes that extracted feature sets from one or more intrinsic images. At the higher level these feature sets were compared with internal models in order to recognise objects and to reason about the physical reality that produced the image(s) and that had been "reconstructed" by the previous processes.

Unfortunately, the task revealed itself as being much more difficult than expected, maybe because vision is so easy and effortless for us that people hoped it would have been almost the same for computers, maybe because vision escapes the domain of introspection, maybe because the researchers did not see some important connections at the system integration level, maybe because the chosen target was misplaced.

As a matter of fact, the reconstructionist school, which has seen its highest expression in the influential work by David Marr[1], has neglected for a long time the importance of the observer as being immersed in the scene and interacting with it in ways much more complex than the simple acquisition of images. Nevertheless an important part of Marr's contribution, perhaps the most important one, is the rigorous computational approach to vision, of which we will speak later, that deserves careful consideration.

In the work by Marr and the other researchers that followed his approach, attention is explicitly focused on WHAT is present in a scene and WHERE it is, and on the PROCESS for extracting this information from images. To know what is present in a scene means to recognise objects, and the most distinctive characteristic that allows to discriminate between different objects is shape. To define what is shape is difficult, but it is easy to accept its relationship with the topological and geometrical properties of the object. In the same way, to know where an object is can mean to know its geometrical relationship with the three-dimensional world in which it is placed. Thus, as obvious consequence, the importance given to the problem of obtaining a good geometrical representation of the viewed scene. This approach has been called *reconstructionist* because it tries to reconstruct the three-dimensional structure of the scene from its two-dimensional projections on images. It has given rise to many sophisticated algorithms sometimes collectively indicated as "Shape from X", where "X" can be for example instantiated by "shading", "contour" or "texture", in which the local three dimensional shape of the surface of an object is recovered from two dimensional image cues.

Problems arise from several factors, some of which are:

- the image formation process itself, that is the projection on a two dimensional surface of a three dimensional scene and that entails loss of depth information;
- the uncertainty about the number and direction of the illuminants;
- the variability of reflectance characteristics of objects surfaces;
- the complex interaction among local reflectance and surface orientation and illumination and observation directions.

Due to these and other aspects many shape from X problems suffer from ill-posedness, non-linearity, instability, or a combination of them.

Active reconstructionist solutions

In a seminal work fist presented in 1987 Aloimonos and others[2,3,4] demonstrated that a controlled motion of the sensor can considerably reduce the complexity of the cited classical vision problems that fall into the "Shape from X" group. For all the three cases cited above the authors showed that adequately controlled observer motion results in a mathematical simplification of the computations.

Let us take for example shape from shading, that is, a process to infer local orientation of object surfaces from apparent luminance values[5,6]. It results that if we want to solve it with a single monocular view, we need to take additional assumptions about the smoothness and uniformity in reflectance characteristics of the surface, and the uniformity of lighting. These often unrealistic assumptions limit the applicability of techniques and algorithms based on them, but they must be taken to regularise a generally underconstrained, ill-posed problem like passive shape from shading. Things become easier with two images in stereo configuration, leaving aside the correspondence problem, and even better with three images. But the problem is still non linear. Moreover stereo images pose two different kinds of problems in short and long baseline configurations: a short baseline makes the correspondence problem easier and allows for a linearization of the equations to be solved by taking only up to the linear term of a Taylor series expansion, but the resulting accuracy is low; a long baseline reaches a good accuracy but suffers from a difficult correspondence problem, that can trap the search for solution in a local minimum.

Aloimonos and colleagues presented an active technique that takes the advantages of both stereo configurations. This technique decouples the process of finding the unknowns (surface orientation and reflectance parameters, plus lighting distribution) in every single point in the image. The adopted reference system is fixed. The camera moves by shifting the position of the optical centre in the XY plane that is parallel to the image plane. For the mathematical details the reader can see one of the works published by the authors[2-4]. The effect of camera motion is that more data are collected in a controlled way such that the equations for the different surface points become coupled along time (subsequent camera positions) and are no more coupled for adjacent points on the surface. This takes rid of the old unrealistic assumptions on maximal surface smoothness and uniform reflectance and lighting. In this way it is possible to separate shape information from reflectance and lighting variations, thus detecting and measuring all of them. Mathematical and computational advantages of an active approach to other Shape from X modules has been similarly shown.

Purposive vision

Object recognition constitute only one of the major visual problems of interest, but other activities build on the information obtained through vision. One of the most important is *navigation*, that is, moving around in the world with the help of our visual sensors. Moreover there are many problems or sub-problems, that must be frequently solved by an intelligent, autonomous, perceiving agent, that need only some specific, task dependent information to be extracted from images. Some of these problems are: obstacle detection and avoidance, detection of independent motion, tracking of a moving object, interception, hand-eye coordination, etc. Trying to solve each of these problems separately can lead to much more economical solutions and to the realisation of effective working modules to be activated on demand when need arises.

The alternative approach now sketched has been called "Purposive vision", referring to the utilitarian exploitation of both system resources and surrounding world information, or "Animate vision", with approximately the same meaning[7].

We have already seen that an active approach can help in the recovery of the three-dimensional structure of the observed scene. And the results presented in the literature until now are only the first ones obtained by following this approach; research in active vision is flourishing all around in the world and are already bringing new insights and practical applications[8-12]. Now, some questions arise. If there is convenience in selecting the information to be extracted from the scene, what kind of information is best suited to the task and to the resources at hand? How to combine perception with action, vision with behaviour? How much explicit representation do we need? And which representations?

Intelligent perception and the active approach

Active vision is actually an approach, a shift in perspective, that considers vision itself not as an isolated functionality, but as a set of tools that an intelligent agent uses at its best to reach a goal. So, active vision is not a part of vision; rather it is vision performed in an active way. And, once for all, it should not be confused with active sensing, in which the sensory system sends some sort of energy onto the observed scene in order to measure, through its use, some physical property. In active vision it is the observer that is active, while the sensors can be either active or passive. In essence, an active observer uses its resources at its best to achieve a goal. And resources include sensors, computing machinery and actuators. Active vision achieves the best conditions for performing a given visual task by adjusting observer and sensors parameters.

It is important to see vision in the context of the overall behaviour of an intelligent system in order to avoid the pitfall of considering this sense, mostly useful to humans and to many other animals, only as a separate functionality providing input to higher processes in an open loop fashion. If vision is not to be considered only as an open ended input channel to the conscious mind, then it is not a self contained problem anymore; and it does not represent the final purpose but a tool, or a set of tools, to be used by an intelligent agent to help it carrying out its own tasks. This means that not every visual information nor every part or object present in a scene should be respectively extracted or analysed in detail, but only the single elements that are currently useful for the completion of the tasks at hand.

By taking a systemistic view of perception and of perceiving agents, we should not overlook any of the "subsystems" that contribute to the perceptual task. We constantly move our eyes, head and body to act in the world, and we have other sensors, in our ears and muscles, that provide information about our *movements*; so maybe the observer motion is not such an obstacle on the road to vision; maybe it is more an advantage. For sure it is an effective way to collect an amazing amount of useful information while dominating the spatiotemporal huge dimension of the input data by sequentially directing the fovea only on small regions of interest in the scene.

There is an aspect of image understanding that deserves careful consideration. It has been shown (Yarbus[13]) that the visual exploration of a picture by a human observer follows different paths according to the particular task he has been assigned with. Not only different global tasks may cause the exploration of different regions in the image, but some regions that are of particular significance for the given task are analysed more carefully; gaze is positioned on them with insistence, often coming back to these points of interest after a round trip to other regions. These experiments suggest two conclusions:

- what is important is to collect, in as short a time as possible, the information that allows to perform the task assigned to the system, leaving aside the rest;
- to answer some questions about the scene one needs to analyse few restricted areas for a long time, meaning perhaps that the extraction of certain kinds of information is difficult compared to the extraction of others.

This view is not completely new, and the idea that a complete internal representation of the world might not be the best approach to the vision problem can be traced for example in the ecological approach of Gibson[14-16]. According to him the world around us acts as a huge external repository of the information necessary to act, and we directly extract from time to time the elements that we need.

Computational vision and representations

The considerations just presented in the previous paragraph introduce us into the other problem that I wanted to address: representations. Do we need them? And, if yes, at what extent? And of which kind? The ecological approach seems to suggest a negative answer to the first of these questions; and the work of Brooks[17] in the track of behaviourism follows this approach. Conversely, in the Marr paradigm representations play a very important role. This is due also to the rigorous computational approach followed by him. In this computational perspective the role of representations is correctly emphasised as that of formal schemes that make explicit some information about an entity. Moreover, formal schemes are the only objects that can be manipulated and transformed, one into the other, by computer programs. It should also be noticed that even the extreme position of Brooks against representation is substantially against the use of certain kinds of them, and even his robots make use of representations. The point is that there exist many form of representation that are influenced by the available hardware, both for information gathering (sensors) and processing (neurons, transistors). Going back to the computational vision approach, the visual process, or at least its prominent part, is seen as transforming an image into a different kind of representation. As image, or input representation, is generally taken a regularly spaced grid of pixels, each of which carrying a numerical measure of the irradiance, taken at the corresponding point on the sensor. In this way what is made explicit by the representation is the measure of light, while other information, such as reflectance of the surface, texture, shape, relative position and motion of the object are left in the background. The effort is then to devise a process, or a collection of processes, that extracts interesting information by transforming the input representation into a new one that make that information explicit. Given the enormous difficulties often encountered on this path, we could ask ourselves if the addition of some different kind of input signal could help.

And how should the output representation be organised? What kind of information, embedded in the input, should be made explicit? The answer depends on the task for which visual processing is performed. The suitability of a representation for a given task depends on whether it makes explicit and easy to use the information relevant to the task itself. The traditional reconstructionist approach gave absolute preference to shape and spatial relations. If shape is the key point, then the final output representation has been selected as a three dimensional model of the observed scene. Model based recognition uses internal geometrical representations, often of the same kind of those used in CAD systems, to be matched against the resulting representation produced by the visual processing.

It is clear that a certain amount of representation is needed, at least because today we are forced to use digital computers and software programs. But it is not necessarily the case that these representations must resemble the geometrical appearance of the observed scene. I agree that the key point is space representation, but it seems reasonable to me to suggest the role of eye movements and gaze control into an internal representation of space. Just think about what happens when somebody throws a ball to you. The key task is visually tracking the ball, foreseeing its trajectory in 3D space and organising and activating muscle commands in order to intercept it. During that time interval CAD-like 3D scene reconstruction has no meaning, although the spatial relationship between you and the ball is really the only important parameter. But the tight integration in the same control loop of ball detection and localisation in the image and oculo motor commands that allow for fixation of the ball represents the most economical solution of the task[18,19]. Vergence commands give the distance of the ball at every instant, but they may also represent it. Gaze control commands give the direction in space from the observer to the ball, and, similarly, they may represent it, along with velocity information that is necessary for interception.

If internal motion commands can play a significant role in "representing" the dynamic situation of tracking and interception[20] there is no apparent reason why they could not take an active role in space representation in general. The spontaneous exploration carried out by any normal individual introduced for the first time into an unknown environment (a room for example) can be reasonably interpreted as the process of building a model of the space in which he is situated. But a satisfying answer to the question if this representation entails 3D surface models or internal exploration movements commands is far from being here. Personally, I am more attracted by the second one.

Conclusions

Active vision is one of the most actual subjects of research in the field of computer vision. This approach has already proved its usefulness in solving some hard problems in computer vision, but it is still in full development and more results are expected. It will not solve all the problems that still confront the researchers in this stimulating area, but it has already reached the result of revitalising interest in the research community. Moreover, two important aspects of visual perception have been put in a different perspective: the position of vision , and perception in general, in the context of an intelligent autonomous agent, and the nature, role and meaning of representations. Vision should not be considered as an isolated functionality; rather it should be viewed as a powerful tool, so strictly integrated with the capabilities of an intelligent system that any artificial separation between vision and intelligence should be banned. This interactions will probably bring new insights and advantages in the study of both disciplines. In the same way active vision has the merit of having made explicit the relationship between perception and action, and the consequent importance of control.

References

1 D. Marr, (1982), *"VISION. A Computational Investigation into the Human Representation and Processing of Visual Information"*, San Francisco, W. H. Freeman & Co.

2 J. Aloimonos, I. Weiss and A. Bandyopadhyay, (1987), *"Active vision"*, Proc. 1st ICCV, London, IEEE, pp.35-54.

3 J. Aloimonos, I. Weiss and A. Bandyopadhyay, (1988), *"Active vision"*, Int. Jou. of Computer Vision, vol.1, pp.333-356.

4 J. Aloimonos and D. Shulman, (1989), *"Integration of Visual Modules: an Extension of the Marr Paradigm"*, Boston, Academic Press.

5 B. K. P. Horn, (1977), *"Understanding image intensities"*, Artificial Intelligence, vol.8, pp.201-231.

6 B. K. P. Horn, (1986), *"Robot Vision"*, Mc Graw Hill.

7 D.H. Ballard and C.M. Brown, (1992), *"Principles of Animate Vision"*, CVGIP: Im. Und., Academic Press, vol.56, n.1, pp.3-21.

8 R. Baicsy, (1988), *"Active Perception"*, Proc. of the IEEE, vol.76, pp.996-1005.

9 J. Aloimonos, ed., (1992), special issue on: *"Purposive, Qualitative, Active Vision"*, CVGIP: Im. Und., Academic Press, vol.56, n.1.

10 E. D. Dickmanns and T. Christians, (1989), *"Relative 3D-state estimation for autonomous visual guidance of road vehicles"*, Int. Aut. Systems, vol.2, pp.683-693.

11 E. D. Dickmanns, B. Mysliwetz and T. Christians, (1990), *"An integrated spatio-temporal approach to automated visual guidance of autonomous vehicles"*, IEEE tr. on SMC, vol.20, pp.1273-1284.

12 A. Blake and A. Yuille eds. (1992), *"Active Vision"*, Cambridge, Mass., MIT Press.

13 A. L. Yarbus, (1967), *"Eye Movements and Vision"*. Plenum Press.

14 J. J. Gibson, (1950), *"The Perception of the Visual World"*, Houghton Mifflin.

15 J. J. Gibson, (1966), *"The Senses considered as Perceptual Systems"*, Houghton Mifflin.

16 J. J. Gibson, (1979), *"The Ecological Approach to Visual Perception"*, Houghton Mifflin.

17 R. Brooks, (1987), *"Intelligence without representation"*, Proc. Workshop on the Foundations of AI.

18 C. M. Brown, (1990), *"Gaze controls with interactions and delays"*, IEEE tr. on SMC 20(3).

19 D. J. Coombs and C. M. Brown, (1991), *"Cooperative gaze holding in binocular robot vision"*, IEEE Control Systems, pp. 24-33.

20 A. Berthoz ed. (1993), *"Multisensory Control of Movement"*, New York, Oxford Univ. Press.

VISUAL EMERGENCE

Massimo Savini

Dipartimento di Informatica e Sistemistica
Universita` di Pavia
via Abbiategrasso, 209
27100 Pavia
savio@ipvvis.unipv.it

Abstract

The term "emergence" has a characteristic meaning in the study of complex phenomena, and is often associated with terms like "population", "evolution", "mutation", "chaos", "catastrophe", and the like; but it can convey significant meanings into the world of vision and artificial vision too. In this writing we will try to first clarify the term "emergence" itself and then its relationship with perception.

Emergence

Let us try to give some definition of emergence: "rise of a system that cannot be predicted from antecedent conditions", "surprise of an individual about a rising phenomenon that he cannot completely understand", "formation of properties, structures and functions on a higher level of being", "something that appears when the observer changes his point of view".

If these definitions convey at least a flavour of the subject, nevertheless their significance still presents rather obscure aspects. For sure there are at least two entities involved: a "system" and an "observer". It seems that the system must be something complex enough to escape complete knowledge and deterministic predictability, like for example social or economic systems, and it must be dynamic. Moreover, the system itself can be regarded as composed of a large number of "simpler" entities, being them humans or micro-organisms, cells or molecules, co-operative agents or spatially arranged signals. But the system results as being more than just the sum or union of its constituents, as it reveals to the observer the presence of structures that can evolve with time under the influence of going on "processes".

Actually, the way the system is perceived by the observer depends on the level of organisation, grouping, resolution, structuring, abstraction that the observer itself cast onto the observables. Think for example of the molecules that make up a fluid: they can be seen as individually distinct entities, as gas or liquid, as a cloud, as a hurricane, as the atmosphere. All these descriptions come from different levels of observation, different points of view. In the visual domain we can see pixels, edges, textures, blobs, objects, scenes; and the aim of computer vision is getting to "objectness" starting from pixels.

Regarding the identification of the processes that cause or influence the emergence of new structures or phenomena, there coexist three approaches among researchers:

- Internalist : emergence has its origins in processes internal to the system;
- Externalist : emergence is attributed to processes that are external to the system;
- Hybrid : both internal and external causes are involved in emergence.

These distinctions arise especially in the field of Evolutionary Computation and Genetic Algorithms, but the identification of the borders of the system can be regarded as somewhat arbitrary. In fact we can reason that the appearance of structure in the system can be observed through measurement of some global, holistic property. And the same choice of which property to measure is equivalent to taking a particular point of view, that excludes others. By taking a point of view we select groupings and break a symmetry, a wholeness; and we cannot thus avoid to overlook some part of the knowledge enclosed into the system. This is accepted because we need to make some information explicit in order to know something, and because it is the way we have always followed. It is the same limitation intrinsic to the sequential paradigm of thought and discourse when confronted with Gestalt perception. This limitation is one of the reasons for the importance of good scientific visualisation tools in the interpretation of huge amounts of data (for example computed by a program that simulates some complex physical phenomenon like those that take place in the atmosphere). But even visual perception does not avoid symmetry breaking because it implies choosing one organisation against alternative ones. And here comes an other important aspect of emergence: the role of the observer. By selecting the parameter to measure into the system and the spatio temporal window of observation the observer builds a space, his own observation space, in which emergence of some structure can occur, while others pass unseen. This active role of the observer in the construction of the space in which emergence can take place includes him among the "processes" that give origin to emergence itself. For this reason we can tendentially agree, in an extensive way, with researchers that adopt an hybrid approach to the interpretation of the phenomenon of emergence. It is probably reasonable to assume the acceptability of the identification of the "system" with the "observation space", at least for what concerns emergence. It is a subjectivist point of view, but an "objectivist" approach of any kind faces the risk of actually being based on an unconscious hiding or underenphasizing of some aspect that can produce a false sense of scientific "objectivity".

Visual emergence

Talking about emergence in the visual domain can make us think of something that was not there before and is now, that we could not see and that we can see now, as if somebody or "something" had put it in front of us with a magic touch. Thinking about this strange phenomenon we could be led to consider ideas like special effects, image synthesis, hallucinations, dreams, etc. Actually the phenomenon of emergence is very common in the everyday visual experience and it should not be regarded as unusual or as a voluntarily produced artifact. Rather, it is perhaps the most common and important phenomenon occurring during the normal activity of "seeing". In the following we will try to highlight the domains, the mechanisms and the phenomenology of what we call Visual Emergence.

In computer vision the starting point is an image, or a sequence of images, that is, an array of numbers. The purpose of computer vision techniques is to identify and extract meaningful structures and information from this array, where "meaningful" indicates both with a *coherent* relation to real objects or events in the scene and *relevant* for the overall task at hand. Coherence saves the observer from hallucinations, errors and misunderstandings. Relevance allows for economic and effective selection of useful information, while discarding an enormous amount of data and avoiding too many heavy computations. This second aspect of visual perception is the central point of the change in perspective carried about by the so called Purposive approach to vision[1].

The dual aspect of this meaningfulness strongly recalls the dualism between the internalist and externalist approach to emergence. If many current researchers agree on the acceptability of an hybrid approach that considers both internal and external causes for emergent phenomena in the domain of genetic and evolutionary computations, this acceptance should be even stronger in the field of vision, either biological or artificial.

Normally, when thinking about vision, we think of images, and if we focus on emergence we are probably led to consider dynamic images or sequences. By this we imply that emergence takes place in time, and we observe it by monitoring the "system" or "space" represented by a greylevel or colour image. But this is an over simplistic view. It is true that something that enters our field of view could be considered as an emergent phenomenon, but this is not the kind of emergence of interest here.

Let us define the two actors of the play: the "system" evidently includes the physical objects present in the scene, while the "observer", either human or computer, draws some conclusions on the contents of the scene. This "drawing conclusions" is a fundamental characteristic of the observer, in the sense that the observer itself must have a purpose in order to perceive anything; and this purpose may change from time to time, and be different from a geometric recovery of the lost tridimensionality of the scene. Even looking around just to kill time entails a purpose: keeping the mind busy with the flow of information extracted from the image sequence. It is not necessary that the purpose be "voluntary", for the human observer; striving for survival among predators, enemies and dangers of any kind, and detecting and catching preys has shaped our visual processing system by cabling inside it the fundamental capabilities for extracting all the basically useful information in real time. So many processes are going on simultaneously in the brain that it requires an effort to keep our eyes open and fixed. And in the same moment in which we succeed we stop perceiving. All the processes that make an observer active (gaze control, vergenge, focusing, etc.) seem thus to be fundamental to perception. The reason is that it is not sufficient to have the transducer flooded with light, we must build an observation space and there detect emergence.

In computer vision there are several spaces that can be built on the basis of input images. We will make here just few examples in order to emphasise the underlying role of emergence in so many processes.

Starting with the raw image, we can detect blobs of uniform grey or colour more or less easily depending on their uniqueness or isolatedness as opposed to being immersed among other regions of similar colour. This can be easily verified, for example by building a "redness" space in which only red points are enhanced while those of different colour receive lower values.

The need for differentiation from a surrounding context is the essence of emergence and its detection. The core of the job is to devise a process that builds an observation space in which the entity that we want to detect/perceive can stand out of the rest of the environment, and where the other entities become its background.

Traditional image processing operations, like template matching through convolution, are designed to build an observation space where the position of the feature of interest is indicated by an outstanding peak. Peak detection becomes the final step of feature detection strategies if the space in which it takes place is built with the purpose of enhancing and letting emerge the features of interest. In this sense the observer is one of the important causes of emergence. Other examples of the same kind are: edge and motion detection, tracking, region of interest detection, probably object recognition. Of course the design of such space construction processes is left to the researcher/programmer in computer vision, whereas it is built-in into biological vision systems. Research in computer vision can profit from the process of emergence detection in several ways, but it is especially true for those researchers that follow an active approach to vision. Here follow some examples.

A usual way for separating an object from the background in monocular views is to perform a small lateral shift and to measure the amount of "change" thus caused in the image. It will be maximum in points inside and along the occluding contours of the object.

A similar effect is that of stereo. We do not realise the real effect of vergence control because of the extremely fast and "natural" stereo fusion performed in everyday life, but it becomes apparent by looking at a stereogram[2] that the space in which the emerging figures live is built by our internal "processing". This construction process can be given strong evidence by trying to fuse an autostereogram, or single-image stereogram[3]. In this case you are given only one image and you can perceive objects floating above and below the image plane only if you consciously diverge your eyes of the correct amount. The advantage in revealing the space construction process comes from the fact that it requires a sensible effort (from seconds to minutes).

In motion perception the construction of an observation space for emergence is easy in the case of a fixed camera. It is sufficient to continuously subtract one frame from the previous one. Objects moving with an angular speed in respect to the viewpoint above a threshold determined by the sampling rate create easily detectable blobs. Things become difficult if the camera is moving, since in this case everything else in the scene appears to move relatively to the observer. But active techniques have been devised to detect independent motion in a robust and reliable way[4] that build an observation space in which the image areas with a different FOE (Focus Of Expansion) from that of the moving camera distinctly differentiate themselves.

Still regarding motion, an important visual capability is that of tracking, that is, keeping an object at the centre of the image in a situation of motion, either of the object, of the camera or both. Some interesting active techniques have been devised in one of which two observation spaces, stereo disparity and optical flow, cooperate for keeping the target centred[5,6].

If emergence holds an important position in the field of navigation, it seems having also an essential role in other basic tasks like that of segmentation and object recognition[7]. Segmentation is the process of dividing an input image into parts implicitly or explicitly recognised as corresponding to projections of real physical objects or surfaces. Segmentation is thus a fundamental step of object recognition and scene interpretation. Segmentation can be explained as the result of different "forces", both coming from the image as emerging features and from the observer as grouping and completion processes.

There is an other way to look at segmentation and object recognition that focuses more on the active role of the observer as constructor of the observation space. Both segmentation and object recognition can be seen as hypothesis and testing cycles in which a set of beliefs is cast by the observer, the visual consequences of these beliefs are foreseen and projected onto the scene, and sequential checks are performed through foveations in order to confirm or deny the hypothesis[8]. Once we have devised a probability space as the place where we can detect emergence, we can regard this activity of selective information gathering as local sources in this space that propagate their influence to the whole through the laws of probability. The winning solution/interpretation is the one that results in a peak of probability. An indirect confirmation to the possibility of this hypothesis comes from the perceptual instabilities that can be artificially produced, of which the Necker cube is the most famous one, by providing the visual system with a problem with two or more equally probable solutions.

Conclusions

The problem of emergence in the visual domain has been only sketched in this short presentation, even if it would deserve a much larger treatment. The reason is that it has not received up to now in the computer vision community the correct attention.

The concept of emergence here presented, somewhat extended in respect to other treatments, may not encounter the favour of everybody, but, especially to researchers in computer vision, it should sound like a stimulus, a different "point of view", from which something new and useful might "emerge".

References

1 J. Aloimonos, ed., (1992), special issue on: *"Purposive, Qualitative, Active Vision"*, CVGIP: Im. Und., Academic Press, vol.56, n.1.

2 C.W. Tyler and M.B. Clarke, (1990), *"The Auto-stereogram"*, SPIE Stereosc. Displ. and Appl., vol.1258, pp.182-196.

3 H.W. Thimbleby, S. Inglis and I.H. Witten, (1994), *"Displaying 3D Images: Algorithms for Single-Image Random-Dot Stereograms"*, IEEE Computer, vol.27, n.10, pp.38-48.

4 R. Sharma and J. Aloimonos, (1991), *"Robust detection of independent motion: an active and purposive solution"*, CAR-TR-534, Univ. of Maryland.

5 D. J. Coombs and C. M. Brown, (1991), *"Cooperative gaze holding in binocular robot vision"*, IEEE Control Systems, pp. 24-33.

6 C. Fermuller and J. Aloimonos, (1992), *"Tracking facilitates 3-D motion estimation"*, CAR-TR-618, Univ. of Maryland.

7 S. Grossberg and E. Mingolla, (1985), *"Neural dynamics of perceptual grouping: Textures, boundaries and emergent segmentations"*, Perception & Psychophysics, vol.38, n.2, pp.141-171.

8 P. Viviani, (1987), *"Fonctions de prise d'information et d'exploration"*, in: "Encyclopedie de la Pleiade: La Psychologie", J. Piaget, P. Mounoud and J.P. Bronckart eds., Paris, Gallimard, pp.1663-1711.

SHAPE PRIMITIVES FROM VISUAL PATTERNS

L.P. Cordella

Dipartimento di Informatica e Sistemistica
Universita' di Napoli "Federico II"
Via Claudio, 21 - 80125 Napoli

Abstract

Shape, as appraisable through visual sense, is a complex property of 2D and 3D patterns which represents one of the main features according to which humans classify objects. Some of the problems to be faced when defining an artificial vision system capable to describe and classify objects according to their shape are discussed and some solutions proposed. An example of application of the introduced concepts to a real case is briefly outlined.

Introduction

Shape is an often complex property of both 2D and 3D patterns (I will alternatively call them objects) which, according to the case, can be perceived not only through visual sense, but through tactile sense too. It is the main property used by man to recognize objects. Visual recognition can be described as the matching of the retinal image of an object, or more generally of a processed version of it, to a description or representation of the object, stored in memory, and adequately reflecting, among others, the shape properties of the object.

Although mechanisms by which shape is evaluated by humans are not completely clear, there is experimental evidence [1-3] that shape is assessed in terms of a set of geometric features, that are often called shape primitives, and of their relations.

From the point of view of an automatic pattern recognition system, it can be said [4] that shape is a property of both a set of objects and a particular method of observation or measurement (e.g., different methods can consider different features or be more or less sensitive to certain details). The cardinality of the considered set may be finite or infinite and the observations are restricted to geometric characteristics of the objects and are appropriate only if they are invariant to traslation, scale and rotation. However, size and orientation can determine the attribution of objects having the same shape to different classes. All the objects that cannot be distinguished by the given method of observation are said to have the same shape. The problem is which are the appropriate measures and method of observation or, in other words, which are the most convenient shape primitives and how they can be accessed.

In the framework of computer vision, the term "visual primitive", or better "primitive component", has a clear meaning for people familiar with the structural approach to visual pattern recognition [5]. This approach assumes that a complex pattern can be decomposed into simpler subpatterns, possibly in a iterative way, and then characterized (i.e., described) in terms of simple components, which are called "primitives", and of their relations. In this way the structure of a pattern can be outlined. Primitive components should not be further structured, or better their structure should not be of interest for the considered purpose.

In the decision-theoretic approach to recognition [6], the term "feature" is certainly more familiar than the term "primitive". In this case, a pattern is characterized by a set of features (e.g., simply a set of measurements performed on the raw data) and a feature vector is assumed to represent the pattern to be recognized. Recognition implies the partition of a feature space into regions each pertaining to a different class. The elements of a feature vector should be such as to characterize a pattern so that it can be attributed to a specific class with high reliability. If this aim is achieved, we can say that the used features effectively represent the pattern or that its description has been given in terms of essential features which should include information about the primitives.

In pattern recognition, the most general meaning of the term feature could be: any entity that can be parameterized. The term could thus be applied uniformly to specify any type of entity assumed to characterize a pattern, from the simplest to the most structured one. However, in practice, a hierarchy can be established, from "local features" directly generated from the raw data, to more or less structured sets of them called "primitive features" whose semantically significant assemblies make up objects.

This terminology may also suggest a strategy for achieving pattern description. However, the determination of a set of shape primitives in terms of which the patterns of interest can be described is influenced by a number of circumstances, such as the nature of the data, the specific application considered and the technology available for implementing the recognition system.

In the following Sections, this problem will be first addressed in its conceptual aspects and then faced from an operative point of view, with reference to a specific application. Differences and relations between description and recognition tasks will be outlined.

Primitive selection

Primitive features selection is a complex problem that has no general solution. There isn't a "universal picture element" nor a simple alphabet of primitives generally valid.

Shape primitives assumed for 2-D pattern description can be line configurations of the boundary or of the skeleton of a figure [7], but can also be two-dimensional parts of this latter[8,9]: decompositions into primary convex subsets have been quite common [5]. As for

3-D objects, they are often represented by a single 2-D projection. In this case the problem is that of recognizing 3-D objects from 2-D images. The majority of the approaches uses for description simple 2-D primitives such as line segments, corners, inflections and 2-D perceptual structures[10-12].

Although the selection of simple primitives makes more tractable the problem of extracting them, it is easy to show that, generally speaking, the more the primitives are simple, the more inefficient and unreliable is the recognition process. Consider for instance the shape of characters (Figure 1). Let us assume that, after suitable processing, a character is represented, for computer purposes, as a digital line: it could be described assuming as a primitive the direction of the unitary segment connecting successive pixels of that line. The character description would be a sequence of codes, each representing one of the eight possible directions of the digital plane. However, it is very likely that, in this way, different specimens of the same character would have quite different descriptions, whilst we would like to say that their shape is essentially the same.

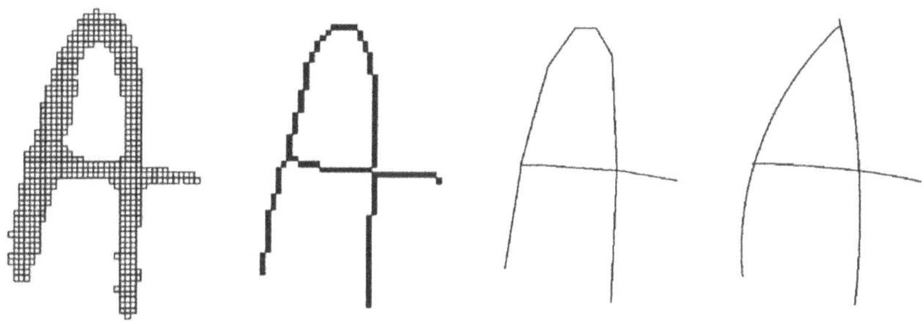

(a) (b) (c) (d)

Figure 1. The bit-map representing the digital version of a character (a) and some alternative representations, obtained after processing, in terms of a digital line (b), a polygonal line (c) and a set of circular arcs (d).

More powerful descriptions (i.e., descriptions less sensitive to not significant shape variations) can be obtained by using as primitives longer straight segments or circular arcs, but this requires that a suitable approximation of the digital line has been preliminarily made. In other words, extracting more complex primitives implies the difficulty of grouping simple features to form them.

For complex images, the most interesting approach to shape description seems to be the structural one. Image description is achieved through a parameterization process according to which a suitable representation of the image is transformed into a data structure made up of primitive features and of relations among them. The process can be implemented with a bottom up method, combining simpler features to form complex primitives and then objects, although, as far as higher abstraction levels are reached, the need for model based knowledge is regarded unavoidable [13]. In a automatic recognition system, a number of successive steps are necessary to pass from the input analog image to the description of its shape and eventually to its classification (recognition).

In early phases, information is processed without any prior knowledge of the viewed image. The goal is to extract relevant information useful for further analysis (e.g., sharp changes in image luminosity are a priori relevant information). The result of early processing is generally still an image, which is a possible initial representation of the original one. Further processing can lead, through successive stages, to obtain more convenient representations.

The aims of the representation phase are manyfold: it serves to reduce noise, to compact information, to decompose the pattern into "meaningful" primitive components, to represent components in such a way that not significant shape variations are hidden. In fact, features actually invariant within a class and peculiar of that class may not be directly detectable from the raw data. Suitable processing is generally needed. The representation phase should be intended as a first step of a process of abstraction that leads from the particular specimen to the prototype.

Although the decomposition process outlines specific features of an object, it leads to a representation that is still susceptible of different descriptions depending on the scope, the wideness of the image class, and so on. It can be said that a meaningful decomposition and description implies to single out primitives (features, component parts) as much as possible invariant with respect to the differences existing among the specimens of a same class, and to describe the components and the structure they form (i.e., what we have called the representation) in such an essential way as not to display the differences between members of the same class, but rather to outline the similarities. Ideally, all the members of a same class should be described in the same way.

Some generally accepted criteria for obtaining effective descriptions are: attention to the scope, conciseness, accessibility of the features to be used, uniqueness of the obtained descriptions. Still, a description must be robust to occlusion and noise, invariant over a range of viewpoints and scales, and computationally efficient. Moreover, descriptions should be stable, while remaining sensitive enough. However, to absorb variability within a same class and not to destroy information necessary to discriminate between different classes are generally conflicting aims.

Indeed, all of the above issues are often competitive, in the sense that using certain features may maximally satisfy some criterion but preclude to maximize the others. This could suggest to introduce, in a recognition scheme, a redundancy of feature types.

As already mentioned, number and nature of the features to be used for describing a pattern depend on the task. If a description serves to the purpose of faithfully reproducing the described object, then it has to be quantitative and very detailed. On the contrary, for recognition purposes, few qualitative features are more convenient especially if, for the application at hand, the number of classes is restricted. In human vision, the dependency of the used features on the task, has been outlined since long time ago by recording eye movements during constrained and unconstrained observation of pictures [14]. These experiments also outline the role of knowledge when looking for primitive features.

If the task is recognition and not merely description, it is finally necessary to compare specimen description with prototype descriptions, in order to obtain classification. As mentioned in the Introduction, it has to be remarked that, to this purpose, different orientation and/or size, so as context and knowledge base, can determine the assignment of objects having the same shape to different recognition classes. Examples are given in Figure 2. This makes the classification phase often very complex.

Figure 2. Two complex shapes which, if rotated by 180°, are differently classified (dog/cat and 6/9).

In any case, it is rarely required to compare descriptions for assessing their perfect identity, rather it is generally necessary to recognize which prototype a description is more similar to. This implies that inexact matching techniques must be employed and a metric must be defined to quantify the similarity or dissimilarity between shape descriptions. Depending on the form of the adopted description (feature vector, attributed relational graph, sentence of a formal language) different recognition schemes can be used, from the classical statistical classifiers, to inexact graph matchers and syntactic parsers. Graphs and graph matchers are the most appealing structure and recognizer respectively in case of complex shapes, even if the computational cost they involve makes their use problematical in some applications.

A real case

An example of object representation, description and recognition scheme according to shape, implementing some of the considerations illustrated in the previous Section, will be now briefly discussed. The considered application is optical character recognition (OCR), one of the oldest, but still widely studied problems in Computer Vision. The method that will be illustrated has been especially devised for the case that a large variability exists among the shapes of characters belonging to a same class, like for unconstrained handprinted and omnifont printed characters. Reliable recognizers for this type of data are not yet commercially available and the matter is still object of research. Let us follow the sequence of processing steps devised for recognizing character shapes.

Given a document at least partially made of printed and/or handprinted text, text regions must, first, be automatically outlined and character strings segmented into components (ideally, single characters). The bit map of each character is then processed in order to obtain a more compact character representation.

Character digital images can be considered as ribbons evolving, joining and intersecting in a 2-D space. According to the semantic held by the ribbon shape, a suitable representation could be achieved by applying to them a thinning transformation, that thins the ribbon to a unit width digital line (often called skeleton) centered with respect to ribbon boundaries. Under certain hypoteses, the transformation achieves the goal of compressing information without loss, but its use for recognition purposes is not straigthforward. In fact, thinning techniques may give place to distorted representations of the shape of the ribbons one would describe. The most important shape distorsions introduced by such techniques are present at the junction between character strokes. This is particularly undesirable because, as it will be seen, correct information about the directions of the lines merging at a junction is necessary to reliably decompose characters into meaningful parts (strokes). Therefore, an acceptable character skeleton should faithfully reflect the shape of the original ribbon.

Since avoiding the generation of distorsions is neither easy nor reliable, the possibility of correcting distorsions after thinning has been considered. It has been shown [15] that this is possible under condition that the skeleton preserves information about the distance of its pixels from background. By using this information, together with information about direction of skeleton lines, measured over a polygonal approximation of the skeleton, the mentioned shape distorsions can be reliably corrected, avoiding side effects and at a contained computational cost. The obtained shape-preserving skeleton, approximated with a polygonal, is stored in a graph-like data structure. A first decomposition of this polygonal into simpler pieces takes now place by keeping together or splitting lines merging at the joints, according to good continuity rules[16]; that is why eliminating shape distorsions at the junction points was considered an essential step.

However, polygonal lines representing characters of the same class, although carefully fit on input data, show an extreme variability in terms of number of sides and shape. In order to reabsorb variability within each class and to make easier the decomposition of a character into meaningful strokes, more general descriptors, obtained by means of an integration process starting from the polygonal lines, were preferred.

Experiments have shown that, in the world of latin characters, the circular arc has enough descriptive power to substitute curves of different shape, but having the same contextual value, without destroying discriminant features. In practice, given a piece of polygonal line (i.e., a set of connected segments), we first determine the circular arc that better approximates the polygonal according to a suitable criterion of shape similarity and then decide whether the found arc can be actually assumed to stand for the original line, according to the purpose[17].

The process is performed by a merging technique. The procedure to find the approximating arc involves a transformation that changes a polygonal line in the (x,y) plane into a set of horizontal segments (a staircase function) in a (l,a) plane, where l is the distance of points along the polygonal from a reference point (curvilinear abscissa), and a is the angle between each segment of the polygonal and a starting segment. This transformation reduces the problem of fitting an arc of circle to a polygonal line to that of approximating a staircase function with a straight line. This implies a reduction of the whole computational complexity since second order curves in (x,y) are changed into first order curves in (l,a).

In order to find the straigth line that better approximates a staircase in (l,a), the value of a suitably defined error parameter is minimized by using the least square method and weighting the contribution given by any staircase segment according to its lenght.

As it is obvious, the above approximation process leads to decompose the lines making up a character into pieces (strokes), each one represented by an arc. In Figure 1, the results of the successive steps of the representation phase are shown in case of a handwritten character.

The developed character representation and decomposition methods have been metodically tested in order to verify their adequacy for representing both omnifont printed and handprinted characters. By the way, testing has shown that, according to the chosen criteria, the best approximation found for a given polygonal is comparable to the one a human subject would suggest. The decompositions demonstrated very stable with respect to character shape variability, allowing a good compromise between the opposite requirements of having very similar representations of characters of the same class and of preserving information necessary to discriminate between characters of different classes.

After the illustrated processing, the obtained character representation is intrinsically size, rotation and traslation invariant and is made of a set of connected arcs. It has to be described in order to achieve classification: this latter actually implies a comparison between specimen description and prototype descriptions.

Although character representations given in terms of sets of arcs appear very simple and allow to absorb much of noise and shape variability, the problem of suitably describing them has not a unique solution and is absolutely not trivial. Different description schemes have been considered, all requiring to parameterize, in a more or less quantitative way, the morfology of each single arc so as some kind of topological relations among them. Critical points are the operative definition of arc contiguity, the quantization intervals chosen for the values of the various parameters, the type of data structure adopted to support the description and, more generally, the choice of the features to be used at the description level, taking into account the devised classification strategies which will intervene in the next step.

A suitable metodology for comparing different description schemes was defined. Two main schemes were considered, essentially differing for the features used to describe arcs, for the type of relationships between them, and for the way of quantifying the chosen parameters. According to one of the tested schemes, the morphology of a generic arc is described by three parameters:

a) the dimension of the arc espressed as a percentage of the dimension of the box surrounding the whole character;

b) the curvature radius of the arc;

c) the direction of the normal to the chord.

In order to reduce the complexity of the description, on the one side, and to take into account the possibility that character strokes may be broken because of different reasons, on the other side, topologic relationships between arcs were established only between "virtually" related arcs. A virtual relation between two arcs is assumed to hold either if they touch each other or if their distance is lower than a threshold depending on the character size. Instead of describing the relations between pairs of arcs in the xy-plane, the relative positions of the projections of the arcs on the x and y axes are described.

The considered description techniques were tuned to the domain of interest (e.g., as regards number of quantization levels most appropriate for each of the parameters used, values of the quantization thresholds, etc.). Experiments have outlined the importance of using information about the relative position of the connection points between arcs. Thus, this information has been included in one of the tested schemes.

The most convenient data structure, providing the most effective coding of the proposed descriptions, would be an Attributed Relational Graph (ARG), i.e., a two dimensional data structure. Consequently, the most suitable type of classifier would be an Inexact Graph Matcher. This approach has been taken into account and a prototype classification stage developed and tested[18]. A less accurate, but computationally less complex classification strategy, based on neural networks, was implemented too.

The neural network experimented for the classification is a feed-forward completely connected net with three layers of neurons. For the learning phase the back propagation algorithm has been used with starting weights randomly set and a constant value for the back propagation factor. For the purposes of the neural network classifier, a coding technique has been developed for representing an ARG with a one-dimensional feature vector of relatively limited size, with a reduced loss of information. The performance of the neural classifier has been enhanced by introducing a set of classification rules allowing to measure the degree of reliability of a decision and to choose to reject the input sample instead of running the risk of misclassifying it [19].

The whole system has been tested by using a very large database including multifont printed text and unconstrained handprinted characters and digits written by people of different age, sex and educational degree. The recognition rates obtained with the neural classifier varied from about 93% to 99%, depending on the type and the quality of the considered data. The first version of the graph matcher did'nt give much better results, but it can be substantially improved. It has to be remarked that these results have been obtained by using only information about the shape of isolated characters, without the contribution of any type of contextual information, and that the considered handprinted characters were completely unconstrained.

On the whole, automatically recognizing characters by describing their shape according to the illustrated approach, revealed to be a powerful method particularly for isolated handprinted characters written within boxes, as in typical forms.

References

1. D.I. Perrett and M.W. Oram, Neurophysiology of shape processing, *Image and Vision Computing*, vol.11, n.6, 317-333, (1993).
2. K. Tanaka, H. Saito, H. Fukada and M. Moriya, Coding visual images of objects in the inferotemporal cortex of the macaque monkey, *J. Neurophysiology*, vol.66, 170-189, (1991).
3. I. Biederman, Human Image Understanding: Recent Research and a Theory, *Computer Vision, Graphics and Image Processing*, 32, 29-73, (1985).
4. G. Nagy, An induced taxonomy of visual form, in *Aspects of Visual Form Processing*, C. Arcelli, L.P. Cordella and G. Sanniti di Baja Eds., World Scientific Publ., Singapore, 410-419, (1994).
5. T. Pavlidis, *Structural Pattern Recognition*, Springer- Verlag, (1977).
6. P. Devijver and J. Kittler, *Pattern Recognition: a Statistical Approach*, Prentice-Hall, Englewood Cliffs, N.J., (1982).
7. K.S.Fu, *Syntactic Methods in Pattern Recognition*, Academic Press, (1974).
8. L.G. Shapiro, A structural model of shape, IEEE Trans. on Pattern Analysis and Machine Intelligence, Vol. 2, 111-126, (1980).

9. L.P. Cordella and G. Sanniti di Baja, An approach to the decomposition of complex figures, in *Digital Image Analysis*, S. Levialdi Ed., Pitman Publ. Inc., London, 155-162, (1984).
10. S.J. Dickinson, A. Rosenfeld, A.P. Pentland, Primitive-based shape modeling and recognition, in *Visual Form: Analysis and Recognition*, C. Arcelli, L.P. Cordella and G. Sanniti di Baja Eds., Plenum Press, New York, 213-229, (1992).
11. M.D. Levine, R. Bergevin, Q. L. Nguyen, Shape description using geons as 3D primitives, in *Visual Form: Analysis and Recognition*, C. Arcelli, L.P. Cordella and G. Sanniti di Baja Eds., Plenum Press, New York, 363-377, (1992).
12. P.J. Besl, R.C. Jain, Three-dimensional object recognition, *ACM Comput. Surveys*, 17, 1, 75-145, (1985).
13. R. T. Chin, C. R. Dyer, Model-based recognition in robot vision, *ACM Comput. Surveys*, 18, 1, 67-108, (1986).
14. A. L. Yarbus, *Eye Movements and Vision*, Plenum Press, New York, (1967).
15. G. Boccignone, A Chianese, L.P. Cordella and A. Marcelli, Using skeletons for OCR, in *Progress in Image analysis and Processing*, V. Cantoni, L.P. Cordella et al. Ed.s, World Scientific Publ. Co., 275-282, (1989).
16. G. Boccignone, A Chianese, L.P. Cordella and A. Marcelli, Recovering Dynamic Information from Static Handwriting, Pattern Recognition, Vol.26, N.3, 409-418, (1993).
17. A Chianese, L.P. Cordella, M. De Santo, M. Vento, Decomposition of ribbon-like shapes, *Proc. 6th Scandinavian Conf. on Image Analysis*, Oulu, Finland, 416-423, (1989).
18. A Chianese, L.P. Cordella, M. De Santo, M. Vento, Classifying character shapes, in *Visual Form: Analysis and Recognition*, C. Arcelli, L.P. Cordella and G. Sanniti di Baja Eds., Plenum Press, New York, 155-164, (1992).
19. L.P. Cordella, C. De Stefano, F. Tortorella and M. Vento, A method for improving the classification reliability of multilayer perceptrons, to appear on *IEEE Trans on Neural Networks*.

COMBINED OPTICAL, NEUROANATOMICAL, ELECTROPHYSIOLOGICAL AND BEHAVIOURAL STUDIES ON SIGNAL PROCESSING IN THE FLY COMPOUND EYE

Nicolas FRANCESCHINI

C.N.R.S., Laboratoire de Neurobiologie, Neurocybernétique
31, Chemin J. Aiguier, 13402, MARSEILLE, Cedex 20 (France)
FAX (33) 91 22 08 75; e-mail: enfranceschini@irlnb.cnrs-mrs.fr

I/ INTRODUCTION

The visual system of the fly is a powerful data processing machine and a fine example of optronics and neuronics. Its orderly optical and neural organisation and its identifiable neurones make it a valuable model system for analysing how messages from the physical world are processed by a natural neural network, in this case to guide a wing on the basis of optical signals. In the eye of the fly, hundreds of thousands of neurones, most of which are arranged in crystalline layers (87,88), form "smart sensors" which are responsible for steering this sophisticated flying machine. Some of them deal specifically with colour vision, motion detection, or polarized light vision, for example. They all take care of the dynamic stability of the aircraft, its collision free navigation in the most unpredictable environments, its chasing of mates or intruders and its automatic landing. Visuomotor behaviour in flies has been analysed from the quantitative point of view over the past decades (e.g. 27,68,8,83,98,28,43,97,82,100,101). Exploring the subtle interplay between vision and action in this humble creature is a valid approach towards discovering how fancy visuomotor control can be achieved by high-flying neural networks.

Over the past ten years, our laboratory has been both analysing and reconstructing visual systems, and has thus been working concomitantly in the fields of visual physiology and mobile robotics. Designing and building mobile creatures equipped with biologically inspired signal processing systems (23) turns out to be a valuable step towards understanding natural creatures themselves because it puts us directly in touch with the real-life problems that animals have successfully solved in the course of evolution. It therefore helps us to ask the relevant physiological questions without having to worry about purely academic questions that never occur to animals.

Here we briefly review our recent attempts to analyse the signal processing which takes place in the fly's visual system. This review is essentially a summary of some papers recently published by our research group.

2/ DISTAL PROCESSING

The housefly's visual system is based on two compound eyes giving a panoramic visual field, and on three ocelli located on the dorsalmost part of the head (Fig.1a,b). Since each compound eye contains about 3000 ommatidia, it samples the environment in 3000 directions. This visual system involves about one million neurones, which have specific shapes and functions. Some of them are small-field columnar elements arranged in a remarkably crystalline lattice, with a retinotopic mapping between the three visual ganglia: the *lamina*, the *medulla*, and the *lobula complex*, which itself comprises the *lobula* and *lobula plate* (Fig.1b). Other neurones, with larger receptive fields, have their neurites running perpendicular to the columnar elements and provide for various kinds of lateral interactions or pool the signals delivered by the columnar elements (87,42,51,81).

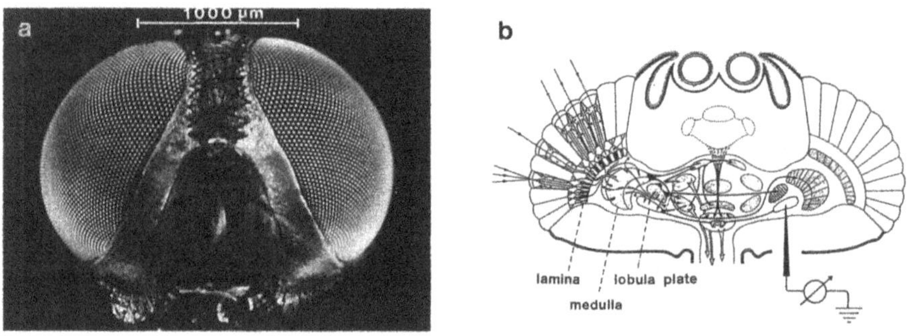

Fig. 1: (a) Head of the fly Musca domestica *(male) as seen through a Lieberkühn microscope (from 20). Each white dot is a reflex from one facet lenslet. b) Schematic horizontal section through the head, showing the layer of receptor cells and three underlying optic ganglia dedicated to neural image processing, with outputs to the brain, the contrateral eye and the descending neurones that run along the neck down to the thoracic interneurones (modified after 59). A microelectrode can be seen penetrating the posterior part of the head capsule to record from an identifiable motion sensitive neurone in the* lobula plate, as discussed in Section III *(From 22)*

This orderly neural network is driven by an array of 3000 x 8 = 24000 micrometer-sized photosensors, the spectral and polarisation sensitivities of which were analysed in detail, by applying three techniques to the living animal, namely microspectrophotometry microspectrofluorimetry and intracellular recording procedures (63,34,16,18,33). The autofluorescence of the photoreceptors (16) provided us with a genuine "colour code" for their spectral sensitivities, each of which can henceforth be individually located on the retinal mosaic, which contains no less than 6 different spectral types (34,17,18,32). Receptor cells R1-6 all display the same dual-peak spectral sensitivity (with maxima in the blue-green and near UV) and all fluoresce red under blue excitation (Fig.2d). They owe their UV peak to an accessory, photostable pigment, which plays the role of a sensitizer (63). In making these cells "panchromatic", the latter boosts their quantum collect, and consequently improves their contrast sensitivity.

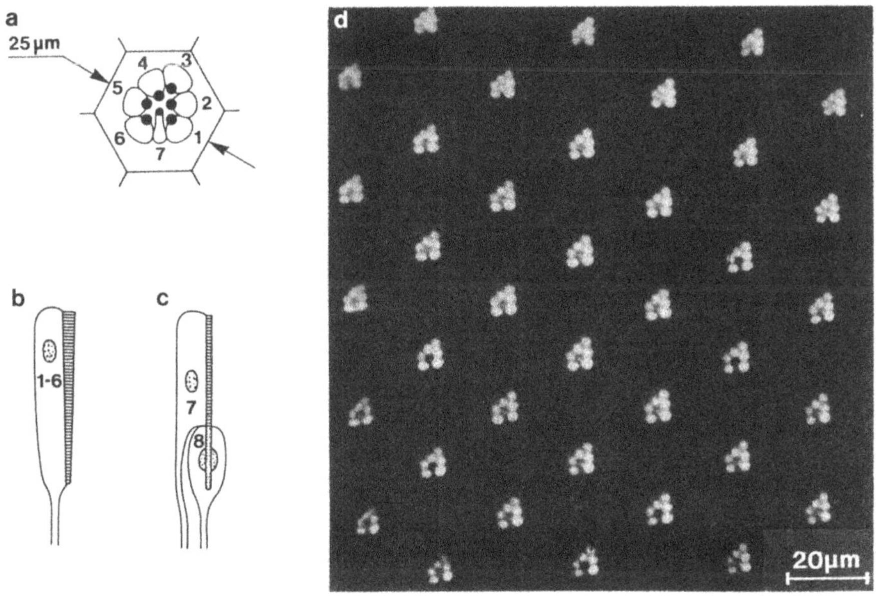

Fig. 2: Photoreceptor cells in the housefly retina. (a,b) Scheme of the organisation and numbering of the set of 8 photoreceptors present in each ommatidium. The rhabdomere *(hatched in b and c) is the light-sensing part of the photoreceptor cell. It contains the visual pigment(s). The rhabdomeres of the two central cells, R7 and R8, are arranged in tandem. (d) part (about 1% only !) of the photoreceptor mosaic observed* in vivo *under epifluorescence microscopy (blue excitation) combined with corneal neutralisation (13). Whereas all rhabdomeres R1-6 fluoresce red throughout the eye, R7 appears either green, or red , or black, depending on the ommatidium* (Musca domestica, *white eye,) (From 16,19)*

All the photoreceptor cells are protected against overstimulation by a fast pigment migration mechanism acting as a kind of "intracellular pupil" that operates independently within each cell in a matter of seconds (62,15).

The projection pattern of photoreceptor axons onto second order neurones has been deciphered (95,58,2). Briefly, six receptor cells of the type R1-6 (from six neighbouring ommatidia) send their axons to a common neural column, called "cartridge", of the *lamina* (Fig.3b) Each of the post-synaptic neurones L1, L2, L3 present within each cartridge is therefore driven by six afferences with virtually the same informational content, as they stem from receptor cells looking in the same direction in space (58, see, however: 80).

The signal summation, and hence the ensemble averaging which takes place within each cartridge (Fig.3b) increases the signal-to-noise ratio by a factor $\sqrt{6}$ in this transmission channel (37), which repeats itself over and over again with each of the 3000 pixels of the eye (56). This neat distal processing ability has earned the fly's eye the name "neural superposition eye" (58).

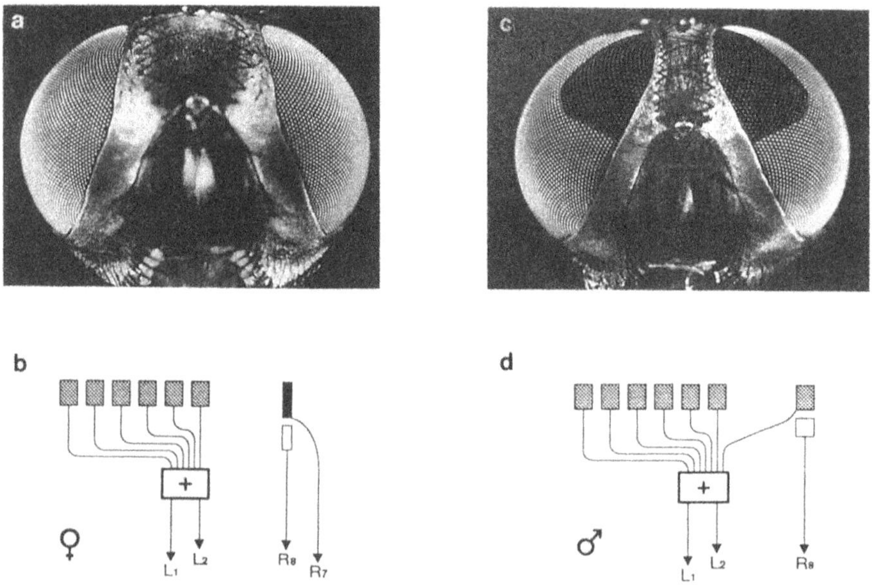

Fig. 3: Principle of "neural superposition" in the fly's eye and sexual dimorphism in the circuit. Convergence of photoreceptor axons onto the lamina second-order neurones L1 and L2 (seen as two dark profiles in the core of each cartridge in Fig. 4d). In the classical scheme (b) found throughout the eye of the female (a), six photoreceptors R1-6 belonging to 6 different ommatidia but with common line of sight converge onto L1 and L2.; whereas, the R7 and R8 axons bypass the underlying cartridge and project directly to the underlying medulla column. A dramatic sexual dimorphism exists, however, in this scheme of neural wiring, since in the frontal-dorsal part of the male retina (c), the R7 axon ends in the underlying lamina cartridge. (after 17,30,35)

As a corollary to this principle, the six peripheral cells (R1-6) from a single ommatidium dispatch their axons to six neighbouring cartridges in the lamina (2), according to the scheme shown in Fig.4a. Novel anatomical techniques such as those involving neuronal "photo-degeneration" (77) and "photo-permeabilisation" (78) have confirmed these facts in a striking manner (Fig4).

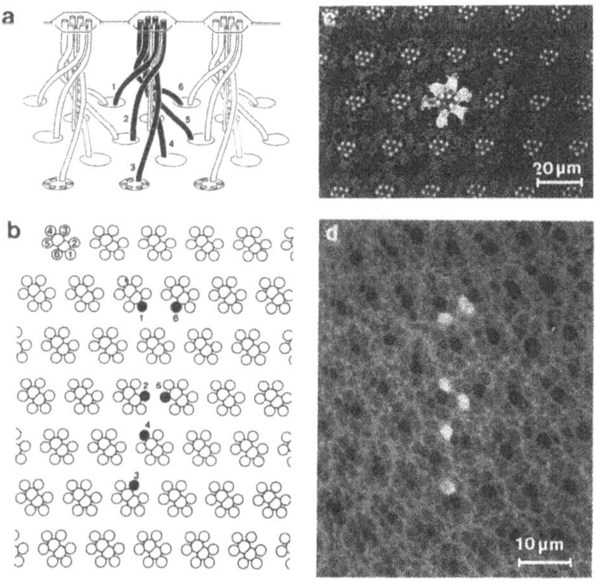

Fig.4: (a,b) projection pattern of the 6 peripheral cells (R1-6) of an ommatidium according to the principle of "neural superposition" (after 58,2). (c,d): Demonstration of this principle using the novel "neuronal photo-permeabilisation" technique, which consists in illuminating one or N neurones in the presence of an extracellularly applied dye. Focal illumination of a single ommatidium achieved in vivo *under corneal neutralisation (13) led to "photo- permeabilization" and the filling of the cells R1-6 with the dye Lucifer Yellow (c), so that their axons could be traced down to their endings in the lamina cartridges (d) (from 78).*

As in the vertebrate retina, the synaptic transmission from photoreceptors to second-order neurones occurs via chemical synapses transmitting graded potentials. Each of the six afferent fibers makes about 200 histaminergic synapses with its target neurones L1 and L2 (72,32), each of which thus receives no less than 1200 synapses. These in turn transmit a high-pass filtered, sign-inverted version of the photoreceptor signal to L1 and L2 (1,69,85,94). The signal bandwidth is extended thanks to the existence of special conductances in the membrane (70). Detailed cable modelling of the signal transmission along the axon of the receptors and post-synaptic neurones accounts well for the observed response kinetics (36).

There exists a second visual system overlapping that described above. It is composed of the two remaining cells R7/R8, which are curiously arranged in tandem in the centre of each ommatidium (Fig.2c). It has long been known that these two cells do not project their axons to the underlying cartridge in the lamina, but convey their "long visual fibers" directly to the *medulla*, in which they form another point-by-point image of the visual world (2,87,88). This distinct projection of cells R7/R8 makes sense because these cells have meanwhile been found to display a gamut of spectral sensitivities (34,35,33,18) as attested by their various autofluorescence colours, which predestines them to act as separate "chromatic channels" (32,92). If these cells also projected onto the cartridge underneath, their spectral signature would get buried in the summation process.

In one extraordinary case, however, the R7 axons were found to end in the lamina (17,35,30), forming synapses with L1 and L2 (Fig.3d), and even profuse synapses with L3 (30). This odd connection is the privilege of the *male* eye, and occurs throughout the "love-spot" (Fig. 3c), i.e. the frontal-dorsal region of the retina (17,35), which is used by the male to detect and track the female in flight (69,99,97). But here again, this unusual axonal projection makes sense because in this strategic part of the male eye, all the R7's exhibit the very same spectral sensitivity as R1-6 (17,35)). The participation of a seventh, spatially and spectrally matched input to the "neural superposition" scheme (Fig.3d) makes an additional improvement to the signal-to-noise ratio (by a factor $\sqrt{7}$ instead of the usual factor $\sqrt{6}$). This can be expected to improve the smoothing of the photon noise at the level of the second-order neurones and hence to also improve the ability to detect a small dark object such as a female fly. It looks just as if the male fly had to sacrifice the luxury of colour vision in favour of a more serious concern, such as the ability to postpone late into dusk the time at which it will have to give up chasing fast moving black spots (60). It is worth noting that underlying the eye region of interest here is a set of male specific neurones such as MLG1 and MLG2 (42,89), which are lying in wait for and actually respond to the expected stimulus: a small moving dark spot (93,25,96).

The fact that we have such detailed knowledge about the early processing of information in the fly's eye makes this animal a unique model for research, particularly when it comes to looking for the logic behind various neural networks. Several methods we originally developed for analysing receptor cells *in vivo*, involving for example *corneal neutralization* (13),*farfield radiation pattern of an ommatidium* (61), *telescopic observation of the eye* (15), and *the deep pseudopupil* (14), were subsequently applied for the reverse purpose of optically stimulating identified cells in the living animal, while simultaneously recording the activity of various neurones along the visual chain using microelectrodes (84,36, 86).These methods, in particular, have provided new means of analysing motion detection (Cf Section III) and retinal movements (Cf Section IV).

III/ DIRECTIONALLY SELECTIVE MOTION DETECTION

The head of insects is full of smart sensors called "motion sensitive neurones", which are used not only to sense the motion of prey or predators, but also to ensure the animal's dynamic stability and course control (28,83,41) and to assess the distance to obstacles based on the motion parallax (6,98,71,64). In particular, about 50 large-field neurones in the *lobula plate* show a clear-cut directional preference in their response to the motion of a pattern in a specific direction (10,81,38,39,51,41). An example is provided in Fig.5, which gives the response of the H1 neurone in the left eye of a fly (*Musca domestica*) to the movement of a striped pattern presented to the ipsilateral eye in two opposite, horizontal directions.

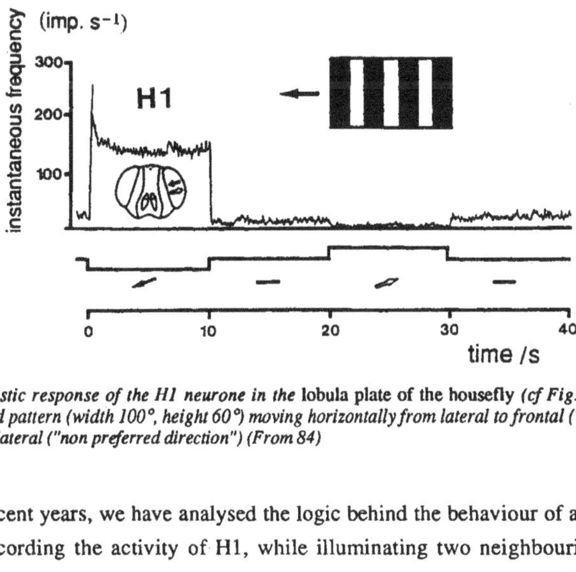

Fig. 5 Characteristic response of the H1 neurone in the lobula plate of the housefly *(cf Fig.1b) to the horizontal motion of a striped pattern (width 100°, height 60°) moving horizontally from lateral to frontal ("preferred direction") or from frontal to lateral ("non preferred direction") (From 84)*

Over recent years, we have analysed the logic behind the behaviour of a motion detecting neurone by recording the activity of H1, while illuminating two neighbouring receptor cells *sequentially* so as to simulate a micro-motion occurring in the animal's visual field. The special optical instrument built for this purpose was a triple-beam incident light "microscope-telescope", the main lens of which was quite simply one of the eye's facets (diameter: 25 µm, focal length: 50 µm). This instrument (Fig.6d) can project *a sequence* of two microspots with a diameter of 1 µm onto two identifed rhabdomeres (diameter 1-2 µm) in the focal plane of one facet (Fig.6a). This "apparent motion" experiment carried out at the single cell level showed that H1 responds with an increase in spike frequency when the two receptors in question are presented with a sequence that mimics motion in the "preferred" direction (Fig.6c, top trace: sequence 1-6), and with a decrease in spike rate in the case of the opposite sequence (Fig. 6c, bottom trace: sequence 6-1).

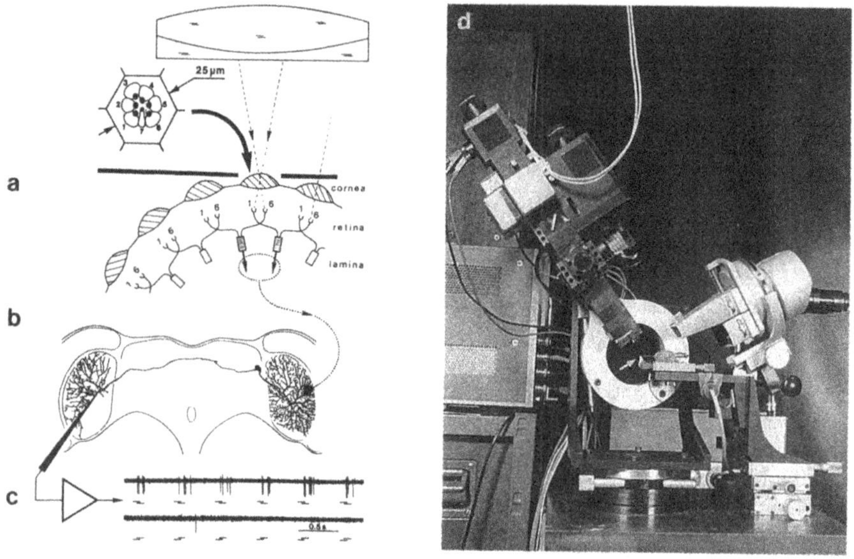

Fig.6: (a) Principle of the microstimulation experiment and scheme of the information flow from the objective of the "microscope-telescope" (shown in d) down to the H1 neurone, the electrical activity of which was recorded contralaterally with a tungsten microelectrode (c).The selected facet lens itself (a) acted as the "primary objective lens" of the instrument. The two selected cells (R1 and R6) drive two adjacent cartridges in the lamina, *as shown in Fig.4a. Sequentially presenting two light spots (each for100 ms) to the receptor pair R1/R6 stimulated a single "Local motion detector " (L.M.D.) and consistently induced a prominent spike discharge in H1, provided the delay between the two flashes was such as to mimick motion in the preferred direction (c, top trace). Note that the opposite sequence induced no response (c, bottom trace). (From 20)*

A series of experiments in which microstimulation was applied to single photoreceptor cells yielded the following findings (20, 22):

- the consistency of the responses obtained when sequentially flashing R1/R6 in *any* ommatidium of the frontal eye region, in *any* given fly (*Musca domestica*) shows that the numerous Local Motion Detectors (L.M.D.'s, cf ref.3) feeding H1 are wired down with a fine grain, making up a *homogeneous structure* underneath the photoreceptor mosaic.

- H1 responds to a sequence of stimuli presented to a *horizontal* pair of photoreceptors such as R1/R6 or R2/R5, but does not respond to the sequential stimulation of a *vertical* pair of receptors such as R1/R3 or R6/R5. This is in agreement with the known property of H1, which responds most conspicuously to horizontal, back-to-front motion of a striped pattern.

- the sequence must start with R1 (or R2) and end with R6 (or R5) in order for a positive response (increase in spike frequency) to emerge. This sequence corresponds in fact to the preferred direction of H1 observed under classical stimulation conditions (Fig.5), provided one takes into account the optical inversion effected by the facet lens.

- the *first* stimulus in the sequence does not in general induce any major changes in the spike activity of H1(Fig.7a,c,d). This first stimulus is essential, however, in that it *facilitates* the response to the *second* one.

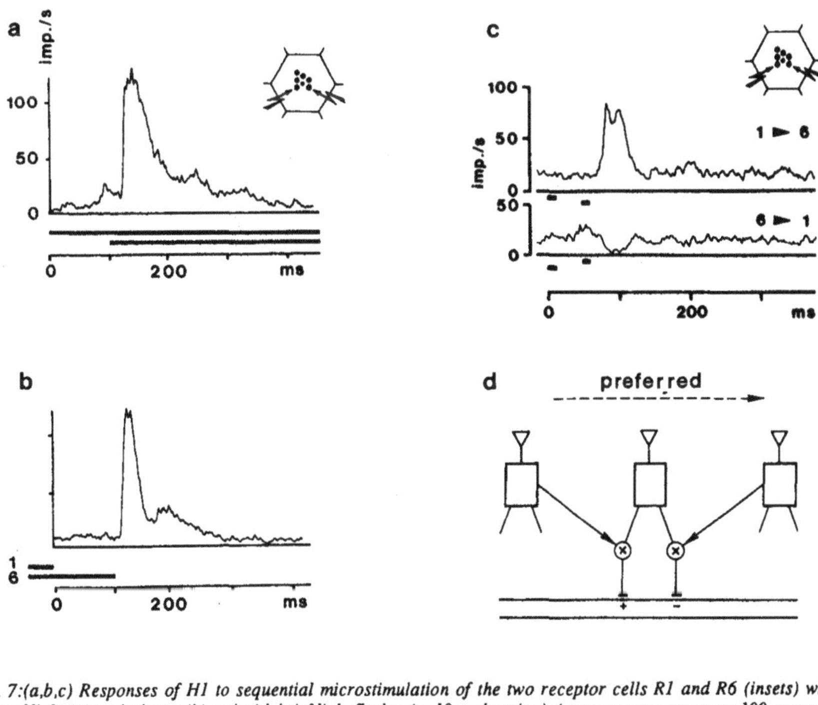

Fig. 7:(a,b,c) Responses of H1 to sequential microstimulation of the two receptor cells R1 and R6 (insets) with steps of light (a) or darkness (b) and with brief light flashes (c: 10ms duration), (average responses over 100 sequence presented with a repetition frequency of 0.5 Hz (a,b) or 2 Hz (c). Note the inhibition of the resting discharge elicited in (c) by a sequence corresponding to the nonpreferred direction. (d) functional structure of an L.M.D.on the basis of which all the responses observed can be accounted for. Two synaptic feet with opposite polarities can be seen driving a piece of H1 dendrite Whether this leads to excitation or inhibition depends on which of these two pathways has been facilitated by a lateral pathway (from 20)

- when a sequence of brief light pulses mimicks motion in the *preferred* direction, the first pulse *facilitates* the response to the second one (Fig.7a) during a "time window" of about 200 ms, which actually governs the speed range of the L.M.D under consideration (84,20).

- when the same sequence mimicks motion in the *nonpreferred* direction, the first pulse again facilitates the response to the second one. This response is "negative", however, this time, corresponding to a decrease in the spike frequency.

- a sequence of two steady light *onsets* mimicking motion in the preferred direction induces only a *transient* frequency increase ("positive response") in H1. The decay of this "step response" has a time constant of the order of 100 ms (Fig. 7a).

- a similar sequence of two steady light *offsets* leads to a qualitatively similar transient *increase* in the spike frequency (Fig. 7d).

- these "positive" transients explain why the response to a sequence of square waves of light or darkness consists of a series of two transients, which are time-locked to the onset and the offset of the second stimulus of the sequence, respectively.

- provided the requirements of (preferred) direction and delay are met, a light onset facilitates only a light onset and a light offset facilitates only a light offset. This means that each LMD is actually split into two coextensive "subunits", each of which is sensitive to the motion of a dark edge or a light edge but not to both (cf. 57,55). One benefit of this division of labour is that it provides H1 with twice as much data about the relative motion of an object in the environment even if this object subtends an angle which is smaller than the sampling basis of the LMD. This feature moreover helps to solve the "correspondence problem" arising in motion detection, since only contrast borders with like signs are brought into correspondence.

Experiments of this kind helped us to elucidate the properties of the single L.M.D.'s that feed the LP motion detecting neurones, to establish a detailed block diagram of an L.M.D., with its various time constants and nonlinearities, and finally to transcribe the fly L.M.D. into another technology (23).

IV/ GAZE SHIFT IN A FIXED COMPOUND EYE

The housefly has a highly mobile head and controls its gaze by moving its head along with the two compound eyes. Head orientation, head stabilization and the multisensory control of compensatory head movements have been the subject of detailed, quantitative behavioural experiments (67,52,47,48,53), and some of the visual interneurones associated with head turning motor neurons have been described (73,29,90,26).

In the early 70's, an enigmatic muscle of the dipteran's head, the MOT (*Musculus Orbito-Tentorialis*, after Schiemenz,1956), was found to fire constantly at a high, regular rate (5,43,75), and to be able to slightly move the photoreceptors (44,76). The observed changes in the spike rate amounted to only a few percent of the average, however, and the MOT muscle virtually fell into oblivion as regards its possible role in the fly's gaze control processes (49,50).

More recently, we established that another muscle of the dipteran head, the MOS (*Musculus Orbito Scapalis,* after Vater, 1962) was able to move the photoreceptor distal endings transversally, shifting the visual axes of the photoreceptor cells by an angle of up to several degrees (21) (Fig.8).

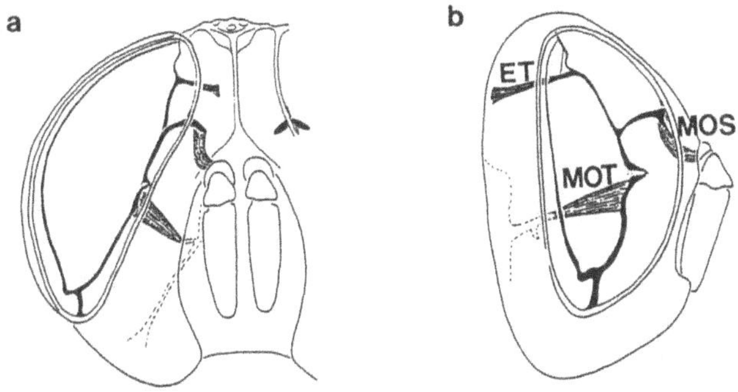

Fig.8: Organization of the eye-muscles (MOT and MOS) and eye tendon (ET) in the head of the fly Musca domestica, based on reconstructions using various muscular and cuticular staining techniques *(from 21 and 74).*

Inserted distally on the scape of the antenna, the MOS muscle in *Musca* is attached proximally to a complex cuticular lever acting as a "bellcrank". When the MOS contracts, the photoreceptors of a frontal ommatidium move laterally, and consequently their visual axes in space shift towards the midsagittal plane. We also found a specialized "eye tendon" (ET in Fig.8), attached to the posterior part of the head capsule, and pulling the retina backwards, like a "return spring"(21). A paired MOS and a paired ET were also found in the male and female of other dipteran species such as *Calliphora, Drosophila, Sarcophaga, Chrysomya* and *Lucilia*..

We have obtained simultaneous recordings from the MOS and the MOT in the walking fly *(Calliphora erythrocephala*), and correlated the spike activity of these two muscles with the micrometric shift of the ipsilateral photoreceptors measured optically on the *deep pseudopupil* (DPP: ref.14,15). At rest, both the MOS and the MOT are characterized by their high and steady spike frequency (which, however, can gradually decrease to OHz in the MOS in darkness, in contrast with the MOT). Whenever the fly walks, however, large changes in the spike frequency occur, not only in the MOS but also in the MOT (Fig.9). The activity of the latter can transiently drop from 120 Hz to 40 Hz (Fig. 9b). The large "negative transients" caused by spike drop-outs (cf 43), often occur synchronously in the two types of muscle (comp.*a* with *b*). They result in photoreceptor shifts, which were assessed microphotometrically by recording the light reflected from the frontal and dorsal borders of the DPP in the frontal-dorsal part of the eye (Fig. 9.c,d).

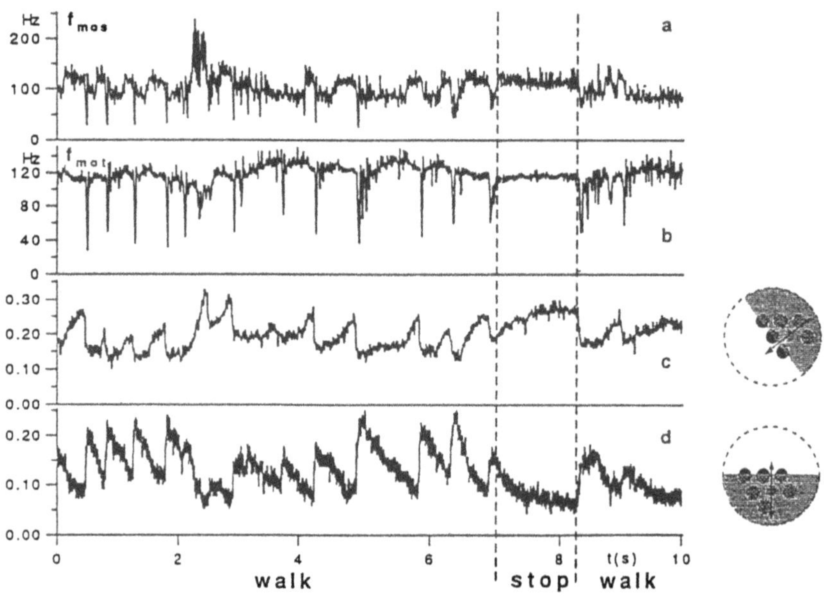

*Fig.9: (a,b) Instantaneous frequency of the MOS (a) and the MOT (b) eye muscles in a tethered blowfly (*Calliphora erythrocephala*) walking on a small, frictionless styrofoam ball (3) under the objective lens of a microscope focused onto a frontal* deep pseudopupil *(azimuth 0°, elevation 25°), all movements of which were recorded optically (c,d).(From 24)*

Rapid flicks were observed, followed by smooth phases during which the DPP - and hence the visual axes of the eye - returned to their original position or a nearby position. These optical recordings (Fig. 9.c,d) show that the movement of the photoreceptors under these conditions is a complex, not purely translational one and that their lines of sight therefore wander about within a solid angle as wide as one interommatidial angle. Frequency transients were observed as often as 6 times per second and seem to be an inherent property of the visual system during walking.

Simultaneous recordings obtained from the two opposite MOS (Fig. 10) showed that their spike rates, whether transient or tonic, grow and decline in parallel. Thus, the two opposite MOS act in concert, shifting the visual axes in symmetrical directions up to several degrees on either side of the midsagittal plane, in the frontal, binocular part of the eyes. In so doing, they provide the eyes with the ability to perform what amounts to slow (Fig.10a) and fast (Fig.10b) *vergence movements* (21).

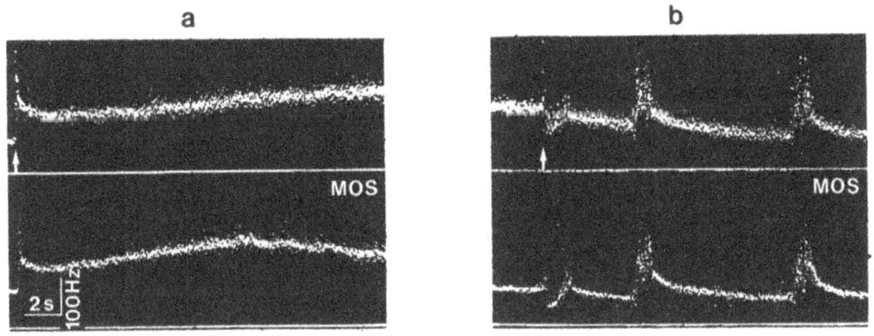

Fig. 10: Instantaneous frequency of the muscle spikes recorded simultaneously from the left and right MOS eye muscles in Calliphora with metal microelectrodes (the white line below each curve is the 0Hz baseline). (a) an airpuff onto the antenna (arrow) induced a kind of startling reaction manifested by a transient increase in spike rate followed by a graded change - which slightly differed in the two muscles. (b) switching the light off (arrow) induced a transient activity depression in the two MOS. After a few seconds in darkness, high frequency spike discharges (transients up to 300 Hz here) occurred spntaneously and repeatedly in the two MOS, with a period of 8-12 seconds. Calibrations in (s) and (Hz) apply to (a) and (b).

CONCLUSION

One of the lessons learned from our studies on "early colour processing" in the fly's eye was that the retina of so-called "simple creatures" may attain a degree of complexity and organisation that dwarfs even that of the human retina (33). The story of the "love-spot" with its specific neural connections illustrates how the structure, mechanism, and biological function of a given neural circuit can suddenly all make sense in the context of animal behaviour. Along the same lines, the fly's compound eye, like other compound eyes, possesses special UV and highly dichroic photoreceptors on its frontal dorsal rim (31), which drive a specialized neural machinery likely to be involved in polarized light navigation (91), possibly via polarization-opponent interneurones like those discovered in the eye of the cricket (66).

Electrophysiological recordings obtained on identifiable neurones while applying optical microstimulation to the retinal mosaic in the living animal have yielded insights into the functional organisation of a given visual neural network. With this approach, the receptor mosaic can be viewed as a huge "keyboard" controlling the underlying neural circuits. As described above, activating two selected "keys" in sequence readily induced a clear-cut response in a large-fielddirectionally selective motion sensitive neurone such as H1. Special sequences of pulses and steps were designed to answer specific questions about the functional principles underlying the individual Local Motion Detectors, in particular as regards the kinetics and the nonlinearities present within each branch of an L.M.D. These L.M.Ds are reduplicated some thousand-fold behind the facet array and drive the huge dendritic field of "collector neurones" such as H1.

These experiments, which yield a description of neural events at the system level, do away with having to worry about whether, for example, the inhibitory pathway in the nonpreferred direction involves GABAA or GABAB receptors. Extracting a functional principle from the neural hardware (cf 82) is a valuable approach for two main reasons. First of all it encourages the experimenter to form hypotheses about the possible involvement of well known neurones in the system under investigation (cf 40,9). Secondly, it is the abstract level *par excellence* at which a system can be described - and therefore transcribed in terms of a man-made technology such as optronics (23).

Although motion detection might seem at first sight to be a rather highly specialized part of the overall activity of a visual system, the neural mechanisms involved do in fact provide a useful general model for naturally occurring spatio-temporal signal processing. These ubiquitous mechanisms perform a well-defined, non-trivial operation called "sequence discrimination", and they use a nonlinear approach to processing two signals occurring at different times and places within a small set of neurones.

In spite of its fixed dioptrics, each eye of the fly can change its gaze rapidly via the co-ordinated action of two eye muscles (MOS and MOT) that act on the orbit and shift the photoreceptors dynamically in different directions. This fast gaze control, which comes as a further asset in addition to the ability to perform rapid head turns, was found in our studies to occur markedly in the walking animal. In retrospect, it seems likely that the puzzling, highly regular,"clock-like" activity of the MOT described in the early days (65,5,43), may have been due simply to the fact that the animal undergoing electrophysiological recordings was immobilized. The complex eye muscle system of the fly - which involves two MOT and two MOS for the whole visual system - is at present being investigated using a combination of optical, neuroanatomical, electrophysiological, behavioural, modelling and robotic approaches.

Acknowledgements:

Supported by Max-Planck Society, C.N.R.S. (Life Sciences and PIR Cognisciences), and Commission of the European Communities (ESPRIT SSS 6961). I thank J. Blanc, M. Boyron and R. Chagneux for their expert technical assistance.

References:

1. H. Autrum, F. Zettler, M. Järvilehto (1970)
   Postsynaptic potentials from a single monopolar neuron of the ganglion opticum 1 of the blowfly *Calliphora*.
   Z. Vergl. Physiol. 70, 414-424.
2. V. Braitenberg, N.J. Strausfeld (1973)
   Principles of the mosaic organisation in the visual system's neuropil of *Musca domestica*
   In: *Handbook of Sensory Physiology*. Vol. VII/3. Central visual information. A. R. Jung (ed.), Berlin, Springer, pp. 631-659.
3. E. Buchner (1976)
   Elementary movement detectors in an insect visual system.
   Biol. Cybernet. 24, 85-111.
4. F.H. Schuling, H.A.K. Masterbroek, B.P.M. Lenting (1989)
   Properties of elementary movement detectors in the fly *Calliphora erythrocephala*
   J. Comp. Physiol. 165, 179-192
5. E.T. Burtt, J.A. Patterson (1970)
   Internal muscle in the eye of an insect.
   Nature 228, 183-184.
6. T.S. Collett (1978)
   Peering - a locust behaviour pattern for obtaining motion parallax information
   J. Exp. Biol. 76, 237-241
7. T.S. Collett (1980)
   Some operating rules for the optomotor system of a hoverfly during voluntary flight.
   J. Comp. Physiol. A. 138, 271-282.
8. T.S. Collett, M.F. Land (1975)
   Visual control of flight behaviour in the hoverfly, *Syritta pipiens L.*
   J. Comp. Physiol. A. 99, 1-66.
9. J.K. Douglass, N.J. Strausfeld (1995)
   Visual motion detection circuits in flies: peripheral motion computation by identified small-field retinotopic neurons.
   J. Neurosci. 15, 5596-5611.
10. D.R. Dvorak, L.G. Bishop, H.E. Eckert (1975)
    Intracellular recording and staining of directionally selective motion detecting neurons in fly optic lobe.
    Vision Res. 15, 451-453.
11. S.Exner (1891)
    Die Physiologie der facettierten Augen von Krebsen und Insekten. Leipzig, Deuticke (translated and annotated by R.C. Hardie, 1991, Berlin, Springer)
12. R. Feiler, W.A. Harris, K. Kirschfeld, C. Wehrhahn, C.S. Zuker (1988)
    Targetted misexpression of a *Drosophila* opsin gene leads to altered visual function.
    Nature 333, 737-741
13. N. Franceschini (1971a)
    Etude optique *in vivo* des éléments photorécepteurs dans l'oeil composé de *Drosophila*
    Kybernetik 8, 1-13
14. N. Franceschini, K. Kirschfeld (1971b)
    Les phénomènes de pseudopupille dans l'oeil composé de *Drosophila*
    Kybernetik 9, 159-182
15. N. Franceschini (1975)
    Sampling of the visual environment by the compound eye of the fly: fundamentals and applications
    in: *Photoreceptor Optics,* A. Snyder, R. Menzel (eds.), Berlin , Springer, pp.98-125

16. N. Franceschini, K. Kirschfeld, B. Minke (1981)
Fluorescence of photoreceptor cells observed *in vivo*
Science 213, 1264-1267
17. N. Franceschini, R. Hardie, W. Ribi, K. Kirchfeld (1981)
Sexual dimorphism in a photoreceptor.
Nature 91, 241-244.
18. N. Franceschini (1984)
Chromatic organization and sexual dimorphism of the fly retinal mosaic.
In: *Photoreceptors*, A. Borsellino, L. Cervetto (eds), New York, Plenum, pp.319-349.
19. N. Franceschini (1985)
Early processing of colour and motion in a mosaic visual system
Neurosc. Res. ,Suppl. 2, 17-49
20. N. Franceschini, A. Riehle, A. Le Nestour (1989)
Directionally selective motion detection by insect neurons
In: *Facets of Vision*, D.G. Stavenga and R.C. Hardie (Eds.), Berlin, Springer, pp. 360-390
21. N. Franceschini, R. Chagneux, K. Kirschfeld, A. Mücke (1991)
Vergence eye movements in flies
in: *Proc. 19th Göttingen Neurobiology Conference 1991*, N. Elsner, H. Penzlin (eds), Stuttgart, G. Thieme, p.275
22. N. Franceschini (1992)
Sequence-discriminating neural network in the eye of the fly
In: *Analysis and Modeling of Neural Systems*. F. Eeckman, Norwell (U.S.A.°), Kluwer, pp.189-197
23. N.Franceschini, J.M. Pichon, C. Blanes (1992)
From insect vision to robot vision
Phil. Trans. Roy. Soc. B 337, 283-294
24. N. Franceschini, R. Chagneux, K. Kirschfeld (1995)
Gaze control in flies by coordinated action of eye muscles
in: *Göttingen Neurobiology Report 1995*, N. Elsner, R. Menzel (Eds.), Stuttgart, G. Thieme Verlag, p.402
25. C. Gilbert, N.J. Strausfeld (1991)
The functional organization of male-specific visual neurons in flies.
J. Comp. Physiol. A. 169, 395-411.
26. C. Gilbert, N.J. Strausfeld (1992)
Small-field neurons associated with oculomotor and optomotor control in muscoid flies: functional organization.
J. Comp. Neurol. 316, 72-86.
27. K.G. Götz (1972)
Principles of optomotor reactions in insects.
Biol. Ophthalmol. 82, 251-259.
28. K.G. Gôtz (1983)
Bewegungssehen und Flugsteuerung bei der Fliege *Drosophila*.
In: *BIONA report 2*. W. Nachtigall (ed), Stuttgart, Fischer, pp. 21-34.
29. W. Gronenberg, N.J. Strausfeld (1991)
Descending pathways connecting the male-specific visual system of flies to the neck and flight motor.
J. Comp. Physiol. A 169, 413-426.
30. R.C. Hardie (1983)
Projection and connectivity of sex-specific photoreceptors in the compound eye of the male housefly (*Musca domestica*).
Cell Tissue Res. 233, 1-21.

31. R.C. Hardie (1984)
Properties of photoreceptors R7 and R8 in dorsal marginal ommatidia in the compound eyes of *Musca* and *Calliphora.*
J. Comp. Physiol. A 154, 157-165.
32. R.C. Hardie (1985)
Functional organization of the fly retina.
In: *Progress in Sensory Physiology.* D. Ottoson (ed.) Berlin, Springer, pp. 1-79.
33. R.C. Hardie (1986)
The photoreceptor array of the dipteran retina.
TINS, 9, 419-423.
34. R.C. Hardie, N. Franceschini, P. Mac Intyre (1979)
Electrophysiological Analysis of the fly retina: II/ Spectral and polarization sensitivity in R7 and R8.
J. Comp. Physiol. 133, 23-39
35. R.C. Hardie, N. Franceschini, W. Ribi, K. Kirschfeld (1981)
Distribution and properties of sex-specific photoreceptors in the fly *Musca domestica.*
J. Comp. Physiol. A, 145, 139-152.
36. J.H. van Hateren (1986)
An efficient algorithm for cable theory, applied to blowfly photoreceptor cells and LMC's.
Biol. Cybern. 54, 301-311.
37. J.H. van Hateren (1987)
Neural superposition and oscillations in the eye of the blowfly
J. Comp. Physiol.A 161, 849-855
38. K. Hausen (1976)
Functional characterization and anatomical identification of motion sensitive neurons in the *lobula* plate of the blowfly *Calliphora erythrocephala.*
Z. Naturforsch. 31c, 629-633.
39. K. Hausen (1981)
Monocular and binocular computation of motion in the *lobula* plate of the fly
Verh. Dtsch. Zool. Ges.1981, 49-70
40. K. Hausen (1993)
Decoding of retinal image flow in insects
in: *Visual Motion and its Role in the Stabilization of Gaze*
F.A. Miles, J. Wallman (Eds), Amsterdam, Elsevier, pp. 203-235
41. K. Hausen, M. Egelhaaf (1989)
Neural mechanisms of visual course control in insects.
In: *Facets of vision.* Chap. 18. D.G. Stavenga, R.C. Hardie (eds), Berlin, Springer, pp. 391-424.
42. K. Hausen, N.J. Strausfeld (1980)
Sexually dimorphic interneuron arrangements in the fly visual system.
Proc. R. Soc. London, Ser. B. 208, 57-71.
43. M. Heisenberg,R.Wolf (1984)
Vision in *drosophila*
Berlin, Springer
44. R. Hengstenberg (1971)
Das Augenmuskelsystem der Stubenfliege *musca domestica.*
Kybernetik 2, 56-77.
45. R. Hengstenberg (1972)
Eye Movements in the housefly *musca domestica.*
In: *Information processing in the visual systems of arthropods.* R. Wehner (ed.), Berlin, Springer, pp.93-96.

46. R. Hengstenberg (1982)
Common visual response properties of giant vertical cells in the lobula plate of the blowfly *Calliphora.*
J. Comp. Physiol. A 149, 179-193.
47. R. Hengstenberg (1988)
Mechanosensory control of compensatory head roll during flight in the blowfly *Calliphora erythrocephala* Meig.
J. Comp. Physiol. A. 163, 151-165.
48. R. Hengstenberg (1989)
Stabilizing head/eye movements in the blowfly *Calliphora erythrocephala.*
IBRO Symposium Head/neck system. Fontainebleau, 1-10.
49. R. Hengstenberg (1991)
Gaze control in the blowfly *Calliphora:* a multisensory, two-stage integration process.
In: *Sensorimotor integration.* Seminars in the Neurosciences, Vol. 3. D. Sparks (ed.), Saunders Scientific Publications, pp. 19-29.
50. R. Hengstenberg (1993)
Multisensory control in insect oculomotor systems.
In: *Visual Motion and its role in the stabilization of gaze.* Reviews in Oculomotor Research. Vol. 5. J. Wallman, F.A. Miles (eds), pp. 1-23.
51. R. Hengstenberg, K. Hausen, B. Hengstenberg (1982)
The number and structure of giant vertical (VS) in the *lobula* plate of the blowfly *Calliphora erythrocephala.*
J. Comp. Physiol. A 149, 163-
52. R. Hengstenberg, D.C. Sandeman, B. Hengstenberg (1986)
Compensatory head roll in the blowfly *Calliphora* during flight.
Proc. R. Soc. Lond. B. 227, 455-482.
53. E. Horn (1982)
Gravity reception in the walking fly, *Calliphora erythrocephala*: tonic and modulatory influences of leg afferents on the head position.
J. Insect Physiol. 28, 713-721.
54. G.A. Horridge (1987)
The evolution of visual processing and the construction of seeing systems.
Proc.R.Soc.Lond. B 230, 279-292.
55. G.A. Horridge , L. Marcelja (1990)
Responses of the H1 neuron of the fly to jumped edges.
Phil. Trans.R. Soc. Lond. B , 329, 65-73.
56. A. Horridge, J. Meinertzhagen (1979)
The accuracy of the pattern of connexions of the first and second-order neurons of the visual system of *Calliphora*
Proc. Roy. Soc. Lond. B175, 69-82
57. J. Kien (1975)
Motion detectors in locusts and grasshoppers
in: *The Compound Eye and Vision of Insects,* G.A. Horridge (ed.), Oxford, Univ. Press, pp. 410-422
58. K. Kirschfeld (1967)
Die Projektion der optischen Umwelt auf das Raster der Rhabdomere im Komplexauge von *Musca.*
Exp. Br. Res. 3: 248-270
59. K. Kirschfeld (1979)
The visual system of the fly: physiological optics and functional anatomy as related to behavior.
In: *The neurosciences IV.* F.O. Schmitt, F.G. Worden (eds), Cambridge Massachusetts, M.I.T. Press, pp. 297-310.

60. K. Kirschfeld (1992)
Wie wird im Verlauf der Evolution optimiert ? "Erfindungen" im Komplexauge der Stubenfliege
In: *Technische Biologie und Bionik 1*, Ist Bionik Kongress, Wiesbaden, 1992, W. Nachtigall (ed.) pp. 39-52
61. K. Kirschfeld, N. Franceschini. (1968)
Optische Eigenschaften der Ommatidien im Komplexauge von *Musca*
Kybernetik 5, 47-52
62. K. Kirschfeld, N. Franceschini (1969)
Ein Mechanismus zur Steuerung des Lichtflusses in den Rhabdomeren des Komplexauges von *Musca*.
Kybernetik 6, 13-22.
63. K. Kirschfeld, N. Franceschini, B. Minke (1977)
Evidence for a sensitizing pigment in fly photoreceptors
Nature 296: 386-390.
64. W.H. Kirschner, M. Srinivasan (1989)
Freely flying honeybees use image motion to estimate object distance.
Naturwisssenschaften 76, 281-282.
65.J. Kuiper , J.T. Leutscher-Hazelhoff (1965)
High precision repetitive firing in the insect optic lobe and a hypothesis for its function in object location
Nature, London 206, 1158-1160
66. T. Labhardt (1988)
Polarization-opponent interneurons in the insect visual system.
Nature, Lond. 331, 435-437.
67. M.F. Land (1975)
Head movement and fly vision.
In: *The compound eye and vision of insects*. G.A. Horridge (ed), Oxford, Clarendon, pp. 469-489.
68. M.F. Land, T.S. Collett (1974)
Chasing behaviour of houseflies (*Fannia canicularis*).
J. Comp. Physiol. 89, 331-357.
69. S. Laughlin (1989)
Coding efficiency and design in visual processing
In: Facets of Vision, D.G. Stavenga, R.C. Hardie (eds.), Berlin, Springer, pp. 212-234
70. S. Laughlin, M. Weckström (1993)
Fast and slow photoreceptors - a comparative study of the functional diversity of coding and conductances in the Diptera.
J. Comp. Physiol. A172, 593-609
71. M. Lehrer, M.V. Srinivasan, S.W. Zhang, G.A. Horridge (1988)
Motion cues provide the bee's visual worl with a third dimension.
Nature. Lond. 332, 356-357.
72. J. Meinertzhagen (1982)
An analysis of the number and composition of synaptic population formed by photoreceptors of the fly
J. Comp. Neurol. 207, 29-44
73. J.J. Milde, N.J. Strausfeld (1986)
Visuo-motor pathways in arthropods. Giant motion-sensitive neurons connect compound eyes directly to neck muscles in blowflies (*Calliphora erytyrocephala*).
Naturwissenschaften 73, 151-153.
74. A. Mücke, N. Franceschini (1991)
Procion staining: novel applications
in: Proc. 19th Göttingen Neurobiology Conference, Stuttgart, G. Thieme, p. 276

75. J. Patterson (1973a)
The eye muscle of *Calliphora vomitoria L.* I. Spontaneous activity and the effects of light and dark adaptation.
J. Exp. Biol. 58, 565-583.
76. J. Patterson (1973b)
The eye muscle of *Calliphora vomitoria L.* II.
J. Exp. Biol. 58, 585-598.
77. S. Picaud, HJ. Wunderer, N. Franceschini (1988)
Photo-degeneration of neurons after extracellular dye application
Neurosci. Lett. 95, 24-30
78. S. Picaud, HJ Wunderer, N. Franceschini (1990)
Dye-induced photo-permeabilization and photo-degeneration: a lesion technique useful for neuronal tracing
J. Neurosci. Meth. 33, 101-112
79. J.M. Pichon, C. Blanes, N. Franceschini (1989)
Visual guidance of a mobile robot equipped with a network of self-motion sensors.
In: *Mobile Robots IV.* W.KJ. Wolfe, W.H. Chun (eds), *Proc. S.P.I.E.* 1195, pp. 44-53.
80. B. Pick (1977)
Specific misalignments of rhabdomere visual axes in the neural superposition eye of Dipteran flies.
Biol. Cybernetics 26, 215-224.
81. R. Pierantoni (1976)
A look into the cockpit of the fly: The architecture of the *lobula* plate.
Cell Tissue Res 171, 101-122.
82. W. Reichardt (1987)
Evaluation of optical motion information by movement detectors.
J. Comp. Physiol. A 161, 533-547.
83. W. Reichardt, T. Poggio (1976)
Visual control of orientation behaviour in the fly. Part I: A quantitative analysis.
Q. Rev. Biophys. 9, 311-375.
84. A. Riehle, N. Franceschini (1984)
Motion detection in flies: parametric control over ON-OFF pathways
Exp. Br. Res. 54, 390-394
85. S.R. Shaw (1984)
Early visual processing in insects.
J. Exp. Biol. 112, 225-251.
86. F.H. Schuling, H.A.K. Mastebroek, R. Bult, B.P.M. Lenting (1989)
Properties of elementary movement detectors in the fly *Calliphora erythrocephala.*
J. Comp. Physiol. A 165, 179-192.
87. N. J. Strausfeld (1976)
Atlas of an insect brain., Berlin, Springer.
88. N.J. Strausfeld (1989)
Beneath the compound eye: Neuroanatomical analysis and physiological correlates in the study of insect vision
In: Facets of Vision, D.G. Stavenga, R.C. Hardie (eds.), Berlin, Springer, pp.317-359
89. N.J. Strausfeld (1991)
Structural organization of male-specific visual neurons in calliphorid optic lobes.
J. Comp. Physiol. A. 169, 379-393.
90. N.J. Strausfeld, C. Gilbert (1992)
Small-field neurons associated with oculomotor control in muscoid flies: cellular organization in the lobula plate.
J. Comp. Neurol. 316, 56-71.

91. N.J. Strausfeld, H. Wunderer (1985)
Optic lobe projections of marginal ommatidia in *Calliphora erythrocephala* specialized for detecting polarized light.
Cell Tissue Res. 242, 163-178

92. N. Troje (1993)
Spectral categories in the learning behaviour of blowflies
Z. Naturforsch. 48c, 96-104

93. K. Umeda, H. Tateda (1985)
Visual interneurons in the *lobula* complex of the fleshfly, *Boettcherisca peregrina.*
J. Comp. Physiol. A 157, 831-836.

94. R.O. Uusitalo, M. Juusola, M. Weckström (1995)
Graded responses and spiking properties of identified first-order visual interneurons of the fly compound eye.
J. Neurophysiol. 73, 1782-1792.

95. P. Vigier (1907)
Mécanisme de la synthèse des impressions lumineuses recueillies par les yeux composés des Diptères.
C.R.Acad.Sci. 6, 122-124.

96. A. Wachenfeld (1994)
Elektrophysiologische Untersuchungen und funktionelle Charakterisierung männchenspezifischer visueller Interneurone der Schmeissfliege *Calliphora erythrocephala* (Meig.). Dr Thesis, University of Cologne.

97. H. Wagner (1986)
Flight performance and visual control of flight of the free-flying housefly *Musca Domestica* II/ Pursuit of targets
Phil. Trans. Roy. Soc. B312

98. R. Wehner (1981)
Spatial vision in arthropods.
In: *Handbook of Sensory Physiology. VII. 6c.* H. Autrum (ed), Berlin, Springer, pp. 288-616.

99. C. Wehrhahn (1979)
Sex-specific differences in the chasing behaviour of free flying houseflies (*Musca*).
Biol. Cybernet. 21, 213-220.

100. R. Wolf, M. Heisenberg (1990)
Visual control of straight flight in *Drosophila melanogaster.*
J. Comp. Physiol. A, 167, 269-283.

101. R. Wolf, M. Heisenberg (1991)
Basic organization of operant behaviour as revealed in *Drosophila* flight orientation.
J. Comp. Physiol. A 169, 699-705.

FROM BIOCYBERNETICS TO BIONICS: ON VISUALLY-GUIDED NAVIGATION IN ANIMALS AND MACHINES

Nicolas FRANCESCHINI, Jean Marc PICHON, Christian BLANES

CNRS, Laboratoire de Neurobiologie, Neurocybernétique
31, Ch. J. Aiguier; 13402 MARSEILLE , Cedex 20 (France)
FAX (33) 91 22 08 75; e:mail: enfranceschini@irlnb.cnrs-mrs.fr

I/ INTRODUCTION

A fact which has to be faced is that in matters of visuomotor skill, i.e., visuomotor intelligence, there exist innumerable lowly earthly creatures which can easily outdo human beings, however hard we may try. For this reason, a strictly anthropomorphic approach may well be out of place when designing tomorrow's *seeing vehicles*, which will have to behave swiftly and efficiently in unstructured, unknown and hazardous environments. Designers of autonomous vehicles wishing to profit from previous successful and thoroughly tested achievements may have much to gain by casting a sidelong glance not only at human cognitive processes but also at the sensori-motor control systems of humble but highly skilled creatures such as insects, arachnids, and crustacea.

In the companion paper (26), we present some recent data on the art of signal processing in the cockpit of the fly. At our laboratory, flies are viewed as perfect miniature, self-propelled and cordless 'vehicles' (7), which are able to use vision to navigate swiftly and land safely in the most unpredictable, obstacle-cluttered environments. These sighted agents take the bearings of obstacles and avoid them without resorting to any sonars or laser range finders. They carry their own energy resources, and process their sensory signals onboard, without having to trail behind them any cumbersone umbilical connecting them to a supercomputer and a power supply. There exist so far no visually guided artificial agents which are able to match the real-time visuomotor prowess of the fly. In the major engineering challenge which consists of designing sighted vehicles capable of steering their way quickly through complex, unstructured environments, the successful fly guidance system may provide some unexpected solutions. Surely it is high time that this little aircraft, which can be so objectionable in many ways, stopped adding insult to injury by doing what it does better than the present-day robots can.

Since 1985, in parallel with our basic research on the neuronal pathways involved in animal vision, our group has been developing sighted, self-guided mobile robots which make use of lessons drawn from the insect world: lessons learned in neuroethology, lessons about the mechanisms underlying signal processing in nervous systems, and lessons about the optical and neural architectures. The aim is to fuse ideas from biology and robotics, with potential benefits to both fields - since many of the tasks to be performed are similar.

Here we report briefly on the design and construction of a fly-sighted demonstrator, i.e., a robot equipped with a compound eye comprising an array of fly-inspired 'Local Motion Detectors' (LMD's). This bionic attempt to incorporate into an 'auto-mobile' the working principles of specific neuronal circuits - those we previously analysed in the visual system of the common housefly (31) - ended up with a ten kilogramme, self-driven terrestrial vehicle capable of moving towards a luminous target at a fairly high speed (50cm/s, i.e., about two body lengths per second) while avoiding any unpredictable obstacles encountered on the way. This account briefly summarizes the philosophy behind our recent work on these lines and readers can refer to the original papers for further details (60,29,30,50,54). The authors of more recent studies have used a similar, biologically inspired approach to the visually-guided, reactive navigation of an artificial agent (15,63,67,10). The physical robot we arrived at is unique, however, in that it operates using the parallel, analog method of signal processing, favoured by Nature.

Our approach fits the general framework of 'active perception', in that an *ad hoc* motion of a sensor is used to constrain the inputs and reduce the computational burden involved in perceptual tasks (4,2,3,1). Readers interested in the general design of - seeing or not seeing - reactive agents ("animats") and behavior-based AI can refer to the proceedings of several recent conferences on this subject (69,52,19,53,34,11).

II/ PRODUCING A SIGHTED MOBILE CREATURE

As early as 1867, von Helmholtz discussed the possibility that humans might use *motion parallax* to assess the three-dimensional layout of the environment. For some reason, however, the idea took about 100 years to mature, perhaps because it took nearly 100 years to realize that the vertebrate visual system indeed includes a battery of smart sensors called "motion sensitive neurones" - as postulated by Exner in 1894, see Fig. 2 in ref.31 - which are able to determine the retinal slip speed at various azimuths and elevations in the visual field, and thus to provide their host with an estimate of the *optic flow field* (cf. 35,36,48,56,44,40,42,45,72).The idea that insects too can use motion parallax and take advantage of the optic flow field to take range of objects and navigate was suggested early on by Exner (22), but actually proved only quite recently in some fine sets of experiments (71,12,16,49,43,66) (see also the reviews: 73,9,41,68,14).

In the companion paper (26), we describe how we analysed the basic working scheme underlying the housefly's directionally selective motion sensitive neurones. This research involved carrying out microelectrode recordings on the electrical activity of a specific motion-detecting neurone, called H1. The sensitivity of this neuron to directional motion relies on the pooling of multiple signals from smallfield units called 'Elementary Motion

Detectors' (EMD's, cf. ref.9), also called 'Local Motion Detectors' (LMD's, cf ref.14), which all project to its dendritic fan in a retinotopic manner. To assess the functional properties of one such LMD, we used a specially designed stimulator, a hybrid between a microscope and a telescope, which served to illuminate two neighbouring photoreceptors cells ***sequentially*** within the same facet, thereby mimicking a micromotion occurring in the visual field (62,25,31).

By 1986, we had built the first optronic replica of a fly LMD, which was capable of processing its two input signals stage by stage, in the analog mode, with similar time constants and nonlinearities to those observed in the live animal. Using a 'surface mounted devices' technology, we had more than a hundred such motion detectors made in the form of minute electronic cards (each weighing as little as 6 grammes), which we used to produce a whole compound eye designed to equip a mobile robot and steer it through a cluttered environment, by making use of the optic flow.

The functional scheme on which this robot was based drew heavily on our knowledge of the fly's visuomotor behaviour. Fast motion picture analyses and experiments in a flight simulator have shown that the cruising fly maintains a near constant angular body- (and probably eye-) orientation, interrupted by saccade-like turns of various amplitudes, which are performed at a high angular velocity (13,47,75,76,38,70,14). We took this to be a strategy adopted by the animal to reduce the optic flow field to its *translational component* (cf.5,58). Here the interesting basic principle involved is that only the translational component of the optic flow field generated by an agent's movement can convey a depth map of the extra-personal space (44,61) (Fig.1)

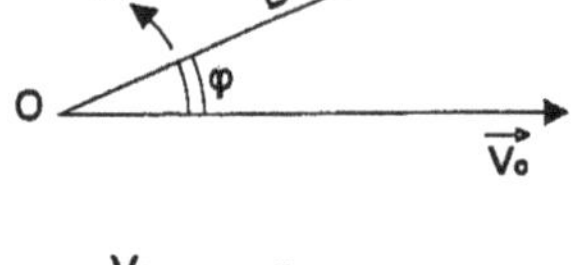

$$\Omega = \frac{V_o}{D} \cdot \sin \varphi$$

Fig. 1: Using the translational optic flow field to avoid obstacles. To an observer O (human, fly, robot, etc.) which translates to the right at speed Vo, a contrast point such as P located at azimuth φ seems to move to the left at speed -Vo. If the visual system of the observer has a means of determining the angular speed Ω at which P seems to be turning, then the observer can gauge the distance D to the contrast point (via the simple formula given, cf 74) - and then produce a suitable steering manoeuvre to avoid it (from 30).

In the same way, our robot proceeds by performing pure translations (during which it collects the relevant visual information 'on the move'), which are interrupted by fast steering actions. The numerous LMD's built into the eye of the moving robot act like so many miniature teledetectors, assessing the distance to any contrasted obstacles encountered as it travels along.

Since we were designing a terrestrial robot capable of moving on a single plane only, we decided to simply equip our automaton with a horizontal ring of facets, the visual axes of which would all cross at a common point: the centre of curvature of the eye - which was also the centre of the (circular) robot. This ring contains about as many facets (around 100) as the number which can be counted on a horizontal cross-section through both eyes of a fly, and the visual field afforded thereby is correspondingly very large (320°). For reasons having to do with the geometry of the optic flow, we provided this eye with a resolution gradient such that the spatial sampling step grows according to a *sine law* as a function of the eccentricity (60,30). This specific non-uniform sampling procedure 'compensates' for the sine law inherent to the optic flow field (cf. Fig.1) and thus ensures that any contrast point located inside a *circle of vision* will be detected upon translation by a step of length ΔL. It should be recalled here that a resolution gradient exists in the fly as well, whose eye is more acute frontally than laterally (6,13,24). The introduction of a sine gradient into this 'centred eye' would lead, however, to an infinitely small interommatidial angle along the heading direction. Therefore a 'deadzone' of +/- 10° was provided around the frontal pole of the optic flow field. This deadzone was covered by two small additional sets of LMD's that were off-centred and symmetrically positioned on each side of the vehicle (60).

Once the photoelectrical signals have been processed in a logarithmic amplification stage, as in the fly, the next stages consist of detecting the contours of obstacles and measuring the time elapsing between the stimulation of any two adjacent facets. Angular velocity thresholds are then set in such a way that nearby contrasts are given priority over those which are farther away from the robot. The motion detecting array thus generates a map of obstacles, which are expressed in polar coordinates in an eye-centred frame of reference at the end of each translatory step. This map is immediately merged with the information about the azimuthal direction of the target (an electric light), before being delivered to a 'motor interface', which then specifies in Volts the new steering angle required to bring the robot one step further towards the target while avoiding the obstacles detected during the previous translation phase.

The trajectory of the robot thus consists of stereotyped translation steps ΔL in length which at the same time serve to acquire visual data. During each translation step, the robot surveys the environment with its numerous local motion detectors, and updates its own picture of the outside world. This picture is immediately obliterated once the obstacle has been left behind. At the completion of each translation step, the next course to be steered is immediately given by the output from the motor interface, and this immediately triggers an (open loop) ocular and body saccade of appropriate amplitude in the appropriate direction. During the actual steering manoeuvres, vision is disconnected because a rotational optic flow is not relevant to the range of obstacles. Vision is not restored until the appropriate translational speed is reached again. The elementary translation steps ΔL are tacked together at a high speed compatible with the parallel, analog mode of processing used. No stop occurs at the end of a translation step if no steering command is issued, i.e., if no obstacle has been detected by the LMD array.

The architecture of our visuomotor control system is reminiscent of Brooks' layered structure called 'subsumption architecture' (8). It differs from the latter in one major respect, however, since it has no enabling or disabling 'switch' deciding which of the two 'behaviours', target pursuit or obstacle avoidance, should be dealt with by the control at any given time. There is no such distinction between two 'behaviours'. Instead, a *graded fusion* permanently occurs between the visual signals giving the bearings of the obstacles and the direction of the target. This kind of 'multisensory fusion', which takes place on a short-lived local 'motor map', ensures that the next heading direction will be the nearest possible to the target direction while avoiding the 'forbidden zone' calculated onboard the robot on the basis of all the obstacles detected during a given translatory step.

Our fly-sighted demonstrator has shown that a compound eye equipped with only a hundred or so motion detecting neurones suffices to steer a vehicle safely and speedily through an unpredictable visual environment. The environment where it can dwell is textured with vertical stripes having random widths (in the range 5cm to 25cm) and random contrast (in the range O.15 to O.75). Alternately it can consists of cylindrical, randomly arranged posts. The collision-free paths taken by this robot between these posts provide evidence that an elementary but fast visual guidance system can be achieved without any symbolic representation and global reasoning about obstacles, without the need to resort to a permanent model of the world, without a central processing unit, without a CCD camera, without signal digitization, without any accurate and precisely matched visual sensors, and without any particularly large memory resources.

In comparison with 'main stream AI' approaches to mobile robotics, one of the most original aspects of this system is the use it makes of the brainlike *parallel and analog* modes of signal processing. The main reason for choosing these modes was the expectation that in practising Nature's way of doing things, and in fighting on the same ground, we might perhaps learn more about the advantages, constraints and adaptability of this mode of computation, and discover how it enabled the winged insects to survive in their highly complex natural environments for 350 million years. We therefore decided to do without any von Neumann-type architecture onboard the robot, and to restrict the use of computers of this kind to the initial simulation phases of the project. By incorporationg a second visual system into the robot, which makes it able to detect the direction of a target, we broke new ground on a territory where few attempts have been made so far in robotics, although animals' nervous systems do it all the time: using analog neural networks to achieve real-time multisensory fusion. Fig. 2 shows the odd routing pattern that connects the thousands of analog devices that blend together the input signals from the compound eye and those from the accessory target-seeking eye, to eventually deliver a single analog output: the steering angle.

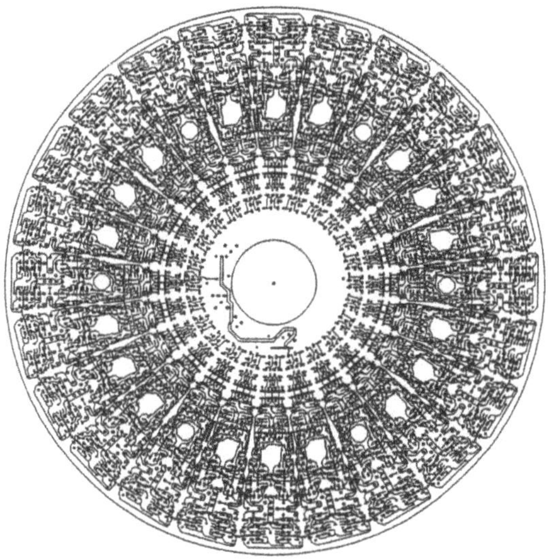

Fig.2: Routing diagram of one face of the printed circuit board that fuses the information about the obstacles and the goal. This six-layered board has about 200 parallel inputs (100 inputs from the LMD's of the compound eye and 100 inputs from the target seeking eye), and only one output, which expresses in Volts the next steering angle required to reach the target while avoiding the obstacles. This side and the reverse side of the board are covered with thousands of analog, surface-mounted devices of only four kinds: resistors, capacitors, diodes, and operational amplifiers. The rose-window-like pattern obtained here is reminiscent of the neural architecture of insect and vertebrate visual systems, and is very different from the architecture of a von Neumann computer. This mosaic organisation results from the numerous repeat units and from the retinotopic projections of the visual signals (from 30)

III/ FROM NEUROSCIENCE TO BIONICS, AND VICE-VERSA

Since this mobile machine was the concrete embodiment of knowledge acquired in neurobiology and neuroethology about insects' visual systems over the last 30 years or so, it incidentally yielded a wealth of physiological insights. The creature we ended up with was simpler of course than its biological forerunner, since we intentionally modelled only one of its aspects (navigation based on optic flow). Owing to its simplificatory nature, our approach shed some new light on sensorimotor processes in general and prompted us to perform novel electrophysiological and behavioural experiments on the fly that we would probably never have thought of otherwise.

One major physiological spin-off from this robotic approach has been the finding that, in spite of their fixed dioptrics, flies can in some cases display 'binocular vergence', a process that is currently being investigated at the laboratory using a variety of techniques (32,28,26). This process itself has generated several hypotheses, some of which have been tested via simulation. Adding a slight amount of vergence gaze control provided the key to having the robot automatically control its cruising speed depending on the density of the obstacles in its way . The keypoint is that a centred eye operating on the basis of motion parallax affords a *radius of vision* that automatically grows larger with speed (50). Now, the visual system described above can be enhanced if its two frontal, off-centred sets of LMD's (dedicated to vision near the heading direction, cf Section II) become off-centred in proportion to the robot's speed, so as to keep sampling the *circle of vision* (50). On the same line, simulation showed that motion parallax can be used to regulate the altitude and speed of a flying agent (54).

This type of research thus provides us with a real state of synergy between two disciplines which might seem *a priori* to be fathoms apart: the study of neurobiological mechanisms, and the production of artificial seeing agents. The notion that both natural and artificial visual creatures have to cope with similar difficulties raised by the complex world which they both inhabit, and the idea that the complex problems they have to solve may have a very limited number of solutions incite one to approach natural and artificial creatures on a common footing and to work towards a really multidisciplinary approach in the study of sensory perception and sensory-motor control.

IV/ CONCLUSION

Issues relating to how vision links up with action in Nature's unmanned aerial vehicles such as insects and birds are really worth being investigated further. Houseflies are perfectly willing and able to show us at a moment's notice exactly what they are able to do. If we have not yet mastered their sophisticated skills, the fact that they can put on such good performances in spite of the low resolution of their eyesight at least shows that the solutions to the underlying problems must exist. These solutions rest on the one hand on specific behavioral strategies that reduce the computational burden, and on the other hand, on dedicated circuits based on exceptionally efficient signal processing architectures.

It was by first exploring the logic behind some specific neural network in the fly's compound eye, using microelectrodes and a specially designed microscope, that we acquired the requisite basic know-how to be able to work on the lines which eventually yielded the robot prototype. This multidisciplinary project obliged us to blend lessons from animal vision, neurobiology, neuroethology, electronics, optics, control theory, information science and robotics.

In the general context of visually-guided navigation of natural or artificial agents, several points are worth mentioning:

1) With just a few pixels, it is possible to obtain skilled visually guided performances. Small numbers of pixels obviously mean considerable savings for designing future robots' eyes with parallel processing. The anthropomorphic notion that the visual guidance of an artificial agent must necessarily start with a pair of high resolution camera eyes needs to be re-considered. At least eighty percent of all the animal species do not do it this way.

2) The plain design of the camera eye with which vertebrates, arachnids and cephalopods are equipped, with its inherently limited field of view, does not necessarily give the right vista when designing robot's eyes. Drawing inspiration from the seven different types of arthropod compound eyes (57) may be a better approach. A compound eye in which each pixel has its own lens turns out to be an efficient means of designing a panoramic visual field, and lends itself more readily to shaping a resolution gradient and a spatial filtering gradient. The interrelationship which exists in compound eyes between spatial sampling and contrast transfer was worked out long ago (37,65).

3) There exist some basic common constraints between the design of a seeing machine and that of a locomoting machine. Constructing a powerful robot by sub-contracting the visual system to one laboratory and the control system to another can therefore lead to a fiasco. Our project brought to the fore the tight interrelationship existing between vision and action (cf. 59). In both our robot and the fly, vision requires locomotion just as much as locomotion is guided by vision.

4) 'Motion blur' does not always constitute a plague in vision and can be put to good use by both natural and artificial creatures, provided their visual systems are equipped with a network of *motion sensors*. The idea that motion blur is detrimental to vision seems to have several origins. On the one hand, it stems from the misconception that an animal's retina may be like a photographic plate or CCD chip, the pixels of which respond to a *number* of absorbed photons (whereas animal photoreceptors respond, as we know, to a *rate* of absorbed photons). On the other hand it stems from the second misconception that animal's and human's visually-guided actions (including navigation) necessarily require a *static image* analysis process and hence are based on a series of stable pictures, each of which is acquired during a given 'exposure time'. If the absence of motion during a 'frame' is required with most man-made optronic devices, this is due to the odd, *sequential* procedure used to read the electrical image - as opposed to the *parallel* reading of the electrical image which takes place in animal focal plane mosaics.

5) Motion detection plays a vital role in animal vision. It serves not only to find preys and mates and to escape predators. Most importantly, it helps animals to perceive their own ongoing bodily movements. This point seems to be still greatly underestimated in general, although several particularly perspicacious authors such as von Helmholtz, Exner and Gibson stressed it a long time ago (cf. 72).

6) Perceiving the optic flow requires a set of N motion detecting neurones, which process the information flowing onto the retina. The fly's visual system, like that of human

beings, contains thousands of motion detecting neurones, each of which carries out a non-trivial task. Computers of the von Neumann type seem to be ill-suited to handling such a large body of data on line, especially when the agent is moving at a high speed. Moreover, digital computers can introduce undesirable effects due to the noncontinuity of the signals (amplitude discretization and temporal sampling), which can adversely affect the motion detection process.

7) The *parallel and analog* mode of processing - that with which animals' sensory systems operate - makes for fast sensorimotor computing. It can work efficiently even when the components are slow, inaccurate, and unmatched. It is likely that this kind of processing will be adopted in tomorrow's vehicles and microvehicles, using low-power analog VLSI circuits, with which massively interconnected, nonlinear processing circuits can be built efficiently and economically (51,18,55,21). A few complete visuomotor systems have already been produced on these lines (20,17)

8) Accuracy is not at a premium when dealing with signal processing in sensorimotor systems. The problem is quite different in fact from those arising in the fields of mathematical computation and telecommunications. In the latter case, sounds or images have to be accurately transmitted or stored to be properly decoded by the fancy human receivers (ears or eyes).

The bionics approach we have adopted yields some valid solutions to the *low-level* visuomotor problem of obstacle avoidance in an uncertain, stationary environment. This problem has in fact been solved by animals throughout the course of evolution, and we have shown how the solution provided by insects can in fact be scaled up to a 10 Kg robot. The animals' own solutions are necessarily both feasible and robust: they have withstood the test of time. Although most roboticists today do not care what solutions may be adopted by animals, they might have much to learn from Nature's ingenious tricks about how to process sensorimotor signals with minimal power, size, time and cost. This is particularly telling in the field of microrobotics (33) where the development of *microvehicles* destined for applications of various kinds depends on the possibility of overcoming the dramatic lack of decent one-cubic-millimeter visual systems. The emergence of flying microvehicles (64,46) and the need to find means of controlling their dynamic stability and visually-guided navigation may well force von Neumann architectures to the wall. Animals are sighted and locomoting agents which not only provide us with eloquent proof that there exist physical solutions to elusive problems such as those involved in visually guided navigation: they also provide us with guidelines for producing efficient robots which are akin to biological systems in spirit. Pursuing investigations on this topic is a really worthwhile scientific challenge, since it seems to work both ways at once, furthering the cause of both neuroscience and robotics.

The market for autonomous mobile vehicles endowed with visual capacities hardly exists yet, but there will be great potential for visually-guided terrestrial, airborne, underwater and space vehicles of all kinds in the centuries to come. It is time for us to realize that even the 'automobiles' of which we have been so proud, from Cugnot's steam truck to the Rolls Royce, are simply blind. These perfidious creatures, as we know, have always cheated by relying completely on the visuomotor skill of a biological operator.

Acknowledgments:

Supported by C.N.R.S., (Life Sciences, Engineering Sciences and PIR Cognisciences), M.R.T., Conseil Régional PACAC, and Commission of the European Communities (ESPRIT SSS 6961, and HCM network). We thank J. Blanc, M. Boyron, R. Chagneux and J. Roumieu for their technical assistance. We are specially indebted to N. Martin and F. Mura for many discussions.

References:

1. Y. Aloimonos (1993)
   Active perception
   Lawrence Erlbaum, Hillsdale (N.J.), U.S.A.
2. Y. Aloimonos, I. Weiss, A. Bandyopadhyay (1987)
   Active vision
   In: *IEEE Ist Intern. Conf. Comp. Vision*, Los Angeles, 35-54
3. D. Ballard (1991)
   Animate vision
   Artif. Intell. 48, 57-86
4. R. Bajcsy (1985)
   Active perception versus passive perception
   In: *Proc. 3rd IEEE Wokshop on Computer vision*, 55-59
5. J. Barnes, H.D. Nalbach (1993)
   Eye movements in freely moving crabs: their sensory basis and possible role in flowfield analysis
   Comp. Biochem. Physiol. 104A, 675-693
6. D.G. Beersma, D.G. Stavenga, J. W. Kuiper (1975)
   Organization of visual axes in the compound eye of the fly *Musca domestica* and the behavioural consequences
   J. Comp. Physiol., 202, 305-320
7. V. Braitenberg (1984)
   Vehicles; Experiments in synthetic Psychology
   Cambridge, U.S.A., M.I.T. Press
8. R. Brooks (1986)
   A robust layered control system for a mobile robot
   IEEE J. Rob. Autom. 1, 14-23
9. E. Buchner (1984)
   Elementary movement detectors in an insect visual system
   Biol. Cyb. 24, 85-111
10. M. Campani, A. Giachetti, V. Torre (1995)
   Optical flow and autonomous navigation
   Perception, vol. 24, 253-267
11. D. Cliff, P. Husbands, J.A. Meyer, S.W. Wilson (1994)
   From animals to animats III
   In: *Proc. Intern. Conf. on Simulation of Adaptive Behavior*
   Cambridge (U.S.A.), M.I.T. Press.

12. T.S. Collett (1978)
Peering: a locust behaviour pattern for obtaining motion parallax information
J. Exp. Biol. 76, 237-241

13. T.S. Collett, M. Land (1975)
Visual control of flight behaviour in the hoverfly, Syritta pipiens
J. Comp. Physiol., 99, 1-66

14. T.S. Collett, H.O. Nalbach, H. Wagner (1993)
Visual Stabilization in arthropods
In: *Visual Motion and its Role in the Stabilization of Gaze*
F.A. Miles, J. Wallman (Eds), Amsterdam, Elsevier

15. D. Coombs, K. Roberts (1992)
Bee-Bot: Using the peripheral optic flow to avoid obstacles
In: *Intelligent robots and computer vision XI*, SPIE 1835, Bellingham, U.S.A., pp. 714-725

16. C.T. David (1982)
Compensation for height in the control of groundspeed by *Drosophila* in a new Barber's pole" wind tunnel
J. Comp. Physiol. 147, 485-493

17. T. Delbrück (1989)
A chip that focuses an image on itself
In: *Analog VLSI Implementation of Neural Systems*,
C. Mead, M. Ismail (Eds), Boston, Kluwer, pp.171-188

18. J.G. Delgado-Frias, W.R.Moore (1991)
VLSI for Artificial Intelligence and Neural Networks
New-York, Plenum

19. J.L. Deneubourg (1993)
Proc. 2nd Inter. Congr. Artificial Life
Cambridge, U.S.A., M.I.T. Press

20. S.P. DeWeerth, C.A. Mead (1988)
A two-dimensional visual tracking array
In: *Advanced Research in VLSI*, J. Allen, F.T. Leighton (Eds), Cambridge (Mass.), M.I.T. Press, pp. 259-275

21. R. Douglas, M. Mahowald, C. Mead (1995)
Neuromorphic analogue VLSI
Ann. Rev. Neurosc. 18, 255-281

22. S. Exner (1891)
Die Physiologie der Facettierten Augen von Krebsen und Insekten. Leipzig, Deuticke
(translated and annotated by R.C. Hardie, 1991, Berlin, Springer)

23. S. Exner (1894)
Entwurf zu einer physiologischen Erklärung der psychischen Erscheinungen
1sten Teil, Leipzig, Deuticke

24. N. Franceschini (1975)
Sampling of the visual environment by the compound eye of the fly: fundamentals and applications.
In: *Photoreceptor Optics*, A. Snyder, R. Menzel (Eds), Berlin, Springer, pp. 98-125

25. N. Franceschini (1985)
Early processing of colour and motion in a mosaic visual system
Neurosc. Res. Suppl. 2, 17-49

26. N. Franceschini (1995)
Combined optical, neuroanatomical, electrophysiological and behavioural studies on signal processing in the fly compound eye
In: *Biocybernetics of Vision: Integrative Mechanisms and Cognitive Processes.*
C. Taddei Ferretti (Ed.), this volume.

27. N. Franceschini (1995)
The fly compound eye: from microelectrode analysis to robot synthesis
In: *IEEE Intern. Conf. Robot. and Autom., Worksh. Bio-mechatronics*, Nagoya,
H. Miura (Ed.)

28. N. Franceschini, R. Chagneux (1994)
Retinal movement in freely walking flies
Proc. 22nd Göttingen Neurobiology conference
29. N. Franceschini, J.M. Pichon, C. Blanes (1991)
Real time visuo-motor control: from flies to robots
Proc. I.E.E.E. Intern. Conf. Advanced Robotics; june 1991, Pisa pp. 91-95
30. N. Franceschini, J.M. Pichon, C. Blanes (1992)
From insect vision to robot vision
Phil. Trans. Roy. Soc. B 337, 283-294
31. N. Franceschini, A. Riehle, A. Le Nestour (1989)
Directionally selective motion detection by insect neurons
In: *Facets of Vision*
D.G. Stavenga, R.C. Hardie (Eds), Berlin, Springer, pp. 360-390
32. N. Franceschini, R. Chagneux, K. Kirschfeld, A. Mücke (1991)
Vergence eye movements in flies.
In: *Proc. 19th Göttingen Neurobiology Conference*
Elsner, N, Penzlin R., (Eds), 275
33. T. Fukuda, T. Tueyama (1994)
Cellular robotics and micro robotic systems
Singapore, World Scientific
34. P. Gaussier, J.D. Nicoud (1994)
From Perception to Action
Los Alamitos (U.S.A.), IEEE Computer Soc. Press
35. J.J. Gibson (1950)
The Perception of the Visual World
Boston, Houghton Mifflin
36. J.J. Gibson, P. Olun, F. Rosenblatt (1955)
Parallax and perspectrive during aircraft landing
Am. J. Psychol. 68, 372-385
37. K.G. Götz (1965)
Die optischen Übertragungseigenschaften der Komplexauge von *Drosophila*
Kybernetik 2, 215-221
38. M. Heisenberg, R. Wolf (1984)
Vision in Drosophila
Berlin, Springer
39. H. von Helmholtz (1867)
Physiologische Optik, Bd 2
Dover, London
40. B.K.P. Horn, B. Schunk (1981)
Determining optical flow
Artif. Intell. 17, 185-303
41. A. Horridge (1987)
The evolution of visual processing and the construction of seeing systems
Proc. R. Soc. London B230, 279-292
42. G. Johansson (1982)
Visual space perception through motion
In: *Tutorial in Motion Perception*, A.H. Wertheim, W.A. Wagenaar, H.W. Leibowitz (Eds), New-York, Plenum Press, pp. 19-39
43. W.H. Kirchner, M. Srinivasan (1989)
Freely flying honeybees use image motion to estimate object distance
Naturwiss. 76, 281-282
44. J.J. Koenderink, A.J. van Doorn (1976)
Local structure of movement parallax in the plane
JOSA 66, 717-723
45. J.J. Koenderink (1986)
Optic flow
Vis. Res. 26,161-180

46. Y. Kubo, I. Shimoyama, T. Kaneda, H. Miura (1994)
Study on wings of flying microrobots
In: *Proc. 1994 IEEE International Conference on Robotics and Automation*, pp. 834-839
47. M. Land (1975)
Head movements and fly vision
In: *The compound Eye and Vision of Insects*
A. Horridge (Ed.), Oxford, Clarendon Press
48. D.N. Lee (1974)
Visual information during locomotion
In: *Perception: Essays in Honor of J.J. Gibson*
Ed; R.B. MacLeod, H. Pick, Ithaca, U.S.A., Cornell Univ; Press, pp. 250-267
49. M. Lehrer, M.V. Srinivasan, S.W. Zhang, A. Horridge (1988)
Motion cues provide the bee's visual world with a third dimension
Nature, London 332, 356-357
50. N. Martin, N. Franceschini (1994)
Obstacle avoidance and speed control in a mobile vehicle equipped with a compound eye.
In: *Intelligent Vehicles*,
I. Masaki (Ed.), Cambridge, U.S.A., M.I.T. Press, pp. 381-386,
51. C. Mead (1989)
Analog VLSI and Neural Systems
Reading (U.S.A.), Addison-Wesley
52. J.A. Meyer, S. Wilson (1991)
From Animals to Animats I
Proc. 1st intern. Conf. on Simulation of Adaptive behavior
Cambridge (U.S.A.), M.I.T. Press
53. J.A. Meyer, H. Roitblat, S. W. Wilson (1993)
From Animals to Animats II
Proc. 2nd Intern. Conf. on Simulation of Adaptive behavior
Cambridge (U.S.A.), M.I.T. Press.
54. F. Mura, N. Franceschini (1994)
Visual control of altitude and speed in a flying agent
In: *From Animals to Animats*, D. Cliff, P. Husbands, J.A. Meyer, S.W. Wilson, (Eds), Cambridge (U.S.A.), M.I.T. Press, pp. 91-99
55. A. Murray, L. Tarassenko (1994)
Analogue Neural VLSI, a Pulse Sream Approach
London, Chapman and Hall
56. K. Nakayama, J.M. Loomis (1974)
Optical velocity patterns, Velocity sensitive neurons and space perception: a hypothesis
Perception 3, 63-80
57. D.E. Nilsson (1989)
Optics and evolution of the compound eye
In: *Facets of Vision*, D.G. Stavenga, R.C. Hardie (Eds.)
Berlin, Springer, pp.30-73
58. H. Paul, H.O. Nalbach, D. Varju (1990)
Eye movements in the rock crab *pachygrapsus marmoratus* walking along straight and curved paths
J. Exp. Biol. 154, 81-97
59. R. Pfeifer, C. Scheier (1994)
From perception to action: the right direction ?
In: *From Perception to Action*. P. Gaussier and J.D. Nicoud (Eds). Los Alamitos (U.S.A.), IEEE Computer Soc. Press, pp. 1-11.
60. J.M. Pichon, C. Blanes, N. Franceschini (1989)
Visual guidance of a mobile robot equipped with a network of self-motion sensors
In: *Mobile Robots*, W. Wolfe, W. Chun (Eds), Bellingham (U.S.A.), S.P.I.E. Vol. 1195, pp 44-53

61. K. Prazny (1980)
Egomotion and relative depth map from optic flow
Biol. Cyb. 57, 1-9
62. A. Riehle, N. Franceschini (1984)
Motion detection in flies: parametric control over ON-OFF pathways
Exp. Br. Res. 54, 390-394
63. G. Sandini, F. Gandolfo, E. Grosso, M. Tistarelli (1993)
Vision during action
In: *Active Perception*,
Y. Aloimonos (Ed.), Hillsdale (U.S.A.), L. Erlbaum Ass., pp. 151-190
64. I. Shimoyama, Y. Kubo, T. Kaneda, H. Miura (1994)
Simple microflight mechanism on silicon wafer
In: *Proc. IEEE Workshop on Microelectromechanical Systems*
pp. 148-152
65. A. Snyder (1981)
Physics of vision in compound eyes
In: *Hdbk of Sens. Physiol., Vision in Invertebrates* Vol. VII/6A
HJ. Autrum (Ed.), Berlin, Springer, pp. 225-313
66. E.C. Sobel (1990)
The locust's use of motion parallax to measure distance
J. Comp. Physiol. A167, 579-588
67. P.J. Sobey (1994)
Active navigation with a monocular robot
Biol. Cyb. 71, 433-440
68. M.V. Srinivasan (1993)
How insects infer range from visual motion ?
In: *Visual Motion and its Role in the Stabilization of Gaze*,
F.A. Miles and J. Wallman (Eds).
69. F.J. Varela, P. Bourgine (1992)
Towards a Practise of Autonomous Systems
Cambridge (U.S.A.), M.I.T. Press
70. H. Wagner (1986)
Flight performance and visual control of flight of the free-flying housefly *Musca domestica*, I/II/III
Phil Trans Roy. Soc. B312
71. G.K. Wallace (1959)
Visual scanning in the desert locust *Schistocerca gregaria*
J. Exp. Biol. 36, 512-525
72. R. Warren, A.H. Wertheim
Perception and Control of Self-motion
Hillsdale (U.S.A.), L. Erlbaum Ass.
73. R. Wehner (1981)
Spatial vision in arthropods
In: *Hdbk Sensory Physiology* VII, 6c, H. Autrum (Ed.), Berlin, Springer, pp. 288-616
74. T.C. Whiteside, D.G. Samuel (1970)
Blur zone
Nature, 225, 94-95
75. R. Wolf, M. Heisenberg (1980)
On the fine structure of yaw torque in visual flight orientation of *Drosophila melanogaster* II/ A temporally and spatially variable weighing function for the visual field ('visual attention')
J. Comp. Physiol. 140, 69-80
76. J. Zeil (1983)
Sexual Dimorphism in the visual system of flies: the free-flight behaviour of male bibionidae (Diptera)
J. Comp. Physiol. 150, 395-412

THE ROLE OF THE BACKGROUND: NON-LOCAL TEXTURE SEGMENTATION AND FIGURE-GROUND PROCESS

Giovanni B. Caputo

Istituto di Psicologia, Università di Pavia, piazza Botta 6, 27100 Pavia, Italy. E-mail gcaputo@coscipv.unipv.it

Abstract- We psychophysically studied the effect of a large surround of uniformly oriented line elements on the orientation discrimination of a single line element (singleton). We found an effect of the surround only if a figure was present. This demonstrated (1) that a background is a unitary surface; (2) that global surface influences local gradient detection leading into question theories of early selection; (3) that lateral long-range interactions among detectors are activated in a all-or-none manner depending on global stimulus configuration.

INTRODUCTION

Visual perception is usually conceived as the result of late structuring of features detected in a early processing stage [1]. This early processing could consist of local filtering of the input image. For instance, it was proposed that texture segmentation could depend on local gradient of some early features called textons [2].

Recent research in neurophysiology challenged the above view: already in the striate cortex the response properties of many cells show considerable dynamics: some V1 cells are able to complete a stimulus behind an occluder [3], the effect of retina lesioning on some V1 cells that lie into the deafferented zone is to shift and enlarge their receptive fields [4]. These findings suggest that the "classical" receptive field could be the apparently local result of non-local lateral connections onto the cell.

Therefore, it is interesting to ask whether also in psychological early processes, that are thought to be local, an influence of the global stimulus configuration is present. In effect, if a local computation is influenced by non-local stimulus properties, the early/late architecture should be reconsidered.

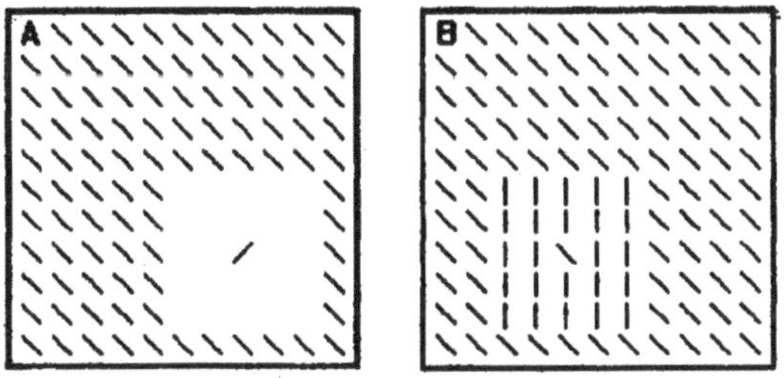

FIGURE 1. Stimuli similar to those used in the experiment. In both figures a single line element (singleton) stays in the middle of a gap in a large surround. **A**, a "gap" stimulus with "orthogonal" orientation between singleton and surround; the gap was empty. **B**, a "figure" stimulus with "same" orientation between singleton and surround; a vertical texture fills in the gap in the surround. In this example the gap size is 5x5.

EXPERIMENT

We used stimuli similar to those shown in figure 1. The subject's task was the orientation discrimination (leftward *versus* rightward) of the singleton that stays in the middle of the gap in the surround. In a first stimulus condition ("gap" stimuli, figure 1A) the gap was empty. In a second stimulus condition ("figure" stimuli, figure 1B) the gap was filled in with a vertical texture. The singleton could have the "same" orientation as the surround or an "orthogonal" orientation. A masking procedure [5] was used: the experimental variable was the stimulus-onset-asynchrony (SOA) between the stimulus (presented for 28 msec) and a masking stimulus (randomly oriented lines presented for 280 msec). The psychophysical method of constant stimuli was used: in each block of trials the stimulus type ("gap" vs "figure") and the SOA were fixed, whereas the trials ("same" vs "orthogonal") were presented in random order. The dependent variable was the percent correct responses for each stimulus condition ("same" *vs* "orthogonal"). The results of one subject are shown in figure 2 for three different gap sizes.

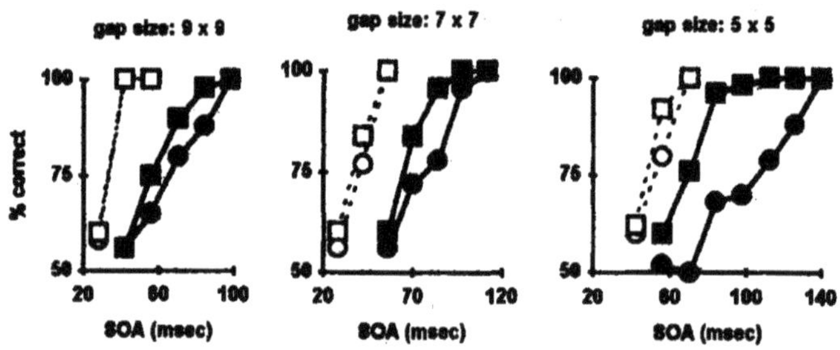

FIGURE 2. One subject's results for a task of orientation discrimination of the singleton: percent correct responses for different SOAs between stimulus and masking. **Open** symbols: "gap" stimuli. **Filled** symbols: "figure" stimuli. **Square** symbols: "orthogonal" orientation between singleton and surround. **Circle** symbols: "same" orientation between singleton and surround. Each plot refers to a different gap size. A surround influence is present only if the gap was filled in with the texture-defined figure; otherwise no-influence was present.

DISCUSSION

The plots of figure 2 indicate that (1) there is an influence of the surround **only if** the gap was filled with the vertical texture; (2) this surround influence decreases the singleton discrimination when it has the "same" orientation as the surround; (3) this surround influence depends on gap size and it is maximal for smaller gaps.

Texture segmentation- The results relative to the "figure" stimuli are critical for theories that conceive texture segmentation as a local filtering. On the contrary, the segmentation of the singleton from the vertical texture is influenced in a non-local manner by the surround and this influence acts only among detectors with the same orientation, thus it is non-linear.

Figure-ground process- The comparison between "gap" and "figure" stimuli indicate that the surround effect is a all-or-none process activated only in a particular ("figure") stimulus condition . We interpret the stimuli of figure 1 in a Gestalt-like manner: in the "gap" stimulus the singleton is perceived as an isolated object, whereas in the "figure" stimulus the vertical texture is perceived as an occluder of the surround and the singleton is perceived either as a part of the surround (in the "same" orientation) or as an object in front of the occluder (in the "orthogonal" orientation). Thus, our results demonstrate that an "active" process [6] makes the background behind the figure in the figure-ground phenomenon.

All two findings criticise against the theoretical assumption of an early computation acting locally on the input image. On the contrary, the early processing stages could be able to make a classification of the visual field into (1) the figure that is a region of interest and (2) the background that is processed as a unitary surface and hence squelched as noise.

REFERENCES

[1] D.H. Hubel, T.N. Wiesel (1962). Receptive fields, binocular interaction and functional architecture in the cat's visual cortex. *Journal of Physiology*, 160, 106-154.

[2] B. Julesz (1986). Texton gradients: the texton theory revisited. *Biol. Cyber.*, 54, 245-251.

[3] M.Jr Fiorani, M.G.P. Rosa, R. Gattass, C.E. Rocha-Miranda (1992). Dynamic surrounds of receptive fields in primate striate cortex: a physiological basis for perceptual completion? *Proc. Natl. Acad. Sci. USA*, 89, 8547-8551.

[4] C.D. Gilbert, T.N. Wiesel (1992). Receptive field dynamics in adult primary visual cortex. *Nature*, 356, 150-152.

[5] J.R. Bergen, B. Julesz (1983). Parallel versus serial processing in rapid pattern discrimination. *Nature*, 303, 696-698.

[6] K. Koffka (1935). *Principles of Gestalt psychology*. New York: Harcourt.

Stochastic resonance in a bistable neural network

Enrico Simonotto and Massimo Riani
Dipartimento di Fisica, Università di Genova
INFM, Unità di Genova
via Dodecaneso 33, 16146 Genova, Italy
email: simonotto@genova.infn.it

Abstract

A neural network model, proposed as a model for the multistability in the perception of ambiguous figures [1], has been studied under the influence of a periodic modulation [2]. The results of the simulations of the neural network are compared with the adiabatic theory of stochastic resonance (McNamara and Wiesenfeld [3]). A good agreement with the theory is found for small enough modulation amplitude values.

1 Introduction

The statistical phenomenon known as stochastic resonance is a property most often observed in periodically modulated, bistable, noise-driven systems. In the absence of the modulation the switching events that carry the system from one stable state to the other one are purely random. However, in the presence of the modulation they become more or less correlated with it. While at both low and high noise intensities, the modulation and the switching events are not well correlated, in a range of intermediate values the degree of correlation grows, hence the name *resonance*.

Stochastic resonance has been demostrated in numerous physical systems (e.g, laser ring, Schmitt trigger); more recently stochastic resonance has been found in simple sensory cells and in a neural network that was introduced as a model of the multistability in the perception of ambiguous figures [2].

Let us consider a two state system with a noisy dynamics, that is, the switching events between the two states are due to stochastic forces acting on the system. In the absence of modulation, the switching rate is given by the well-known Kramers formula:

$$R = \frac{a}{\pi\sqrt{2}} \exp\left(-\frac{2\Delta U_0}{D}\right) \tag{1}$$

where ΔU_0 is the height of the potential barrier between the two stable states and D is the variance of the stochastic forces.

We can introduce a periodic modulation that tilts the potential in such a way that each potential minimum is alternatively raised and lowered relative to the barrier height; in a semi-period one of the minima is raised and the other one is lowered, in the following semi-period the effect is reverted and so the minima that was previously lowered now is raised and so on. The potential height ΔU becomes a time-dependent quantity: $\Delta U \to \Delta U_0 + \epsilon \sin(\omega t)$; also the switching rates become time-periodic quantities.

The periodic modulation effect can be studied by measuring the power spectrum of the output of the system. Due to the modulation of the transition rates a narrow peak at the modulation frequency (and at its higher harmonics if the modulation amplitude is high enough) can be observed in the power spectra, superimposed to the broad band (tipically lorentzian) noisy component of the spectrum. By measuring the height of the peak at the modulation frequency with respect to the amplitude of the noise background at the same frequency, one can measure the Signal-to-Noise Ratio (SNR) of the output.

If the noise level, that is the variance D of the stochastic forces, is very low, the system remains confined in one of the two minima of the energy, for a long period of time, much longer than the period of the modulation. So, the effect of the modulation on the system is small and, as a consequence, SNR is small. Also at very high values of the noise intensity the SNR is small, because the transition rate is much higher than the modulation frequency. Measuring the power spectra for intermediate values of the noise intensity one can observe an increase of the SNR that reaches its maximum for a proper value of such intensity.

By using the adiabatic approximation, that is, assuming that the modulation period is much longer than the relaxation times of the system in the two wells of the potential, McNamara and Wiesenfeld [3] were able to evaluate both the main characteristics of the power spectra of the system and the analytical dependence of the SNR vs. the noise intensity.

The power spectra of the system, as previously stated, are composed of two components: δ peaks at the modulation harmonics and a broad band spectrum. McNamara and Wiesenfeld were also able to evalute the SNR vs. D relationship for a unidimensional system with a continuous two-well potential:

$$SNR = \frac{Q}{D^2} \exp\left(-\frac{2\Delta U_0}{D}\right) \qquad (2)$$

where Q is a parameter which depends on some of the potential parameters and on the modulation amplitude. Such a formula was derived in the the regime of low modulation amplitude values.

2 Neural Network Model

We have studied a neural network model of the phenomenon of multistability in visual perception [1] under the effect of a periodic modulation of its energy landascape [2].

The neural network is made up of two neuron populations, each one representing one of the two interpretations of the visual ambiguous stimulus. The network neurons are binary units with activation levels 0 and 1; the connection matrix, that is the matrix representing the synaptic connections of the neurons, is symmetric. The cognitive phenomenon we model gives some hints on the connection matrix: in fact, we know that the two perceptions exclude each other so the activation of one population must exhert an inhibitory effect on the opponent one. Then, the connection matrix can be subdivided into four blocks: two representing positive (excitatory) connections between neurons of the same popolation, two representing the inhibitory connections between neurons of different populations. We have decided to assign the same value to all the connections in each block, so only three parameters are necessary to describe the connection matrix: p (q) for the connections between neurons of the A (B) population and neurons of the same population and r for the inhibitory connections among neurons of different populations.

Hopfield [4] has shown that in a network with symmetric connection matrix an energy function can be introduced. The restrictions we have posed on the connection matrix allow us to greatly semplify the energy function which, in our model, depends only on the number of active neurons in each of the two populations:

$$E = -\frac{1}{2}a(a-1)p - \frac{1}{2}b(b-1)q + abr \tag{3}$$

where a (b) is the number of neurons with activation level 1 in the A (B) population and p, q, r are the connection matrix parameters. We have found conditions to ensure that the energy has only two minima, namely the states in which one of the two populations is fully activated and the opponent one is inhibited.

Oscillations between the two basins of attraction of the minima of the energy can be observed introducing a thermally activated dynamics in the network. The pseudotemperature is then linked to the noise in the system: for low temperature values we get low intensity noise levels and so on.

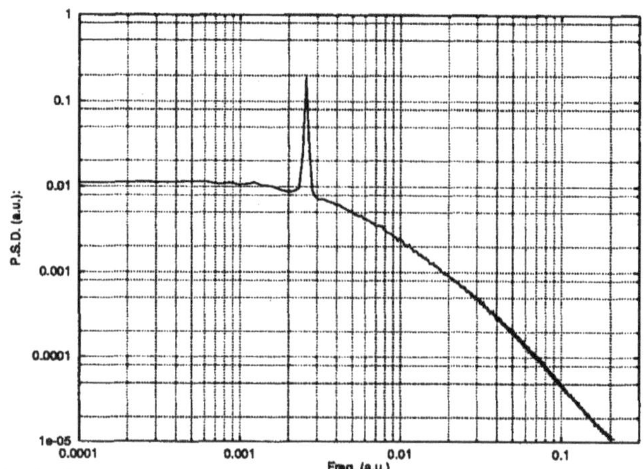

Figure 1: Typical power spectra: a peak is superimposed to the broad band lorentzian spectra.

The periodic modulation can be introduced as a perturbation of the energy:

$$E = -\frac{1}{2}a(a-1)p - \frac{1}{2}b(b-1)q + abr + \mathrm{H}_m \sin(\omega_m t)\,(a-b) \tag{4}$$

where H_m and ω_m are the modulation amplitude and frequency. This perturbation can be interpreted, by exploting the analogy between magnetic systems and neural networks with binary units, as an external time-periodic magnetic field acting in opposite directions on the two populations [2].

We have measured the power spectra of the output of the network for different temperature values (i.e., noise intensities). As you can see in fig. 1, the peak at the modulation frequency is very pronounced. If the modulaton amplitude is high enough, other peaks may appear at higher harmonics.

From the power spectra we have measured the SNR vs. Temperature relationship. We have tried a comparison between the measured curve and the McNamara and Wiesenfeld

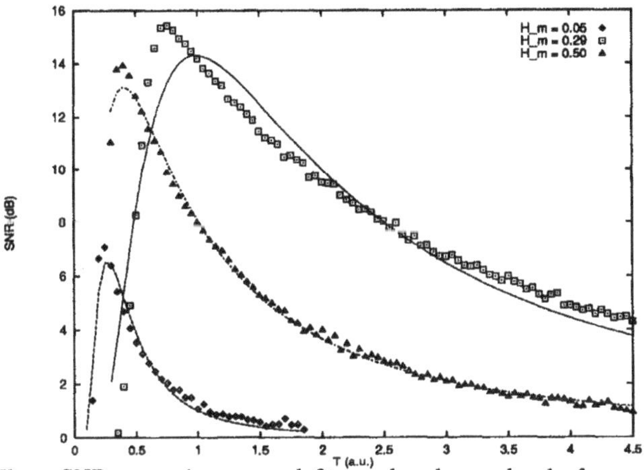

Figure 2: Three SNR vs. noise: a good fit can be observed only for small modulation amplitude.

theoretical curve eqn. 2. Results are shown in fig. 2 for three different values of the modulation amplitude H_m. There is a good fit for small modulation amplitudes ($H_m = 0.05$), that is, when only one peak is evident in the power spectra. In this case, we have found that the noise intensity D is linked to the temperature value and follows the law: $D \propto T^{\frac{3}{2}}$. For increasing modulation amplitudes H_m the theory does not describe the data well.

We have to remark that the theoretical relationship eqn. 2 was derived, in the approximation of only one δ in the power spectra, that is, for small H_m values, for a continuous, unidimensional, two well potential. The neural network is quite a different system: its energy landscape is bidimensional and defined on a grid. Nethertheless, the theory still describes the SNR measured in the neural network simulations when the requirements on the modulation amplitudes are satisfied.

References

[1] F. Masulli, M. Riani, E. Simonotto, and F. Vannucci. Boltzmann ditributions and neural networks: models of unbalanced interpretations of reversible patterns. In *Science of Artificial Neural Networks*, page 267. SPIE, 1992.

[2] M. Riani and E. Simonotto. Stochastic resonance in the perceptual interpretation of ambiguous figures: A neural network model. *Physical Review Letters*, 72(19):3120–3123, 1994.

[3] B. McNamara and K. Wiesenfeld. Theory of stochastic resonance. *Physical Review A*, 39(9):4854–4869, 1989.

[4] J. J. Hopfield. Neural networks and physical systems with emergent collective computation abilities. *Proceedings of the National Academy of Science*, 79:2554–2558, 1982.

PARTICIPANTS

Members of the Scientific Committee, Lecturers and Members of the Organizing Committee are indicated respectively by three, two and one asterisks.

Arcelli Carlo
Istituto di Cibernetica, CNR
Via Toiano, 6
I-80072 Arco Felice (Napoli)
Italy
tel: 0039 81 8534208
fax: 0039 81 5267654
e-mail: car@imagm.na.cnr.it

Beghi Luigi
Dipartimento di Matematica pura e Applicata
Via Belzoni, 7
I-35131 - Padova, Italy
tel: 0039 49 831905
fax.: 0039 49 39351

**Buizza Angelo
Dipartimento di Informatica e Sistemistica
Università di Pavia
Via Abbiategrasso, 209
I-27100 Pavia, Italy
tel: 0039 382 505359
fax: 0039 382 505373
e-mail: angelo@bioing.unipv.it

**Burattini Ernesto
Istituto di Cibernetica, CNR
Via Toiano 6
I-80072 Arco Felice (Napoli)
Italy
tel: 0039 81 8534204
fax: 0039 81 5267654
e-mail: ernb@sole.cib.na.cnr.it

Caputo Giovanni
Istituto di Psicologia
Università di Pavia
Piazza Botta, 6
I-27100 Pavia, Italy
tel: 0039 382 386271
fax: 0039 382 386272

Cesaretti Giovanni
Dipartimento di Fisiologia e Biochimica
Università di Pisa
Via San Zeno, 31
I-56127 Pisa, Italy
tel: 0039 50 553517
fax: 0039 50 552183

Chillemi Santi
Istituto di Biofisica, CNR
Via San Lorenzo, 26
I-56127 Pisa, Italy
tel: 0039 50 513111
fax: 0039 50 553501
e-mail: chillemi@ib.pi.cnr.it

Chukova Svetlana
Laboratory of Physiology of Vision
Pavlov Institute of Physiology
CSE-St. Petersburg
Russia
fax: 007 812 2180501
chi@physiology.spb.su

***,**Cordella Luigi
Dipartimento di Informatica e Sistemistica
Università di Napoli
Via Claudio, 21
I-80125 Napoli, Italy
tel: 0039 81 7683185/158
fax: 0039 81 7683186
e-mail: cordel@nadis.dis.unina.it

*Cotugno Antonio
Istituto di Cibernetica, CNR
Via Toiano, 6
I-80072 Arco Felice (Napoli)
Italy
tel: 0039 81 8534113
fax: 0039 81 5267654
e-mail: neuros@mail.irtemp.na.cnr.it

***De Lotto
Dipartimento di Informatica e Sistemistica
Università di Pavia
Via Abbiategrasso, 209
I-27100 Pavia, Italy
tel: 0039 382 505350
fax: 0039 382 505373

Di Maio Vito
Istituto di Cibernetica, CNR
Via Toiano 6
I-80072 Arco Felice (Napoli), Italy
tel: 0039 81 8534107
fax: 0039 81 5267654
e-mail:vdm@biocib.cib.na.cnr.it

***,**Fiorentini Adriana
Istituto di Neurofisiologia, CNR
Via San Zeno, 51
I-56127 Pisa, Italy
tel: 0039 50 559716
fax: 0039 50 559725
e-mail: adri@neuro.in.pi.cnr.it

***,**Franceschini Nicolas
Laboratoire de Neurobiologie, CNRS
Dept. de Neurocybernétique
31 Chemin Joseph Aiguier
F-13402 Marseille
France
tel: 0033 91 164129
fax: 0033 91 220875
e-mail: ENFranceschini@irlnb.cnrs-mrs.fr

**García-Pérez Miguel
Laboratorio de Visión
Facultad de Psicología
Universitad Complutense
Campus de Somosaguas
E-28023 Madrid
Spain
e-mail: psmet04@sis.ucm.es

**Gerbino Walter
Dipartimento di Psicologia
Via dell' Università, 7
Università di Trieste
I-34123 Trieste, Italy
tel: 0039 40 6767357/29/53
fax: 0039 40 312272
e-mail: gerbino@univ.trieste.it

**Giulio Ludovico Filippo
Università di Torino
Dipartimento di Morfofisiologia Veterinaria
Settore Fisiologia ed Etologia
Viale Mattioli, 25
I-10125 Torino
Italy
tel/fax: 0039 11 6687887

***Götz Karl Georg
Max-Planck-Institut für biologische
Kybernetik
Spemannstrasse, 38
D-72076 Tübingen, Germany
tel: 0049 7071 601560
fax: 0049 7071 601575
e-mail: goetz@sunwan.mpik-tueb.mpg.de

Guglielmotti Vittorio
Istituto di Cibernetica, CNR
Via Toiano, 6
I-80072 Arco Felice (Napoli)
Italy
tel: 0039 81 8534114/117
fax: 0039 81 5267654

**Hengstenberg Roland
Max-Planck-Institut für biologische
Kybernetik
Spemannstrasse, 38
D-72076 Tübingen
Germany
tel: 0049 7071 601566
fax: 0049 7071 601575
e-mail: hbg@sunwan.mpik-tueb.mpg.de

***,**Hochstein Shaul
Department of Neurobioloy
Institute of Life Science
The Hebrew Univesity
IL-91010 Jerusalem
Israel
tel: 00972 2 585193
fax: 00972 2 586296
e-mail: shaul@vms.huji.ac.il

*Izzo Marcella
Istituto di Cibernetica, CNR
Via Toiano, 6
I-80072 Arco Felice (Napoli)
Italy
tel: 0039 81 8534124
fax: 0039 81 5267654
e-mail: secymarc@.na.cnr.it

Kauhsk Neeme
Tallin Pedagogical University
Narva Road 25
EE-2400, Tallinn, Estonia
tel.: 00372 2 426626
fax: 00372 2 425339//00372 6 312588
e-mail: neeme@lin.tpu.ee

Liquori Gianluca
Dipartimento di Ricerche Filosofiche
Università di Roma "Tor Vergata"
I-00133 Roma, Italy
tel: 0039 6 72595120
Mailing Address:
tel: 0039 6 5037787
fax: 0039 6 5037787

**Magnani Lorenzo
Università di Pavia
Dipartimento di Filosofia, Sez. Filosofico-Teorica
Laboratorio di Filosofia Computazionale
Palazzo San Felice
Piazza Botta, 6
I-27100 Pavia
Italy
tel: 0039 382 386341
fax: 0039 382 23215
e-mail: lorenzo@ipvaim.unipv.it

Maguire Greg
Sensory Sciences Center
University of Texas
6420 Lamar Fleming
TX 77030, Houston
USA
tel: 713 745 0757
fax:713 792 4513
e-mail: gmaguire@gsbs.gs.uth.tmc.edu

Maldonato Mauro
Dipartimento di Psichiatria
Università di Napoli
Via Largo Madonna delle Grazie, 4
I-80136 Napoli
Italy
tel: 0039 81 5666505
fax: 0039 81 5666523

**Mallot Hanspeter
Max-Planck-Institut für
biologische Kybernetik
Spemannstrasse, 38
D-72076 Tübingen
Germany
tel: 0049 7071 601603
fax: 0049 7071 601575
e-mail: ham@pea.mpik-tueb.mpg.de

**Marzi Carlo Alberto
Università di Verona
Dipartimento di Scienze Neurologiche e della Visione
Strada Le Grazie
I-37134 Verona
Italy
tel: 0039 45 8098143
fax: 0039 45 580881
e-mail: marzic@borgoroma.univr.it

*Mazzarella Annamaria
Istituto di Cibernetica, CNR
Via Toiano, 6
I-80072 Arco Felice (Napoli), Italy
tel: 0039 81 8534123
fax: 0039 81 5267654
e-mail: cib@secyAnn.na.cnr.it

***,**Mingolla Ennio
Department of Cognitive and Neural System
Center for Adaptive Systems
Boston University
111 Cummington Street, Boston
MA 02215, USA
tel: 001 617 3539485
fax: 001 617 3537755
e-mail: ennio@cns.bu.edu

*Musio Carlo
Istituto di Cibernetica, CNR
Via Toiano, 6
I-80072 Arco Felice (Napoli)
Italy
tel: 0039 81 8534131
fax: 0039 81 5267654
e-mail: camus@biocib.cib.na.cnr.it

Neugebauer Anna
IIGB-CNR
Via Marconi, 10
I-80100 Napoli, Italy
tel: 0039 81 7257111
fax: 0039 81 5936123
e-mail. neugebauer@iigbna.iigb.na.cnr.it

**Patterson John
Qeen Mary and West Field College
Faculty of Basic Medical Sciences
University of London
Department of Physiology
Mile End Road
London E14NS, UK
tel: 0044 71 9826378
fax: 0044 81 9830467
e-mail: ugha085@alpha.qmw.ac.uk

*Piantedosi Salvatore
Istituto di Cibernetica, CNR
Via Toiano, 6
80072 Arco Felice (Napoli), Italy
tel: 0039 81 8534212
fax: 0039 81 5267654
e-mail: sp@imagm.na.cnr.it

**Pizzamiglio Luigi
Università "La Sapienza"
Dipartimento di Psicologia
Via dei Marsi, 78
I-00100 Roma
Italy
tel: 0039 6 49917530/4451667
fax: 0039 6 4451667

Pizzuti Etta
PFTIM
Via Petrarca, 115
I-80100 Napoli, Italy
tel: 0039 81 5750015
fax: 0039 81 5752319
Mailing Address:
Via Manzoni, 257
I-80123 Napoli, Italy
tel: 0039 81 7146731/7691274

**Radilova Irina
Department of Psychology
General Psychology Division
University of Helsinki
Ritarikatu 5
SF-00014 Helsinki
Finland
tel 00358 0 1913496
fax 00358 0 1913443
Mailing Address:
Praha
Czech Republic
tel: 0042 2 4752409

***Radil-Weiss Tomáš
Czechoslovakia Academy of Sciences
Institute of Physiology
1083 Vídenská
142 20 Praha 4 - KRC
Czech Republic
tel: 0042 2 4719741
fax: 0042 2 4719517
e-mail: bohdanec@biomed.cas.cz

**Riani Massimo
Dipartimento di Fisica
Università di Genova
Via Dodecaneso, 33
I-16146 Genova
Italy
tel: 0039 10 3536201
fax: 0039 10 314218
e-mail: riani@genova.infn.it

*Santillo Silvia
Istituto di Cibernetica, CNR
Via Toiano, 6
I-80072 Arco Felice (Napoli)
Italy
tel: 0039 81 8534113
fax: 0039 81 5267654
e-mail: neuros@mail.irtemp.na.cnr.it

**Savini Massimo
Dipartimento di Informatica e Sistemistica
Università di Pavia
Via Abbiategrasso, 209
I-27100 Pavia
Italy
tel: 0039 382 505372/50
fax: 0039 382 391373
e-mail:savio@ipvvis.unipv.it

Scherbakova Natalia
Laboratory of Neurophysiology Sensory-Motor Function
Pavlov Institute of Physiology
CSE-St. Petersburg
Russia
fax: 007 812 2180501
chi@physiology.spb.su

Sepe Gennaro
Clinica Oculistica
Università Federico II
Via Pansini, 5
I-80100 Napoli
Italy

Simonotto Enrico
Dipartimento di Fisica
Università di Genova
Via Dodecaneso, 33
I-16146 Genova
Italy
tel: 0039 10 3536314
fax: 0039 10 314218
e-mail: simonotto@genova.infn.it

**,*Taddei-Ferretti Cloe
Istituto di Cibernetica, CNR
Via Toiano, 6
I-80072 Arco Felice (Napoli)
Italy
tel: 0039 81 8534113
fax: 0039 81 5267654
e-mail: neuros@mail.irtemp.na.cnr.it

Unzicker Alexander
Institut für Medizinische Psychologie
Goethestrasse, 31
D-80336 München
Germany
tel: 0049 89 5996202
e-mail: sascha@space.imp.med.uni-muenchen.de

**Vallerga Silvana
Istituto di Cibernetica e Biofisica, CNR
Via Dodecaneso, 33
I-16146 Genova
Italy
tel: 0039 10 335626
fax: 0039 10 6475800
e-mail: vallerga@ced.ice.ge.cnr.it
International Marine Center
Lungomare d'Arborea, 22
I-09072 Torre Grande (Oristano)
Italy
tel 0039 783 22027/32
fax 0039 783 22002

Xausa Elisabetta
Dipartimento di Matematica Pura ed Applicata
Università di Padova
Via Belzoni, 7
I-35131 Padova
Italy
tel: 0039 49 8275111/905
fax: 0039 49 8758596
e-mail: xausa@pdmat1.math.unipd.it

Zanforlin Mario
Dipartimento di Psicologia Generale
Università di Padova
Piazza Capitaniato, 3
I-35139 - Padova
Italy

www.ingramcontent.com/pod-product-compliance
Lightning Source LLC
LaVergne TN
LVHW051925120926
847902LV00009B/75
* 9 7 8 9 8 1 0 2 3 2 2 9 0 *